T0293438

New Insights into Horticulture

New Insights into Horticulture

Editor: Thelma Bosso

www.callistoreference.com

Callisto Reference,
118-35 Queens Blvd., Suite 400,
Forest Hills, NY 11375, USA

Visit us on the World Wide Web at:
www.callistoreference.com

ISBN: 978-1-64116-765-9 (Hardback)

Cataloging-in-Publication Data

New insights into horticulture / edited by Thelma Bosso.
p. cm.
Includes bibliographical references and index.
ISBN 978-1-64116-765-9
1. Horticulture. 2. Gardening. 3. Agriculture. I. Bosso, Thelma.
SB318 .F86 2023
635--dc23

Table of Contents

Preface

The world is advancing at a fast pace like never before. Therefore, the need is to keep up with the latest developments. This book was an idea that came to fruition when the specialists in the area realized the need to coordinate together and document essential themes in the subject. That's when I was requested to be the editor. Editing this book has been an honour as it brings together diverse authors researching on different streams of the field. The book collates essential materials contributed by veterans in the area which can be utilized by students and researchers alike.

Horticulture refers to the art of cultivating plants in gardens for producing food and medicinal materials, as well as for decorative and comfort purposes. It is classified into multiple groups that focus on cultivating and processing various plants and foods for particular purposes. Horticulture includes plant propagation and cultivation for improving quality, plant growth, nutritional value, yields, and resistance to diseases, insects and environmental challenges. It also encompasses aspects such as soil management, plant conservation, arboriculture, landscape restoration along with garden and landscape design, construction and maintenance. It utilizes small plots of land with a large variety of crops and can be classified into floriculture, olericulture, viticulture, arboriculture, and pomology. This book explores all the important aspects of horticulture in the present day scenario. It presents researches and studies performed by experts across the globe. The extensive content of this book provides the readers with a thorough understanding of the horticulture.

Each chapter is a sole-standing publication that reflects each author's interpretation. Thus, the book displays a multi-facetted picture of our current understanding of application, resources and aspects of the field. I would like to thank the contributors of this book and my family for their endless support.

Editor

Preface

Editor

Anaerobically-Digested Brewery Wastewater as a Nutrient Solution for Substrate-Based Food Production

Ignasi Riera-Vila, Neil O. Anderson, Claire Flavin Hodge and Mary Rogers *

Department of Horticultural Sciences, University of Minnesota, Saint Paul, MN 55108, USA; i.riera.vila@gmail.com (I.R.-V.); ander044@umn.edu (N.O.A.); flavi010@umn.edu (C.F.H.)

* Correspondence: roge0168@umn.edu

Abstract: Urban agriculture, due to its location, can play a key role in recycling urban waste streams, promoting nutrient recycling, and increasing sustainability of food systems. This research investigated the integration of brewery wastewater treatment through anaerobic digestion with substrate-based soilless agriculture. An experiment was conducted to study the performance of three different crops (mustard greens (*Brassica juncea*), basil (*Ocimum basilicum*), and lettuce (*Lactuca sativa*) grown with digested and raw brewery wastewater as fertilizer treatments. Mustard greens and lettuce grown in digested wastewater produced similar yields as the inorganic fertilizer control treatment, while basil had slightly lower yields. In all cases, crops in the digested wastewater treatments produced higher yields than raw wastewater or the no fertilizer control, indicating that nutrients in the brewery wastewater can be recovered for food production and diverted from typical urban waste treatment facilities.

Keywords: urban agriculture; reclaimed wastewater; controlled environment agriculture; soilless production; brewery; *Brassica juncea*; *Lactuca sativa*; *Ocimum basilicum*

---

1. Introduction

Urban agriculture is experiencing a resurgence in popularity in many parts of the world. Beyond the social benefits urban agriculture can provide, such as creating opportunities for community building, jobs, and education, as well as increased access to healthy food [1], urban agriculture is also gaining interest due to the possible environmental benefits it can provide to municipalities. Some of those benefits include increased green spaces, reduction of food imports, and nutrient recycling [2–5]; the latter is the focus of this paper. Modern cities function largely as nutrient sinks, with nutrients shipped in from rural locations through food and transformed to waste once consumed. Historically, and presently in the Global South, these nutrient-rich waste streams were and are prized for fertility in both rural and urban agriculture [5–7]. However, social stigma, concerns over sanitation, and potential presence of pathogens, nitrates, heavy metals, or pharmaceuticals [8] in urban wastewater make this integration a challenge today. Still, advocates argue that finding ways to recycle the organic fraction of waste streams to agricultural production in urban areas will not only reduce soil and water pollution, but also prove central to both urban waste management and agricultural production [9].

Integrating agricultural production with industries that produce wastewater suitable for irrigation may increase sustainability in both sectors [9,10] and reduce the challenge of reusing wastewater [11]. The brewing industry can serve as a model for wastewater and urban agriculture integration. The brewing process creates large amounts of wastewater that, if treated aerobically like most of the wastewater in the US, requires large amounts of energy for water treatment. This usually translates into substantial surcharges to the brewery [11]. The large energy requirement for treatment is due to processing the high number of organic compounds in the water. These organic compounds have the potential to produce hydrogen or methane energy if treated anaerobically, and the resulting nutrients could be a source of fertility for agricultural production [12–14].

Traditional soil-based agriculture may not be a feasible solution in urban environments due to soil contamination and the lack of available land. Soilless production, such as substrate-based, aquaponics (hydroponics) could offer a flexible solution adaptable to different settings [15]. Research has explored how to recycle wastewater for soilless production, mostly for hydroponic production, and found that if the wastewater or organic waste is mineralized and the optimal nutrient content is achieved, comparable yields to plants grown using inorganic fertilizer may be obtained. This has been shown using brewery wastewater [12]. In all cases, researchers promoted nutrient mineralization through either algae ponds [12], bioreactors [16], or introduction of desired bacteria in the system [17]. Part of the rationale of using organic substrates in this work is to enhance mineralizing and heterotroph bacteria and fungi populations [18] in order to promote organic matter breakdown and mineralization [19,20].

The objective of this research was to test the performance of both digested and raw brewery wastewater as a fertility source for substrate-based vegetable production. The use of both digested and raw wastewater seeks to explore the feasibility of using brewery wastewater for substrate production as well as to determine the potential benefits of digesting the wastewater. Three genetically diverse crops were used to study the adaptability of the brewery wastewater to grow high value crops suitable for controlled environment production in soilless media. Mustard greens (*Brassica juncea*) can be easily grown in soilless production as a leafy vegetable and are increasing in consumer demand, particularly in fresh salad mixes [21,22]. Lettuce (*Lactuca sativa*) is widely grown and studied in soilless production [23–26] and basil (*Ocimum basilicum*) is a popular herb that is also widely cultivated and studied in soilless systems [27,28].

Our hypothesis was that if the treatments with wastewater have similar nutrient profiles to the inorganic control treatment, crops should produce similar yields. We hypothesized that differences between the digested and raw wastewater treatments would be due to the higher nitrogen and lower organic carbon content of the digested wastewater as opposed to the raw wastewater treatment.

2. Materials and Methods

This research is part of a larger project that seeks to create an anaerobic wastewater treatment process to reduce carbon load of wastewater combined with urban food production via soilless agriculture. Specifically, we are interested in modeling a decentralized approach to this integration that could happen at various locations in an urban environment, unlike a centralized wastewater treatment facility. For this project, the anaerobic digestion process addresses two objectives: (1) to reduce organic load of the wastewater while producing hydrogen and methane energy and (2) to create a final water solution more suitable for plant uptake in soilless vegetable production, thereby closing a water usage loop.

Three experiments corresponding to three different crops—'Green Wave' mustard greens, 'Nufar' basil, and 'Salanova Green Butter' lettuce (Johnny's Selected Seeds, Waterville, ME)—were conducted between May and September 2018 at the University of Minnesota Plant Growth Facilities in St. Paul, MN, USA (44°59′17.8″ N lat., -93°10′51.6″ W long.). The experiments were the same in methods, only differing in crop type. Each experiment was set up as a completely randomized design, with $n = 5$ replicates of each of the four fertility treatments, for a total of 20 plants in each experiment.

Four different fertility treatments were evaluated: (1) an unfertilized control (i.e., only water); (2) an industry standard inorganic hydroponic fertilizer with the following concentrations: 150 ppm N, 52 ppm P, 215 ppm K, 116 ppm Ca, 53 ppm Mg, 246 ppm SO_4, 3 ppm Fe, 0.5 ppm Mn, 0.15 ppm Zn, 0.15 ppm Cu, 0.5 ppm B, and 0.1 ppm Mo, obtained by mixing 4 L of water with 2.56 g of $CaNO_3$ and 3.88 g of 5–12–26 (Jack's hydroponics fertilizers, JR Peters Inc., Allentown, PA, USA); (3) raw wastewater from a local brewery (Fulton Beer, Minneapolis, MN, USA); and (4) digested wastewater from the same brewery, diluted at 50% with deionized water after digestion (Dr. Paige Novak's laboratory, Dept. of Civil, Environmental and Geo-engineering, University of Minnesota, Minneapolis, MN, USA). All of the plants were watered with 100 mL of well water every morning, along with 50 mL of their respective fertility treatments six afternoons per week (Sunday to Friday), with rates based on previous experience; one day per week had only water applied. Fertility treatments were at the same greenhouse temperature when irrigating.

All raw and digested wastewater was collected before the experiments began and stored in 26.5 L polyethylene containers (Aqua-Tainer, Reliance Products, Winnepeg, Manitoba, Canada) in a cooler at 5 °C (Vollrath Inc., Sheboygan, WI, USA). A subsample of the digested and raw wastewater was submitted for analytical testing (Research Analytical Lab, University of Minnesota, St. Paul, MN, USA) for ammonium-N [29], nitrate/nitrite-N [30], and total phosphorus [31] at the beginning of the experiment. In the early stages of the project, a low nitrogen concentration in the digestate but a high electrical conductivity [32] was detected. Therefore, 150 mg/L of ammonium hydroxide was used for pH adjustment of the brewery wastewater instead of calcium carbonate as a means to obtain a digestate with higher nitrogen content but lower conductivity.

The three different crops—mustard greens, basil, and lettuce—were grown for six weeks in plastic containers with a 10.16 cm diameter and an 8.5 cm depth (Belden Plastics, Saint Paul, MN, USA) filled with a peat-based substrate (Professional Growing Mix #8, Sun Gro Horticulture, MA, USA). This substrate was chosen based on performance in preliminary experiments. Plants were started in a 128 plug tray (TO Plastics, Clearwater, MN, USA) in a peat-based propagating mix (Sunshine Propagation Mix, Sungro, MA, USA), and kept in a mist chamber, misted at intervals of 5 min for 7 days to encourage germination. The mean temperature was 25.78 ± 2.14 °C with 66.94 ± 16.29% relative humidity (RH). After germination, seedlings were moved to the greenhouse and placed on top of a flat 25.4 × 50.8 cm tray filled with 5 cm of water for 7–10 days. The photoperiod was set for long days (16:8 h day:night) with supplemental lighting supplied by high pressure sodium high intensity discharge (HID) lamps at a maxima of 1377 μmol m^{-2} s^{-1}. Mean temperature ±SD was 24.39 ± 4.24 °C. All environmental settings were controlled via an Argus Control Systems Ltd. computer (Surrey, BC, Canada).

Plant biomass, or final yield, was calculated for each plant at the end of the six weeks. Mustard green leaves longer than 10 cm were harvested weekly. Fresh weight of the leaves was recorded and then leaves were dried for 72 h at 70 °C in a hot air oven (Hatchpack, PI, USA) and dry weights were measured. The total above-ground biomass of mustard greens was calculated by adding the cumulative harvested yield to the dry mass of the above-ground plant material at the termination of this experiment. In the case of lettuce and basil, total above-ground biomass present at the end of the experiment was used as the measurement of yield.

Plant growth measurements were recorded weekly. The number of leaves on the terminal stem and stem width (mm) were measured on each experimental unit, or each plant, while chlorophyll level was measured using a soil-plant analyses development (SPAD) meter (SPAD 502, Spectrum technologies, IL, USA) for each leaf on each plant. The SPAD meter provides an output value between −9.9 to 199.9 in SPAD units; the higher the value, the greater the chlorophyll content [33]. The meter measures relative light absorbance at two different wavelengths (650 nm and 940 nm) [34].

Water quality of the leachate (collected after watering with 100 mL of well water) was measured weekly for pH, EC, infiltration, and nitrates. Three out of the five replications of each treatment were sampled, for a total of 12 samples from each crop, which were randomly selected at the beginning of the experiment. Infiltration time was measured as the duration of time from pouring the water until there was no standing water on the substrate surface. Electrical conductivity (EC) and pH of the collected leachate was measured using an electronic pH and EC meter (Milwaukee mw802, Milwaukee Electronics, WI, USA). Nitrate levels of the leachate were measured using a LAQUA twin NO3-11C pocket nitrate reader (Horiba scientific, Minami-ku Kyoto, Japan).

Statistical analyses were conducted using R software (The R foundation, v.3.4.1) to determine significance using Analysis of Variance (ANOVA). If the null hypothesis of no significant differences was rejected, mean separations were accomplished using Tukey's honestly significant difference (HSD) test at a significance level of $\alpha = 0.05$. Plant chlorophyll content and growth parameters were also tested for correlation with yield, using Pearson's correlation test at a significance level of $\alpha = 0.05$.

3. Results

3.1. Biomass and Yield

Fertility treatment had a significant effect on the number of harvested leaves (data not shown) and total above-ground biomass of mustard greens (Table 1). Fertility treatment also had a significant effect on total above ground biomass of lettuce and basil (Table 1). Mustard greens and lettuce plants grown using digested wastewater produced a similar amount of total above ground biomass when compared to plants in the inorganic fertilizer treatment, while raw wastewater plants performed at the same level as those in the no fertilizer treatment (Table 1). For basil, the highest biomass was observed in the inorganic fertilizer followed by digested wastewater treatments (Table 1). Mustard greens in both the inorganic fertilizer and digested wastewater treatments reached harvestable size the second week of the experiment and produced until week 6, while mustard greens grown in the raw wastewater only were harvestable on week 6, and no plants reached maturity in the unfertilized control treatment.

Table 1. Mean ± SE total above-ground biomass, or per plant dry weight (g), for mustard greens, lettuce, and basil grown with different fertility treatments in a greenhouse in St. Paul, Minnesota in 2018. One-way ANOVA results for fertility treatments were performed on each parameter and crop. Values within the same column with different letters are significantly different (Tukey's HSD, $\alpha = 0.05$).

Fertility Treatment	Total Dry Weight Per Plant (g)		
	Mustard Greens	Lettuce	Basil
No fertilizer	0.19 ± 0.04 b	0.19 ± 0.03 b	0.46 ± 0.02 c
Inorganic fertilizer	0.58 ± 0.02 a	2.83 ± 0.11 a	3.57 ± 0.09 a
Raw wastewater	0.29 ± 0.08 b	0.46 ± 0.09 b	0.73 ± 0.13 c
Digested wastewater	0.56 ± 0.04 a	2.09 ± 0.46 a	1.95 ± 0.36 b
ANOVA	$F(3,16) = 16.33$ $p < 0.01$	$F(3,16) = 27.51$ $p < 0.01$	$F(3,16) = 50.71$ $p < 0.01$

3.2. Plant Growth

Due to the leaf removal from mustard green plants, plant growth was monitored using stem width at the substrate level. No significant fertility treatment effects were found for mustard stem width during any of the weeks (data not shown). Weekly stem width measurements were moderately correlated with weekly harvest ($r = 0.57$, $p < 0.01$).

For lettuce and basil, the number of leaves per week was used as a measurement of plant growth. Significantly different numbers of leaves per lettuce plant were counted in weeks 2–6 ($p < 0.05$) between treatments. In weeks 2–5, lettuce grown in inorganic fertilizer treatments had the highest number of leaves, while in week 6, lettuce from both inorganic fertilizer and digested wastewater treatments had a similar number of leaves. Total number of lettuce leaves at the end of the experiment was positively correlated to final above-ground biomass ($r = 0.89$, $p < 0.01$). Basil plants, too, had a similar number of leaves through the first week, but subsequent significant differences were found in weeks 2–6 ($p < 0.01$) between treatments. In later weeks, basil from the inorganic fertility treatment had the highest number of leaves. Similar to the case of lettuce, at the end of the experiment, the total number of basil leaves was highly correlated to final above-ground biomass ($r = 0.97$, $p < 0.01$). Overall, plants from both inorganic and digested wastewater treatments appeared marketable for all crops (Figures 1–3).

Figure 1. Image of 'Green Wave' mustard green plants (day 42) grown in University of Minnesota's Plant Growth Facilities greenhouse (St. Paul, MN). Plants were grouped by fertility treatment in rows; from left to right: digested wastewater, raw wastewater, inorganic fertilizer, and no fertilizer.

Figure 2. Image of 'Salanova Green Butter' lettuce plants (day 42) grown in University of Minnesota's Plant Growth Facilities greenhouse (St. Paul, MN). Plants were grouped by fertility treatment in rows; from left to right: digested wastewater, raw wastewater, inorganic fertilizer, and no fertilizer.

Figure 3. Image of 'Nufar' basil plants (day 42) grown in University of Minnesota's Plant Growth Facilities greenhouse (St. Paul, MN). Plants were grouped by fertility treatment in rows; from left to right: digested wastewater, raw wastewater, inorganic fertilizer, and no fertilizer.

3.3. Chlorophyll Content

In all crops, the chlorophyll content of the digested wastewater treatments stayed at a similar level to inorganic fertilizer treatments throughout the 6 weeks (Table 2). Significant differences in SPAD values of mustard greens were not found until weeks 5 and 6 ($p = 0.03$), where digested wastewater treatments had the highest chlorophyll content and raw wastewater the lowest (data not shown). SPAD values for mustard greens in were not correlated with weekly harvest ($r = 0.05, p < 0.56$).

Table 2. Mean chlorophyll content (SPAD values) averaged across 6 weeks for mustard greens, lettuce, and basil grown with different fertility treatments in a greenhouse in St. Paul, Minnesota in 2018. Values within the same column with different letters are significantly different (Tukey's HSD, $\alpha = 0.05$).

Fertility Treatment	Chlorophyll Content, SPAD Units		
	Mustard Greens	**Lettuce**	**Basil**
No fertilizer	22.23 ab	13.33 c	17.79 c
Inorganic fertilizer	21.01 a	20.76 a	25.93 a
Raw wastewater	20.40 b	17.21 b	21.50 b
Digested wastewater	24.13 a	20.42 a	26.22 a
ANOVA			
	$F(3,133) = 8.32$	$F(3,116) = 22.72$	$F(3,116) = 43.85$
	$p < 0.001$	$p < 0.001$	$p < 0.001$

Significant differences between SPAD measurements became apparent in the fourth week, where digested wastewater and inorganic fertilizer treatments had the highest chlorophyll content; the six-week average is shown in Table 2. SPAD values at the end of the experiment were correlated with lettuce total biomass ($r = 0.78, p < 0.05$).

We observed significant differences in the SPAD values of basil plants grown in the different fertility treatments as early as week 2 and continuing through week 6 ($p < 0.01$). In all cases, basil leaves from the digested wastewater and inorganic fertilizer treatments had the highest chlorophyll content, followed by raw wastewater treatment (Table 2). Basil SPAD values were correlated with total dry biomass at the end of the experiment ($r = 0.77, p < 0.01$).

3.4. Water Test

Comparing the total nitrogen ($T_N = NH_4 - N + NO_2 - N + NO_3 - N$) between the digested wastewater (50% diluted) and the raw wastewater, there was higher ammonium-N content in the digested wastewater, 171.50 ppm compared to 7.23 ppm (Table 3). The ammonium content of the digested wastewater was similar to that used in the inorganic fertilizer control and within the 100–250 ppm recommended range [35,36]. The only concerning factor was that the N ratio ($R_N = NH_4 - N/NO_3 - N$) [37] was extremely high: >1000, and since all the N was in the form of ammonium, this could generate problems of ammonia toxicity and yield reduction [38,39], though this was not observed. Phosphorus tests showed an orthophosphate concentration as low as 22 ppm in the pure digested wastewater (Table 3), a level similar to our previous hydroponic experiments [32] and lower than the raw wastewater, which can be explained by the 50% dilution of the digester leachate before use.

Table 3. Solution test results (pH, electrical conductivity (EC), ammonium-N, nitrate/nitrite N, total phosphorus) for each fertility treatment. Nutrient levels of digested and raw wastewater were obtained from laboratory testing while nutrient levels of inorganic fertilizer obtained from the fertilizer formulation (see Methods).

Fertility Treatment	pH	EC	Ammonium-N	Nitrate/Nitrite-N	Total Phosphorus
Digested wastewater	8.40	1.87	171.50	<0.1	22.00
Raw wastewater	4.40	0.84	7.23	<0.1	79.88
Inorganic fertilizer	6.10	1.72	0	150	52
No fertilizer [1]	7.90	0.24	-	-	-

[1] Tests for ammonium, nitrate/nitrite or phosphorus were not applicable for the unfertilized control.

3.5. Leachate Monitoring

Water infiltration time increased over time in raw wastewater treatments, while digested wastewater showed similar monitoring trends compared to the inorganic fertilizer treatments (data not shown). Nitrate, EC, and pH data of these leachate samples are reported in Table 4 averaged across weeks and crops, as fertility treatment effects were greater than crop effects. Overall, the highest nitrate levels were observed in the leachate from the beginning of the experiment (data not shown), and on average, N levels were highest in leachates from the digested wastewater and inorganic fertilizer treatments (Table 4). At the beginning of the experiment, the leachate was acidic (pH = 5.5–6), but turned more neutral through the experiment (data not shown); raw wastewater treatment leachate was the most basic (Table 4). Electrical conductivity also declined over time, with highest levels measured in the digested wastewater and inorganic fertilizer treatment leachate.

Table 4. Leachate measurements (Nitrate-N, EC, and pH) for each fertility treatment averaged across 6 weeks and crops (mustard greens, lettuce, and basil) grown with different fertility treatments in a greenhouse in St. Paul, MN in 2018.Values within the same column with different letters are significantly different (Tukey's HSD, $\alpha = 0.05$).

Fertility Treatment	Leachate Characteristics		
	Nitrate-N (ppm)	EC (DS/cm)	pH
No fertilizer	340 b	0.51 b	6.52 b
Inorganic fertilizer	604 a	0.90 a	6.36 bc
Raw wastewater	440 ab	0.65 ab	6.85 a
Digested wastewater	653 a	0.97 a	6.22 c
ANOVA			
	$F(3,248) = 5.343$ $p < 0.01$	$F(3,248) = 5.283$ $p < 0.01$	$F(3,248) = 35.93$ $p < 0.001$

4. Discussion

Fertilizing with digested wastewater produced similar yields for lettuce and mustard greens compared to plants grown using commercial inorganic fertilizer. For basil, yields were lower when using the digested wastewater, but still significantly higher than the treatments that received only raw wastewater or well water. Therefore we suggest that digested brewery wastewater has the potential to provide plants with the required nutrients to obtain high yields, at least with certain crops. This is consistent with a similar study that grew tomato, *Solanum lycopersicum*, using digested brewery wastewater [12] and other research that successfully used nutrient solutions partially or totally made with wastewater or organic waste for soilless production [10,39–42].

Lettuce and basil plants that were only harvested once, compared to the multiple harvests of mustard leaves, exhibited the highest chlorophyll content in the inorganic fertilizer and digested wastewater treatments. Moreover, the higher chlorophyll contents at the end of the experiment were strongly positively correlated with yield. This reinforces the conclusion that digested brewery wastewater was able to provide enough nutrients, particularly N, for the plants to produce chlorophyll at the same level as the inorganic nitrogen of the synthetic fertilizer [23,28]. In contrast, SPAD values were not different among treatments for mustard greens. This could be due to the effect leaf removal had on plant health, since stress is a factor known to reduce SPAD measurements [33]. SPAD values for inorganic and digested wastewater were similar to those reported by other authors on basil grown in hydroponic and aquaponic systems [28]. Lettuce in inorganic and digested wastewater treatments obtained much higher SPAD values than those reported for plants hydroponically grown with reclaimed organic wastes [43], and similar to lettuce plants hydroponically grown with 150–200 ppm of total N [23]. The additional ammonium-N we added as a buffer and additional nutrient source is likely the reason for this finding.

Notably, most of the nitrogen present in the digested brewery wastewater came from the addition of ammonia in the digester. This addition may seem to contradict our research objective of growing horticultural crops with the digested wastewater, but due to the low nitrogen concentration found in the wastewater, increasing nitrogen concentration was needed to successfully grow plants, similar to what other studies found [39–42]. Ammonia can increase the risk of ammonium toxicity and nitrogen deficiency [44,45] in some crops, while in other species, like lettuce, it may be the preferred nitrogen source [17,38]. This highlights the need for further research on crop suitability for digested wastewater production in order to better understand which crops may produce acceptable yields, as well as the factors driving those differences and how to enhance nitrification for those crops where nitrate is preferentially taken up. Many factors affect nitrification rates including: pH of the substrate, substrate material [46], temperature [47], ammonia to nitrate ratio, carbon to nitrogen ratio [16], original bacteria population levels, and inoculation [19,37]. Additionally, future studies should test how to optimize this ammonia addition to minimize nitrogen concentration in the final effluent. A possible way to increase this nutrient removal could be through the recirculating of the nutrient solution in the system [48,49].

This study did not address the nutrient composition of the mature plants, nor of the wastewater nutrient solutions, besides nitrogen and phosphorus. No evident nutrient deficiencies were observed in any of the plants, though the plants used reached harvestable maturity relatively quickly and a different response might occur when growing longer maturing or fruiting crops [50]. Micronutrient deficiencies could also occur due to the alkaline pH of the digested wastewater [51]. A nutrient solution could also affect nutrient profile and taste of the produce [23], as the lack of certain elements like silicon or sodium do not lead to deficiencies but can effect quality and yield [25,52,53]. The high levels of sodium detected, around 130 ppm, did not exhibit salinity effects on the plants, but this should be considered in future analyses due to the issues with sodium accumulation in both the nutrient solutions and substrates [54,55]. Further understanding how the brewing process affects those nutrient fluctuations may help anticipate them.

5. Conclusions

This research shows that it is possible to use anaerobically digested brewery wastewater to grow different crops and obtain commercially acceptable yields. Anaerobic digestion is not only a possible way to produce energy [13,56], but also to adjust nutrient profiles and create a better nutrient solution for soilless production. Thus, it is possible to integrate brewery wastewater treatment with soilless urban agriculture. This integration, and energy generation through anaerobic digestion, could help reduce the high environmental footprint that many soilless urban farms have [57] and be a good way to increase food system sustainability by promoting nutrient reuse and reducing waste treatment energy requirements [5,10]. This integration of sectors opens the door to further synergies; for example, if the soilless production happened in protected environment, CO_2 from the brewing process could be used to enhance plant growth [58], the high temperature of the wastewater [14] could heat the growing space, or spent grains could be a large component of the substrate mix instead of peat [59]. Still, in order to promote the implementation of those solutions, there is a need to develop the knowledge and technologies. Important practical considerations will also need to be made, such as synchronizing digestate production and availability with crop needs, as well as scalability of this model.

The best method to integrate both soilless urban agriculture and brewery wastewater treatment still remains unclear. In this article, we have researched how to solve some of the agronomic and technological challenges of this integration, but additional complementary research is needed pertaining to the economic, legal, and social challenges of this decentralized urban system. Urban agriculture can provide food close to home, improve water use efficiency, and utilize locally available sources of nutrients, if other support systems exist. The economic viability of this integration would likely depend, in great measure, on successfully creating a marketing strategy that demonstrates value in sustainably managing urban waste streams and producing food locally.

Author Contributions: I.R.-V., N.O.A. and M.R. equally contributed to this work; C.F.H. provided review and edits.

Acknowledgments: We would like to acknowledge the researchers from the department of civil engineering (Kuang Zhu, Paige Novak and William Arnold) for all their help and knowledge as well as all the members of the Rogers and Anderson labs for their support.

References

1. DeLind, L.B. Where have all the houses (among other things) gone? Some critical reflections on urban agriculture. *Renew. Agr. Food Syst.* **2014**, *30*, 3–7. [CrossRef]
2. Mougeot, L.J.A. *Agropolis: The Social, Political and Environmental Dimensions of Urban Agriculture*; Mougeot, L.J.A., Ed.; Earthscan: London, UK, 2005; ISBN 1844072320.
3. Specht, K.; Siebert, R.; Hartmann, I.; Freisinger, U.B.; Sawicka, M.; Werner, A.; Thomaier, S.; Henckel, D.; Walk, H.; Dierich, A. Urban agriculture of the future: An overview of sustainability aspects of food production in and on buildings. *Agric. Hum. Values* **2014**, *31*, 33–51. [CrossRef]
4. Goldstein, B.; Hauschild, M.; Fernández, J.; Birkved, M. Urban versus conventional agriculture, taxonomy of resource profiles: A review. *Agron. Sustain. Dev.* **2016**, *36*, 1–19. [CrossRef]
5. Smit, J.; Nasr, J. Urban agriculture for sustainable cities: Using wastes and idle land and water bodies as resources. *Environ. Urban.* **1992**, *4*, 141–152. [CrossRef]
6. Metson, G.S.; Bennett, E.M. Phosphorus cycling in Montreal's food and urban agriculture systems. *PLoS ONE* **2015**, *10*, 1–18. [CrossRef] [PubMed]
7. Cohen, N. and Reynolds, K. Resource needs for a socially just and sustainable urban agriculture system: Lessons from New York City. *Renew. Agr. Food Syst.* **2014**, *30*, 103–114. [CrossRef]
8. Kretschmer, N.; Ribbe, L.; Gaese, H. Wastewater Reuse for Agriculture. *Technol. Resour. Dev.* **2000**, *2*, 37–64.

9. McClintock, N. Why farm the city? Theorizing urban agriculture through a lens of metabolic rift. *Camb. J. Reg. Econ. Soc.* **2010**, *3*, 191–207. [CrossRef]
10. Mohareb, E.; Heller, M.; Novak, P.; Goldstein, B.; Fonoll, X.; Raskin, L. Considerations for reducing food system energy demand while scaling up urban agriculture. *Environ. Res. Lett.* **2017**, *12*. [CrossRef]
11. Modic, W.; Kruger, P.; Mercer, J.; Webster, T.; Swersey, C.; Skypeck, C. *Brewers Association Wastewater Management Guidance Manual*; Brewers Association: Boulder, CO, USA, 2015; pp. 1–31.
12. Power, S.D.; Jones, C.L.W. Anaerobically digested brewery effluent as a medium for hydroponic crop production—The influence of algal ponds and pH. *J. Clean. Prod.* **2016**, *139*, 167–174. [CrossRef]
13. Weber, B.; Stadlbauer, E.A. Sustainable paths for managing solid and liquid waste from distilleries and breweries. *J. Clean. Prod.* **2017**, *149*, 38–48. [CrossRef]
14. Enitan, A.; Adeyemo, J.; Kumari, S. Characterization of Brewery Wastewater composition. *World Acad.* **2015**, *9*, 1043–1046.
15. Lind, O. *Organic Hydroponics—Efficient Hydroponic Production from Organic Waste Streams: Introductory Research Essay*; 2016; Available online: http://www.3rmovement.com/wp-content/uploads/2017/04/Hydroponic-produciton-organic-Olle-Lind.pdf (accessed on 30 January 2019).
16. Norström, A. Treatment of Domestic Wastewater Using Microbiological Processes and Hydroponics in Sweden. PhD Thesis, KTH Royal Institute of Technology, Stockholm, Sweden, 2005; 62p.
17. Shinohara, M.; Aoyama, C.; Fujiwara, K.; Watanabe, A.; Ohmori, H.; Uehara, Y.; Takano, M. Microbial mineralization of organic nitrogen into nitrate to allow the use of organic fertilizer in hydroponics. *Soil Sci. Plant Nutr.* **2011**, *57*, 190–203. [CrossRef]
18. Koohakan, P.; Ikeda, H.; Jeanaksorn, T.; Tojo, M.; Kusakari, S.I.; Okada, K.; Sato, S. Evaluation of the indigenous microorganisms in soilless culture: Occurrence and quantitative characteristics in the different growing systems. *Sci. Hortic. (Amsterdam)* **2004**, *101*, 179–188. [CrossRef]
19. Lang, H.J.; Elliott, G.C. Enumeration and inoculation of nitrifying bacteria in soilless potting media. *J. Amer. Soc. Hort. Sci.* **1997**, *122*, 709–714. [CrossRef]
20. Montagu, K.D.; Goh, K.M. Effects of forms and rates of organic and inorganic nitrogen fertilisers on the yield and some quality indices of tomatoes (*Lycopersicon esculentum* Miller). *N. Z. J. Crop Hortic. Sci.* **1990**, *18*, 31–37. [CrossRef]
21. Kansas State University Extension. *Mustard Greens*; Kansas State University: Manhattan, KS, USA, 2016.
22. Parkell, N.B.; Hochmuth, R.C.; Laughlin, W.L. *Leafy Greens in Hydroponics and Protected Culture for Florida*; University of Florida, IFAS Extension: Gainesville, FL, USA, 2016; pp. 1–7.
23. Mahlangu, R.I.S.; Maboko, M.M.; Sivakumar, D.; Soundy, P.; Jifon, J. Lettuce (*Lactuca sativa* L.) growth, yield and quality response to nitrogen fertilization in a non-circulating hydroponic system. *J. Plant Nutr.* **2016**, *39*, 1766–1775. [CrossRef]
24. Brechner, M.; Both, A.J. *Hydroponic Lettuce Handbook*; Cornell University CEA Program: Ithaca, NY, USA, 1996; p. 48.
25. Scuderi, D.; Restuccia, C.; Chisari, M.; Barbagallo, R.N.; Caggia, C.; Giuffrida, F. Salinity of nutrient solution influences the shelf-life of fresh-cut lettuce grown in floating system. *Postharvest Biol. Technol.* **2011**, *59*, 132–137. [CrossRef]
26. Samarakoon, U.C.; Weerasinghe, P.A.; Weerakkody, W.A.P. Effect of electrical conductivity [EC] of the nutrient solution on nutrient uptake, growth and yield of leaf lettuce (*Lactuca sativa* L.) in stationary culture. *Trop. Agric. Res.* **2006**, *18*, 13–21.
27. Treadwell, D.D.; Hochmuth, G.J.; Hochmuth, R.C.; Simonne, E.H.; Davis, L.L.; Laughlin, W.L.; Li, Y.; Olczyk, T.; Sprenkel, R.K.; Osborne, L.S. Nutrient management in organic greenhouse herb production: Where are we now? *Horttechnology* **2007**, *17*, 461–466. [CrossRef]
28. Saha, S.; Monroe, A.; Day, M.R. Growth, yield, plant quality and nutrition of basil (*Ocimum basilicum* L.) under soilless agricultural systems. *Ann. Agric. Sci.* **2016**, *61*, 181–186. [CrossRef]
29. RFA. *Methodology Ammonia Nitrogen A303–S171*; Astoria-Pacific Int.: Clackamas, OR, USA, 1986.
30. RFA. *Methodology Nitrate/Nitrite A303–S170*; Astoria-Pacific Int.: Clackamas, OR, USA, 1985.
31. RFA. *Methodology Total Phosphorus A303–S050*; Astoria-Pacific Int.: Clackamas, OR, USA, 1986.
32. Riera-Vila, I.; Anderson, N.O.; Rogers, M. Anaerobically digested brewery wastewater as a nutrient solution for non-circulating hydroponics. Unpublished work. 2018.

33. Ling, Q.; Huang, W.; Jarvis, P. Use of a SPAD-502 m to measure leaf chlorophyll concentration in *Arabidopsis thaliana*. *Photosynth. Res.* **2011**, *107*, 209–214. [CrossRef] [PubMed]
34. Spectrum technologies inc SPAD 502 plus chlorophyll meter product manual. *SPAD Man.* **2009**, *1–23*, 32.
35. Liu, W.K.; Yang, Q.C.; du Lian, F.; Cheng, R.F.; Zhou, W.L. Nutrient supplementation increased growth and nitrate concentration of lettuce cultivated hydroponically with biogas slurry. *Acta Agric. Scand. Sect. B Soil Plant. Sci.* **2011**, *61*, 391–394. [CrossRef]
36. Resh, H.M. *Hydroponic Food Production*, 7th ed.; CRC Press: Boca Raton, FL, USA, 2013.
37. Trejo-tellez, L.; Gomez-Merino, F. Nutrient Solutions for Hydroponic Systems- A standard methodology for plant biological researchers. *Toshiki Asao IntechOpen* **2012**. [CrossRef]
38. Savvas, D.; Passam, H.C.; Olympios, C.; Nasi, E.; Moustaka, E.; Mantzos, N.; Barouchas, P. Effects of ammonium nitrogen on lettuce grown on pumice in a closed hydroponic system. *HortScience* **2006**, *41*, 1667–1673. [CrossRef]
39. Guo, S.; Brück, H.; Sattelmacher, B. Effects of supplied nitrogen form on growth and water uptake of French bean (*Phaseolus vulgaris* L.) plants: Nitrogen form and water uptake. *Plant. Soil.* **2002**, *239*, 267–275. [CrossRef]
40. Dos Santos, J.D.; Lopes da Silva, A.L.; da Luz Costa, J.; Scheidt, G.N.; Novak, A.C.; Sydney, E.B.; Soccol, C.R. Development of a vinasse nutritive solution for hydroponics. *J. Environ. Manag.* **2013**, *114*, 8–12. [CrossRef]
41. Kawamura-Aoyama, C.; Fujiwara, K.; Shinohara, M.; Takano, M. Study on the hydroponic culture of lettuce with microbially degraded solid food waste as a nitrate source. *Jpn. Agric. Res. Q.* **2014**, *48*, 71–76. [CrossRef]
42. Eregno, F.E.; Moges, M.E.; Heistad, A. Treated greywater reuse for hydroponic lettuce production in a green wall system: Quantitative health risk assessment. *Water* **2017**, *9*, 454. [CrossRef]
43. da Silva Cuba Carvalho, R.; Bastos, R.G.; Souza, C.F. Influence of the use of wastewater on nutrient absorption and production of lettuce grown in a hydroponic system. *Agric. Water Manag.* **2018**, *203*, 311–321. [CrossRef]
44. Silber, A.; Bar-Tal, A. *Nutrition of Substrate-Grown Plants*, 1st ed.; Elsevier Ltd.: Amsterdam, The Netherlands, 2008; ISBN 9780444529756.
45. Ikeda, H.; Tan, X. Urea as an organic nitrogen source for hydroponically grown tomatoes in comparison with inorganic nitrogen sources. *Soil Sci. Plant. Nutr.* **1998**, *44*, 609–615. [CrossRef]
46. Hashida, S.-N.; Johkan, M.; Kitazaki, K.; Shoji, K.; Goto, F.; Yoshihara, T. Management of nitrogen fertilizer application, rather than functional gene abundance, governs nitrous oxide fluxes in hydroponics with rockwool. *Plant. Soil.* **2014**, *374*, 715–725. [CrossRef]
47. Russell, C.A.; Fillery, I.R.P.; Bootsma, N.; McInnes, K.J. Effect of temperature and nitrogen source on nitrification in a sandy soil. *Commun. Soil Sci. Plant. Anal.* **2002**, *33*, 1975–1989. [CrossRef]
48. Bar-Yosef, B. *Fertigation Management and Crops Response to Solution Recycling in Semi-Closed Greenhouses*, 1st ed.; Elsevier Ltd.: Amsterdam, The Netherlands, 2008; ISBN 9780444529756.
49. Bugbee, B. Nutrient management in recirculating hydroponic culture. *Acta Hortic.* **2004**, *648*, 99–112. [CrossRef]
50. Wortman, S.E.; Douglass, M.S.; Kindhart, J.D. Cultivar, growing media, and nutrient source influence strawberry yield in a vertical, hydroponic, high tunnel system. *Horttechnology* **2016**, *26*, 466–473.
51. Tyson, R.V.; Simonne, E.H.; Treadwell, D.D.; Davis, M.; White, J.M. Effect of water pH on yield and nutritional status of greenhouse cucumber grown in recirculating hydroponics. *J. Plant. Nutr.* **2008**, *31*, 2018–2030. [CrossRef]
52. Manzocco, L.; Foschia, M.; Tomasi, N.; Maifreni, M.; Dalla Costa, L.; Marino, M.; Cortella, G.; Cesco, S. Influence of hydroponic and soil cultivation on quality and shelf life of ready-to-eat lamb's lettuce (*Valerianella locusta* L. Laterr). *J. Sci. Food Agric.* **2011**, *91*, 1373–1380. [CrossRef]
53. Petersen, K.K.; Willumsen, J.; Kaack, K. Composition and taste of tomatoes as affected by increased salinity and different salinity sources. *J. Hortic. Sci. Biotechnol.* **1998**, *73*, 205–215. [CrossRef]
54. Riera-Vila, I.; Anderson, N.O.; Rogers, M. Wastewater test of 4 local breweries. Unpublished work.
55. Raviv, M.; Lieth, J.H.; Bar-Tal, A.; Silber, A. *Growing Plants in Soilless Culture: Operational Conclusions*, 1st ed.; Elsevier Ltd.: Amsterdam, The Netherlands, 2008; ISBN 9780444529756.
56. Simate, G.S.; Cluett, J.; Iyuke, S.E.; Musapatika, E.T.; Ndlovu, S.; Walubita, L.F.; Alvarez, A.E. The treatment of brewery wastewater for reuse: State of the art. *Desalination* **2011**, *273*, 235–247. [CrossRef]
57. Goldstein, B.; Hauschild, M.; Fernández, J.; Birkved, M. Testing the environmental performance of urban agriculture as a food supply in northern climates. *J. Clean. Prod.* **2016**, *135*, 984–994. [CrossRef]

58. Tremblay, N.; Gosselin, A. Effect of carbon dioxide enrichment and light. *Horttechnology* **1998**, *8*, 524–528. [CrossRef]
59. NiChualain, D.; Prasad, M. Evaluation of three methods for determination of stability of composted material destined for use as a component of growing media. *Acta Hortic.* **2009**, *819*, 303–310. [CrossRef]

2

Response of Mediterranean Ornamental Plants to Drought Stress

Stefania Toscano [1], Antonio Ferrante [2] and Daniela Romano [1,*]

[1] Department of Agriculture, Food and Environment (Di3A), Università degli Studi di Catania, Via Valdisavoia 5, 95123 Catania, Italy; stefania.toscano@unict.it

[2] Department of Agricultural and Environmental Sciences, Università degli Studi di Milano, Via Celoria 2, 1-20133 Milano, Italy; antonio.ferrante@unimi.it

* Correspondence: dromano@unict.it.

Abstract: Ornamental plants use unique adaptive mechanisms to overcome the negative effects of drought stress. A large number of species grown in the Mediterranean area offer the opportunity to select some for ornamental purposes with the ability to adapt to drought conditions. The plants tolerant to drought stress show different adaptation mechanisms to overcome drought stress, including morphological, physiological, and biochemical modifications. These responses include increasing root/shoot ratio, growth reduction, leaf anatomy change, and reduction of leaf size and total leaf area to limit water loss and guarantee photosynthesis. In this review, the effect of drought stress on photosynthesis and chlorophyll *a* fluorescence is discussed. Recent information on the mechanisms of signal transduction and the development of drought tolerance in ornamental plants is provided. Finally, drought-induced oxidative stress is analyzed and discussed. The purpose of this review is to deepen our knowledge of how drought may modify the morphological and physiological characteristics of plants and reduce their aesthetic value—that is, the key parameter of assessment of ornamental plants.

Keywords: growth; gas exchange; chlorophyll fluorescence; oxidative stress; signal transduction; plant choice; green areas

---

1. Introduction

Drought stress strongly limits the growth of plants in Mediterranean regions. In the world, there are five Mediterranean-climate regions (i.e., areas surrounding the Mediterranean Sea, parts of western North America, parts of western and southern Australia, southwestern South Africa, and parts of central Chile) located between 32°–40° N and S of the Equator [1]. The Mediterranean climate is defined by precipitation and temperature, and it is characterized by a high seasonality summarized as hot and dry summers and cool and wet winters [2]. Despite the fact that these territories occupy less than 5% of the earth's surface, they harbor almost 20% of the world's vascular plant species [3]. The primary aspect that influences plant characteristics and natural vegetation is the extensive dry season. For this reason, plant growth and survival are endangered by long periods lacking rainfall and higher temperatures in the summer that impose more or less intense stress conditions [4]. The global climate changes that are occurring currently will worsen the availability of water, especially in arid and semi-arid environments. The availability of fresh and good quality water will decrease, especially in large cities [2,5]. This will entail difficulties in keeping green areas because the competition for water will be a critical issue.

For these reasons, great attention has recently been placed on the use and management of water to improve the sustainability of ornamental plant maintenance in semi-arid environments, such as the Mediterranean basin. Water scarcity led to the diffusion of techniques for creating green spaces that are able to save water (xeriscaping), favoring the use of species tolerant of water stress, which are native species like the carob tree, a species that is highly tolerant to high temperature and to low soil water efficiency [6]. This attention to water saving depends on the fact that even if the water in the urban environment is widely used for purposes other than irrigation (for example industrial and residential uses), "a landscape may serve as a visual indicator of water use to the public due to its visual exposure" [7]. The water saving can be maximized by utilizing different strategies such as making a suitable choice of ornamental plant species—one that has a high tolerance to drought stress without compromising the ornamental value and/or reducing the effects of drought stress through innovative cultivation methods.

Ornamental plants are not only species and/or cultivars that offer aesthetic pleasure, but they can also improve the environment and the quality of our lives [8]. Thus, ornamental plants can be used to restore disturbed landscapes, control erosion, reduce energy for climatization and water consumption, and improve the aesthetic quality of urban, peri-urban, and rural landscapes, as well as recreational areas, interiors, and commercial sites. In consideration of the many contexts in which plant species can be used for ornamental purposes, the number is very large. The wide number of the ornamental or potentially ornamental species increases the possibility of finding suitable genotypes that are able to cope with drought stress and that can be used for landscaping planning.

For landscaping, plant choice can be based on a very large number of species from a wide geographical range and with different functions [8]. Unlike in agriculture, performance of an amenity landscape is not measured with a quantifiable yield, but rather how well it meets the expectations of the user or the individual paying for installation and maintenance. These expectations include aesthetic appearance and/or utility such as shading, ground cover, and recreation [9]. Sometimes, in degraded environments, plant survival is the only purpose of cultivation. Furthermore, for ornamental plants used in landscaping, fast growth is not always desirable because the excessive shoot vigor often requires frequent pruning with higher management costs. To maintain a compact growth habit, ornamental plants may have to be pruned or treated with plant growth regulators [10].

A reduced quantity of water may have positive benefits on growth control, therefore moderate drought stress can be a useful tool to provide plants with compact habit and slower growth—both parameters required for easier landscape management [11]. Plant drought stress is difficult to study because the sensitivities and response times to water deficit vary among different plant species and are related to the intensity and length of the water stress. Plant response to drought stress involves the interaction of various physiological and biochemical parameters that can be exploited as markers for the identification of tolerant species [12].

2. Ornamental Plant Response to Drought Stress

2.1. Growth and Morpho-Anatomical Modification

Plant responses to drought are different and interconnected. Plant plasticity to drought stress adaptation varies within genera, species [13,14], and even cultivars. The main morphological changes under drought conditions are shoot and leaf growth reduction. These negatively affect the ornamental value and the visual appearance, which are particularly important key factors (from the ornamental point of view) that must be used along with markers for selecting tolerant genotypes [15]. Several experimental studies on ornamental plants showed that plant quality decreased in response to severe drought stress [16–18].

The effect of drought stress on plant growth and dry matter has been noticed in numerous ornamental species—for example, *Pistacia* [12], *Spiraea, Pittosporum* [19], *Bougainvillea* [20], *Callistemon* [21], *Laurus,* and *Thunbergia* [22] (Table 1). Since the photosynthetic pathways strongly influence the response to water stress, only the responses of C3 plants are presented in Table 1.

The reduction of leaf area is another typical response observed in plants subjected to water stress, as confirmed by several authors. Indeed, as reported by Toscano et al. [23], the total leaf area and the leaf number showed the widest variations in *Lantana* between control and severe deficit irrigation, while in *Ligustrum,* the differences were more marked for the total leaf area and not significant for the leaf number. The reduction of the leaf area is a consequence of a reduction in the leaf number [24] or the leaf size (unit leaf area) [22]. Thus, plants counteract the water limitation by reducing the transpiration area. One of the avoidance mechanisms that minimizes water loss when the stomata are closed is, in fact, the reduction of the canopy area. In callistemon plants, drought stress increases the root-to-shoot ratio, causing the reduction of aerial tissues rather than the roots [25–27]. This reduction also occurs when the plants are grown in pots, a frequent condition in the nursery phase.

Table 1. Major effects of drought stress on ornamental plants [1].

Species	Plant Habit	Treatments	Growth stage	Modified Parameter by Drought Stress	Ref.
Rudbeckia hirta, Callistephus chinensis, Althaea rosea, Malva sylvestris	forbs	4 levels of irrigation treatments: 25%, 50%, 75% and 100% of the reference evapotranspiration (ET0)	Seedling one month after transplanting	Plant fresh weight (−); SLA (−); Stomatal Conductance (−); Δ Canopy Temperature (+); water use efficiency index (WUEi) (+); water use efficiency biomass (WUEb) (+)	[28]
Periploca angustifolia	bushy-branched shrub	Full irrigation (FI), Water Deficit (WD), and Rehydrated (R)	11-month-old seedlings	Relative water content (RWC) (−); osmotic potential ($\psi\pi$) (−); water potential (ψw) (−); transpiration rate (−); net CO_2 assimilation rate (ACO_2) (−); stomatal conductance (g_s) (−); water use efficiency (WUE) (+); Proline (+); MDA (+); chlorophyll (a, b, total and a/b) and carotenoid content (−);	[29]
Pistacia lentiscus	bushy shrub	C = 100% water holding capacity; Moderate Water irrigation (MW, 60% of the control) and Severe Water deficit (SW, 40% of the C)	1-year-old seedlings	Dry weight (−), plant height (−), pre-dawn leaf water potential (Ψl) (−); RWC (−) in SW	[12]
Lantana camara, Ligustrum lucidum	bushy shrubs	C = container capacity, or irrigated at 100% of water container capacity (WCC); light deficit irrigation (LDI), irrigated at 75% of WCC; moderate deficit irrigation (MDI), irrigated at 50% of WCC; and severe deficit irrigation (SDI), irrigated at 25% of WCC.	Two-month-old rooted cuttings	Dry weight (−); leaf number (−); total leaf area (−); leaf thickness (−); photosynthesis (−); stomatal conductance (−); variable to maximal fluorescence (Fv/Fm) (−); water potential (−).	[23]
Bougainvillea buttiana 'Rosenka' and *B.* 'Lindleyana'	shrubby vines	C = substrate moisture close to container capacity and irrigation applied when 20% of the water was leached; deficit irrigation (DI), 25% of the amount of water supplied in C.	Two-year-old plants	Leaf, flower, total biomass dry weight, total leaf area (−); stomatal resistance (+); Ψl and Ψp (+); Stomatal length and width (−)	[20]
Spirea nipponica (S), *Pittosporum eugenioides* (P), *Viburnum nudum* (V)	bushy shrubs	4 irrigation levels (100, 70, 50, and 25% of container capacity) and Trinexapac-ethyl (TE) treatments (0.1, 0.2, and 0.3 L ha^{-1})	Plant heights 10 (S and V) and 40 cm (P)	Leaf number and area (−), plant dry weight and height (−), root dry weight (+). A, E, and gs (−). The application of 0.2 and 0.3 L ha^{-1} TE enhanced S, P and V tolerance to drought stress	[19]
Acacia tortilis subsp. *raddiana*	medium-sized tree	C = 80% of field capacity; Stress = withholding irrigation for 25 d.	6-week-old seedlings	Leaf number (−), dry mass (−), shoot length and total leaf area (−), water potential (−), stomatal conductance (−); transpiration rates (−); chlorophyll fluorescence (−) only when soil WC was < 40%, soluble sugars (+).	[30]
Viburnum opulus and *Photinia X fraseri* 'Red Robin'	bushy shrubs	C = 100% ET; Moderate Water Deficit plants (MWD) received 60% ET and Severe Water Deficit (SWD) received 30% ET.	Plants grown in pots (24 cm in diameter)	Water potentials (−); Pn and g_s (−) in SWD in *P. x fraseri*; g_s and leaf transpiration (Tr) (−) in *V. opulus*	[31]

Table 1. *Cont.*

Species	Plant Habit	Treatments	Growth stage	Modified Parameter by Drought Stress	Ref.
Callistemon laevis	bushy shrub	Control (0.8 dS m^{-1}, 100% water holding capacity), WD (0.8 dS m^{-1}, 50% of the amount of water supplied in control), saline (4.0 dS m^{-1}, same amount of water supplied as control) and saline water deficit (4.0 dS m^{-1}, 50% of the water supplied in the control).	2-year-old rooted cuttings	Total biomass (−); plant height (−); osmotic adjustment (−), leaf tissue elasticity (−)	[21]
Viburnum opulus L. and *Photinia x fraseri* 'Red Robin'	bushy shrubs	Control with 600 $mLday^{-1}$ (C), moderate WD (MWD) 66% of C and severe water deficit (SWD) received 33% of C.	One-year-old plants	Stem diameter (−). Modulus of elasticity (−) only in *Photinia*	[32]
B. glabra 'Sanderiana', *B. xbuttiana* 'Rosenka', *B.* 'Lindleyana'	shrubby vine	Three irrigation levels based on the daily water use 100% (C), 50% (MDI) or 25% (SDI)	Rooted cuttings	SDW (−), total DW (−), leaf number (−), leaf area (−), macronutrient concentration (−) in SDI. Stomatal resistance (+), leaf water potential (−), leaf osmotic potential. (−)	[33]
Nerium oleander	bushy shrub	C (field capacity); WD (withholding irrigation)	One-year-old plants	Stem elongation (−); Leaf FW (−); Leaf WC (−); Chl (a, b and total) (−); Proline (+); Glycine betaine (+); Total soluble sugar (+); Total phenolic compounds (+); Total flavonoids (+); ascorbate peroxidase (+); glutathione reductase (+).	[34]
Callistemon citrinus 'Firebrand'	bushy shrub	C (substrate moisture close to container capacity); moderate deficit irrigation (MDI) by applying 50% of the amount of C and severe deficit irrigation (SDI) by applying 25% of the C irrigation	2-year-old rooted cuttings	RGR (−) in MDI; R/S ratio (+); WUE (+); g_s in MDI and SDI (−); Pn/g_s ratios (+); Stem water potential (−); Pn (−) in SDI	[24]
Pelargonium x hortorum	forb	C (100% of water field capacity = WFC); sustainable deficit irrigation (SDI), irrigated at 75% WFC throughout the experiment; regulated deficit irrigation I (RDI I), irrigated at 75% throughout the experiment, except during the flowering phase when plants were irrigated at 100%; regulated deficit irrigation II (RDI II), irrigated at 100% throughout the experiment, except during the flowering phase when plants were irrigated at 75%.	Rooted cuttings (4- to 5-cm tall and with 6–7 leaves)	Height (−), Flowering (−) RDI II; SDW (−), Number of leaves (−); Total leaf area (−).	[35]
Eugenia uniflora 'Etna Fire', *P. x fraseri* 'Red Robin'	bushy shrubs	Well-watered (WW), moderate drought stress (MD, 75%), severe drought stress (35%, SD).	Three months old rooted cuttings	A, g_s and E (−); RWC (−); Fv/Fm (−); Proline and MDA (+) in Eugenia; MDA (+) in SD.	[36]

Table 1. *Cont.*

Species	Plant Habit	Treatments	Growth stage	Modified Parameter by Drought Stress	Ref.
Myrtus communis	bushy shrub	Control (C), 100% water holding capacity [leaching 15% (v/v) of the applied water;]; moderate water deficit; MWD, 60% of the C; severe water deficit; SWD,40% of the C.	Seedlings of 2-year-old	SDW (−); root dry weights (−), leaf numbers (−), Total leaf area (−), plant height (−) in SWD; plant height (−) in MDW. Root hydraulic resistance (+); leaf water potential pre-dawn (−); Pn (−).	[37]
Pelargonium x hortorum	forb	Control, C, container capacity; Moderate deficit irrigation, MDI, 60% of the C; Severe deficit irrigation, SDI 40% of C. After 2 months, all the plants were exposed to a recovery period of 15 days with the same irrigation regime applied to control plants, until the end of the experiment.	Rooted cuttings	SDW (−); leaf area (−); R/S ratio (+); Height (−); Width (−); g_s (−); Pn (−).	[38]
Callistemon citrinus, Laurus nobilis, Pittosporum tobira, Thunbergia erecta	bushy shrubs	Two consecutive cycles of suspension/rewatering (S-R) compared with plants that were watered daily (C).	Six-month-old plants	SDW (−); R/S ratio (+); RWC (−); Leaf water potential (−), g_s (−); Pn (−).	[22]
Passiflora alata, P edulis, P. gibertii, P. setacea, P. cincinnata	climbing vines	Two soil water regimes: soil field capacity and interruption of irrigation until the stomatal closure and apparent wilting of the whole plant.	Six month after sowing	Height (−); Dry weight of leaves, branches, roots (−); g_s (−); palisade parenchyma thickness (+); leaf limb and spongy parenchyma thicknesses (+); stomatal diameter (+).	[39]

[1] C = control; ET = Evapotranspiration; WD = Water deficit; SDW = shoot dry weight; (−) reduction due to WD; (+) increase due to WD.

Increased root-to-shoot ratios are frequently observed in plants under drought conditions which reduces water consumption [40] and increases water absorption [41]. This parameter is also suggested as a screening factor for grading plants with different stress tolerances. In addition to the reduction of the leaf area during drought stress, the modification of the leaf size and the cuticle thickness are also observed.

In a study conducted by Toscano et al. [23] on two ornamental shrubs (*Lantana* and *Ligustrum*) in a Mediterranean area, the analysis of leaf anatomical traits allowed the identification of the different strategies used during water stress conditions. During severe deficit irrigation, *Lantana* plants increased the spongy tissue rather than the palisade tissue; this anatomical modification facilitated the diffusion of CO_2 toward the fixation sites in order to increase the concentration gradient between internal air space and the atmosphere, thus enhancing the competition among cells for CO_2 and light [42]. In both species, an increase in the thickness of the spongy tissue and the palisade tissue was observed. The reduction of the specific leaf area could be a way to improve water use efficiency (WUE). In fact, thicker leaves usually have higher concentrations of chlorophylls and proteins per unit leaf area and thus have greater photosynthetic capacities per unit leaf area than thinner leaves [43].

The leaf anatomical modifications are species-specific. Thus, in *Polygala* and *Viburnum* plants subjected to four levels of irrigation treatments through the use of dielectric sensors (EC 5TE, Decagon Devices, Pullman, Washington, USA) to maintain the substrate water content (WS) equal to 10% (WC10%), 20% (WC20%), 30% (WC30%), and 40% (WC40% = control) of the pot volume, the leaf anatomical modifications were linked to spongy tissue in *Polygala* and palisade tissue in *Viburnum* (Figure 1).

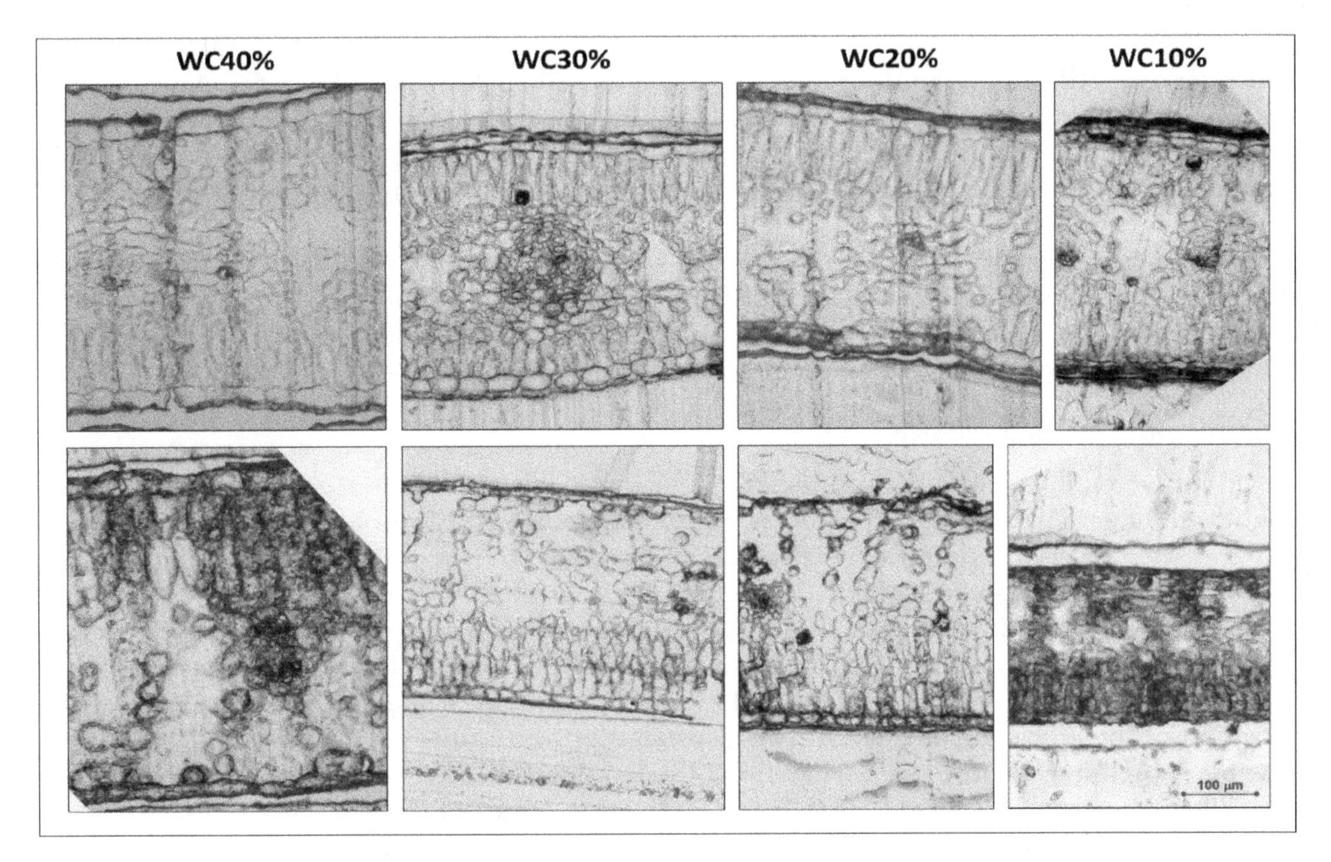

Figure 1. Light microscopy of blade cross-sections in *Polygala* (above) and *Viburnum* (below) at different water regimes (source: Toscano et al., unpublished data).

Acquiring greater knowledge of the morphological, physiological, and biochemical responses of the species in adverse environmental conditions, whether they are occasional, temporary, or long-term, allows us to choose the correct ornamental plants in relation to the interested area.

This information is useful in identifying the mechanisms of the adaptation of plants to adverse conditions such as drought stress [44], allowing us to select the most suitable species without compromising their aesthetic value.

2.2. *Physiological Parameters*

2.2.1. Leaf Gas Exchange

The main consequences of drought stress in plants are stomatal closure, reduction of gas exchange, the slowing down of photosynthetic activity, and the death of the plant [45,46]. Drought stress conditions mainly affect the photosynthetic system and ratio. In particular, they compromise the elements that are involved in the process, such as the electron transport to the thylakoids, the carbon cycle, and the stomatal control of CO_2 supply. Different published papers demonstrated that the reduction of photosynthetic activity is related to the mechanisms of stomatal conductance [47–50]. In fact, the first response of plants to water stress is stomatal closure and the subsequent reduction of the assimilation of the photosynthetic carbon necessary for the photosynthetic activity. As a consequence of the stomatal closure, there is not only a reduction in water loss, but also a reduction in nutrient uptake, consequently altering the metabolic pathways [51]. During drought stress, most species show a reduction in photosynthetic activity and a fast stomatal closure in relation to water potential adjustment [52,53]. The reduction in growth is also related to the reduction in the water potential of the leaves. Upon stomatal closure, a reduction in photosynthetic activity is achieved, which in turn leads to a decrease in plant growth and production [54,55]. The levels of carbon dioxide inside the stomatal chamber, and therefore in the cells, decrease, causing a reduction in photosynthesis. A decrease in the rate of CO_2 fixation is also observed and is associated with a reduction in the stomatal opening [56].

Under drought stress conditions, high conductivity ratios (A_N)/stomatal conductance (g_s) (also expressed as intrinsic WUE) indicate that leaves (the chloroplasts in particular)—even if there is an immediate stomatal closure—try to maintain high photosynthetic performance. As reported by Álvarez et al. [12], the decrease in g_s in *Pistacia lentiscus* subjected to drought stress limited water losses through transpiration control.

In order to estimate the tolerance to drought stress in plants, the transpiration ratio is essential. In fact, it has been observed that species that can retain a greater quantity of water and therefore lose less water through the stomata are more tolerant to drought. [57]. As reported by Galmes et al. [58], shrubs have a better ability to regulate transpiration compared to herbaceous plants.

2.2.2. Chlorophyll *a* Fluorescence

Under water stress conditions, one parameter that is commonly used to identify the presence of photosynthetic plant damage in plants is the measurement of chlorophyll *a* fluorescence. In fact, this parameter is very useful for analyzing the influence of environmental factors on the efficiency of the photosynthetic apparatus [59]. Down regulation of photosystem II (PSII) activity results in an imbalance between the generation and utilization of electrons, apparently resulting in changes in quantum yield [60]. The ratio variable to maximal fluorescence ratio (Fv/Fm) (i.e., the maximum primary photochemical efficiency of the PSII in a sample of leaves adapted in the dark) allows the evaluation of the efficiency of the PSII photosystem, indirectly measuring the physiological state of the plant [61]. Several authors have defined the Fv/Fm threshold values to indicate if a plant is more or less stressed. Values between 0.78-0.85 indicate that the plant is not stressed [62]. In a study conducted by Álvarez et al. [63] on *Callistemon* plants maintained at different levels of drought stress, the Fv/Fm values remained constant at 0.80. The drought stress was not compromised by the PSII. Therefore, the *Callistemon* is a species resistant to drought. Álvarez et al. [12] reported that in *Pistacia lentiscus* plants subjected to different levels of water stress (from May to October), low Fv/Fm values were found in stressed plants during the warmer months. At the end of the trial when the conditions were less stressful, the plants recovered from these values. This shows that the plants did not cause irreversible damage to the foliar tissues, indicating that PSII was not permanently damaged by stressful conditions. This affirms that the chloroplasts of Mediterranean species have different strategies during stress conditions for avoiding photo-inhibitory processes, such as the mechanism to consume the reducing power produced by the PSII [64,65].

2.3. Oxidative Stress

When photosynthetic activity is reduced and light excitation energy is in excess of that used or required by photosynthesis, over-excitation of the photosynthetic pigments in the antenna can occur, leading to the accumulation of reactive oxygen species (ROS) in chloroplasts [66]. During drought stress in plants, there are different biochemical changes. The main response is the accumulation of ROS, which causes the destruction of the cell membranes and results in oxidative damage to plants [67,68]. The plants, in order to oppose this accumulation, have developed many antioxidant activities and a series of secondary metabolites that counteract the generation of ROS and scavenge ROS once they are formed [69–71].

ROS are chemically active free radicals of oxygen. When unpaired electrons are present in the valence shell of these molecules, they become highly reactive and damage the cell structure and function. ROS production takes place within the compartments of different organelles, such as chloroplasts, mitochondria, and peroxisomes [60].

ROS include superoxide anion (O^{2-}), hydrogen peroxide (H_2O_2), hydroxyl radical (OH^-), singlet oxygen ($^1O_2$), and ozone (O_3). ROS are produced by plants continuously because they also have the role of cellular signaling, while excessive production involves oxidative stress [72].

Plants have mechanisms that protect them from the destructive action of oxidative reactions [73]. A mechanism put in place as a defense from stress relates to the production of antioxidant enzymes that protect the plants from ROS.

Garratt et al. [74] highlighted some enzymes among the main natural "detoxifiers" present in plants, such as superoxide dismutase (SOD; EC 1.15.1.1), catalase (CAT; EC 1.11.1.6), glutathione peroxidase (GPX; EC 1.11. 1.7), and ascorbate peroxidase (APX; EC 1.11.1.11). These enzymes are located in different compartments of the plant cells, while the CAT is instead located in the peroxisomes [75].

A type of ROS can be transformed into another type; for example, O_3 is decomposed into H_2O_2, O^{2-}, and $^1O^2$. The O^{2-} is also transformed spontaneously or enzymatically into H_2O_2 through SOD activity [76], which can react further with Fe^{2+} to form OH.

Controlling the production and action of ROS allows a better understanding of the effects of various abiotic stresses on plants. The study of protective mechanisms such as the antioxidant enzymes could allow the identification of processes that are the basis for the response of plants to stress.

When the plants are not stressed, the ROS level is kept low by the scavenger activity of the antioxidant enzymes. In the presence of abiotic or biotic stress (such as water, saline, or ozone stress), these balances are broken and there is an increase in the intracellular ROS levels. About 1%–3% of the oxygen that is consumed by plants leads to the formation of ROS [77,78]. The main changes that occur in plants are the increase in lipid peroxidation, protein degradation, DNA fragmentation, and finally cell death. All of this occurs because ROS are highly reactive [50]. Reacting with proteins and lipids, they modify structure, cellular metabolism, and, in particular, those that are linked to the photosynthetic process [79].

As a defense mechanism, the activity of these antioxidant enzymes increases under abiotic stress conditions such as drought [80–82], salinity [83,84], and ozone [85]. There are also non-enzymatic antioxidants: tocopherol, ascorbate, glutathione, phenols, alkaloids, flavonoids, and proline [60,72,86–91]. A decomposition product of poly-to-fatty acids of polyunsaturated fatty acids is malondialdehyde (MDA). It is considered a marker of membrane lipid peroxidation, which is an effect of oxidative damage. During the various drought stress conditions, some adapted species modify their antioxidant activities, increasing, for example, the activity of SOD and peroxidase (POD) [92]. SOD is the primary defense against ROS because it eliminates superoxide radicals. Specifically, it dismutates two O_2^-

radicals into H_2O_2 and O_2^{-}, which are precursors to other ROS and are generated in different subcellular compartments [93].

3. Mechanism of Signal Transduction and Development of Drought Tolerance

Drought stress is sensed by the roots of plants and the reduction of water availability slowly occurs depending on the soil physical properties. The limitation of water induces in plants several physiological, biochemical, and molecular changes that lead to increased plant tolerance (Figure 2). Since plants cannot escape from adverse weather conditions, survival depends on their ability to develop efficient adaptation strategies. The plant responses start from the activation of specific regulatory genes that lead to the modification of the physiology and the metabolism of the plants. Currently, transcriptional changes are widely studied in different species and under different drought stress conditions. Pioneer studies have been carried out on model plants, such as *Arabidopsis thaliana*, identifying the transcription profiles and transcription factors involved in responses to drought stress [94,95]. Among the different genes, dehydrin was found to be an indicator of the entire transcriptome response under drought stress. The increase in stress intensity induces the activation of genes associated with stress responses [96]. The most important genes involved belong to abscisic acid (ABA) perception and biosynthesis as well as the ethylene pathway. Among the transcription factors involved, the most important are abscisic acid-responsive element (ABRE), ABRE-binding (AREB) proteins, ABRE-binding factors (AREB/ABFs), drought-responsive *cis*-element binding protein/C-repeat-binding factor (DREB/CBF), ABF/AREB, NAC, WRKY transcription factors, Apetala 2 (AP2), and ethylene response elements [97,98]. The ABF/AREB are under ABA regulation involving SnRK2. These transcription factors are able to provide rapid gene activation under different abiotic stresses, including drought. Other transcription factors belong to the MYB family (such as MYB2 and MYC2) and are inducible by ABA [99]. Therefore, this plant hormone has a pivotal role under water stress in the activation of secondary gene networks, which leads to plant adaptation to stress. Mutants lacking ABA biosynthesis or action are very sensitive to drought stress [100].

The genes induced under drought encode for different proteins that are directly or indirectly involved in plant adaptation. Specific genes induced by water stress increase the accumulation of late embryogenesis abundant (LEA) proteins [101]. These proteins are accumulated in tissues under dehydration or desiccation, such as seeds. In plants, the LEA proteins are considered important in plant drought tolerance [102]. Water stress induces gene expression of membrane proteins. Among these, the most important are the aquaporins, i.e., the water channels.

At a biochemical level, plants increase the biosynthesis of osmolytes to lower the cell water potential and increase the water uptake ability of roots. These molecules are responsible for plant osmotic adaptation and include glycine, glycine betaine, proline, sugars, γ-aminobutyric acid, alcohols, sugar alcohols, trehalose, mannitol, polyamines, etc. [103–105]. The accumulation of these substances allows for the improvement of crop tolerance against drought stress, and the visual appearance of the plants does not change. Plants do not seem to be under stress conditions, but the biosynthesis of protectant molecules requires energy that is not exploited for the growth or the yield in agricultural crops. The energy used for the biosynthesis of osmolytes is also known as "fitness cost", which represents the energy costs for the plant to defend itself. The plants reduce photosynthetic activity, and ribulose-1,5-bisphosphate carboxylase/oxygenase (RUBISCO) efficiency declines with the increase in water reduction [106]. Since photosynthesis is a biochemical process that requires water, carbon dioxide, and light, the lack of water directly reduces photosynthesis. The quantum efficiency of PSII at the initial water stress transiently increases and then declines. The light received by the leaves must be dissipated to avoid photo-oxidative damage, and the energy dissipated can be estimated by chlorophyll *a* fluorescence. Gas exchange at the leaf level is regulated by stomatal opening. Under drought, water loss can be reduced by stomatal closure and a reduction in carbon dioxide concentration [107]. The reduction of light use can lead to an excess of excitation energy in leaves with ROS accumulation [108]. The increase in radicals stimulates the plant to activate the antioxidant systems, such as the enzymes involved in the detoxification of cells. The most important enzymes are SOD, CAT, APX, POD, glutathione reductase (GR), and monodehydroascorbate reductase (MDHAR). These enzymes are able to reduce the ROS accumulation and enhance plant tolerance to drought [70]. Drought stress is a common stress in plants grown in the Mediterranean area, and several ornamental shrubs subjected to water availability increase the activity of these enzymes [36].

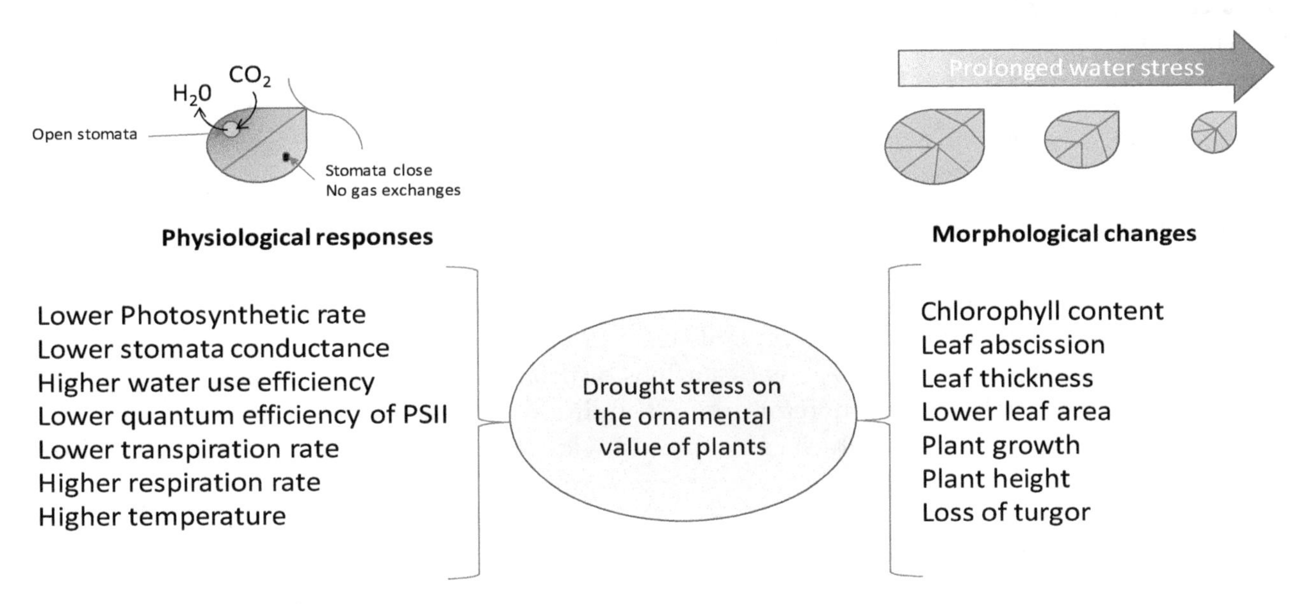

Figure 2. Physiological and morphological changes of plants exposed to reduced water availability. The magnitude of changes depends on the intensity of the stress.

Reduced photosynthetic activity also affects sugar concentration since respiration under drought increases because the plant temperature increases [109]. Plants under normal conditions are able to maintain the leaf temperature in the optimal range for photosynthesis by their thermoregulation ability, which is due to the evaporation of water at the leaf level through transpiration. The water passing from the aqueous state to the gas absorbs the heat from the plants and lowers the temperature. Under drought conditions, the closure of the stomata reduces transpiration and leads to a temperature increase, inducing a higher respiration rate. The lower photosynthesis and the higher respiration rate collectively reduce plant growth [110]. The reduction of plant growth in ornamental plants under water stress has been reported in several species, such as *Eugenia uniflora, Passiflora incarnata, and Photinia x* fraseri [36,111]. Ornamental plants can adopt different strategies under water stress. The study of plant responses to drought can be simulated by reducing water availability. In a study focused on drought responses, it was found that *Penstemon barbatus* was able to counteract drought by increasing root biomass and reducing stomatal conductance [112]. The gas exchange parameters, such as photosynthesis and stomatal conductance, can be considered good parameters for ornamental plant selection for tolerance to drought stress.

ABA is one of the most important plant hormones because it can regulate stomata opening in relation to potassium ions in guard cells [113]. An increase in ABA is crucial for reducing water loss through the stomata. Exogenous applications of ABA demonstrated that treated plants have a higher tolerance to drought. Another plant hormone that is induced by water stress is ethylene. It is also known as a senescence hormone because it is involved in leaf and flower senescence. Several ornamental plants are sensitive to ethylene, and it causes leaf abscission and yellowing, and petal rolling or desiccation [114]. Therefore, water stress can be detrimental for the ornamental plants used in the garden or other urban or peri-urban areas. Ethylene can be produced from endothermic engines. Therefore, in urban areas, plants exposed to ethylene and drought stress accelerate their senescence. Another important plant hormone that can have a positive role in the mitigation of drought stress is represented by the cytokinins. It has been demonstrated that *Arabidopsis* plants overexpressing genes involved in cytokinin biosynthesis showed higher drought stress tolerance [115]. These plant hormones have a preferential site of biosynthesis in roots, and drought stress seems to reduce the concentration of cytokinins with an increase in root growth [116]. The increase in root biomass is considered a first response of the plant to drought stress. However, the application of some plant growth promoting bacteria (PGPB) also induced drought tolerance by increasing their cytokinin concentration and ABA [117].

Therefore, plant adaptation to drought stress is due to plant hormone equilibrium, and the plant responses are consequences of the cross-talk among them [118].

4. Effects of Drought Stress on the Ornamental Value of Plants

Plants under water stress modify their morphology and physiology to survive under stressful environments. These changes can also have a direct effect on the visual appearance and subsequently the ornamental value of the plants. Morphological changes can be observed on the leaves and the plant habit. The most common changes that are observed are leaves that are smaller and have different orientations on the branches. Ornamental plants used in drought-prone environments must be able to adapt to the utilization area, such as private gardens or urban or peri-urban areas without irrigation systems. At nursery levels, the selection for drought environment can be carried out by considering the size and architecture of the roots, which can explore a wide volume of soil. Unfortunately, evaluation of root systems is not easy to perform.

In nursery cultivation, the generalized use of pots, often of small volume, cause root restriction effects. Yong et al. [119] analyzed the influence of substrate volume reduction on cotton plants under conditions where water and nitrogen supplies were not limited. The root-restriction lowered the rate of photosynthesis due to lower stomatal conductance. Root restriction increased the shoot-to-root ratios and reduced the total whole-plant leaf area by 20%.

The critical step for many ornamental plants is transplanting. Therefore, the hardening of plants is important for xerophytic environments [120]. After transplanting, the survival of plants can be guaranteed from their ability to reduce water losses through transpiration and gas exchange. The adapted plants must reduce stomatal conductance, maintain their water balance, and have high WUE [121].

The effects of drought have direct impacts on the habit of plants, and the ornamental quality can be observed at the leaf level. Leaves can drop, change color, or show necrosis from the action of ethylene. Flower life and turnover are also affected in many ornamental plants. The presence of flowers on plants greatly enhances their visual appearance. Therefore, tolerant plants should be able to have a high number of flowers with longer lives because, under water deficits, the turnover of flowers is reduced [122]. Flower turnover or new flower production depends on plant energy availability. Under prolonged drought stress, reduced photosynthesis and fewer carbohydrates are available for flowering.

However, reduced growth can have positive effects for urban green areas and the maintenance of public and private gardens due to lower management costs. Reduced growth is particularly important with ornamental plants that are shaped by pruning. Slower growth contributes to a longer preservation of shape with delayed pruning activities.

5. Use of Different Tools in Mitigating Drought-Induced Damages

A solution to overcoming the problems associated with drought stress is making an appropriate plant choice. The response to drought varies greatly among the plants that can be used in landscaping. In green areas, often a combination of woody and herbaceous ornamental plants is used with various manufactured elements (generally referred to as 'hardscape') [123]. The plant choice can refer to a very large number of species in different environments that are able to assure different functions in the landscape [8]. Plant adaptability to drought stress changes within genera, species, or cultivars [13,14].

Where drought stress is frequent, the ornamental plant choice can favor plants that grow in desert areas (like xerophytes or succulents), which are especially capable of surviving water shortages.

Arbuscular mycorrhizal (AM) symbiosis can also increase host resistance to drought stress, although the effect is not always predictable. Since drought stress is frequent in drying soils, the AM influence on plant drought response can be the result of AM influence on salt stress. With this aim, Cho et al. [124] determined if the AM-induced effects on drought responses would be more accentuated when plants of similar sizes were exposed to drought in salinized soils, rather than only when drought was applied. In the trial, using two greenhouses, different water relations characteristics were measured in sorghum (*Sorghum bicolor*) plants colonized by *Glomus intraradices*, *Gigaspora margarita*, or a mixture of AM species during a sustained drought following exposure to salinity treatments (NaCl stress, osmotic stress via concentrated macronutrients, or soil leaching). The findings confirmed that AM fungi can alter the host response to drought, but they did not lend much support to the idea that AM induced salt resistance. The beneficial effects of AM were related to the improved ability of the roots to adsorb water by increasing the active root surface. The increase in the root adsorption ability was also due to gibberellin- and cytokinin-induced production by AM [125].

Direct and indirect positive roles of PGPB in plants under stress have been reported [126]. The positive effects of PGPB are through the activation of 1-aminocyclopropane-1-carboxylate deaminase enzyme that reduces ethylene production and increases auxin concentration in roots [127]. In recent years, there has been an increase in biostimulants used in agriculture and horticulture to enhance crop abiotic stress tolerance [128]. Alleviation of abiotic stress is perhaps the most frequently cited benefit of biostimulant formulations [129]. Biostimulants are derived from organic substances through different industrial processes. They can be composed of microorganisms such as fungi or bacteria [128] and help to improve plant abiotic adaptation by acting on the physiology and biochemistry of plants [130]. The cytokinin-producing bacteria under drought conditions are of relevant interest [131]. Some microbial inoculants known to have a positive effect on plant development can also help plants overcome or tolerate abiotic stress conditions. In ornamental plants, production can be improved by biostimulant application. Hibiscus (*Hibiscus* spp.) treated with commercial biostimulants showed an increase in gas exchange with higher photosynthetic activities [132]. In a pot experiment with bedding plants, a seaweed extract of *Ascophyllum nodosum* revealed positive effects on the growth and development of petunias (*Petunia* spp.), pansies (*Viola tricolor*), and cosmos (*Cosmos* spp.) exposed to drought [133]. Some reported positive effects of biostimulants are the induction of early flowering, a higher number of flowers, and higher biomass accumulation [134]. With the aim of evaluating the differences in the mechanisms involved in ornamental species' resistance to drought stress resulting from a regular suspension and recovery of the water supply, Toscano et al. [22] subjected plants of five ornamental shrubs (*Callistemon citrinus*, *Laurus nobilis*, *Pittosporum tobira*, *Thunbergia erecta*, and *Viburnum tinus* 'Lucidum') to two consecutive cycles of suspension/rewatering (S-R) and compared them with plants that were watered daily (C). The five species exhibited different responses to drought stress. At the end of the experimental period, S-R treatment had no effect on the dry weight of any species except *Pittosporum*. In *Pittosporum*, drought stress reduced total plant biomass by 19%. Drought stress induced alterations in shrubs, including decreases in shoot dry matter and increases in the root-to-shoot ratio, strongly affecting *Callistemon* and *Pittosporum*. All species adapted to water shortages using physiological mechanisms (RWC and water potential adjustment, stomatal closure, and reductions in photosynthesis). Following re-watering, the species fully recovered. Therefore, they can be considered as suitable for landscaping in the Mediterranean environment. However, *Laurus* and *Thunbergia* seemed to be less sensitive to drought stress than the other species.

Light drought stress can be adopted to control the growth of pot plants. Davies et al. [135] used deficit irrigation in comparison to conventional overhead irrigation in two crops of different canopy structure (*Cornus alba* and *Lonicera periclymenum*). In a subsequent experiment, *Forsythia* × *intermedia* was grown in two substrates with contrasting quantities of peat (60 and 100%). Deficit irrigation was found to be mainly effective in controlling vegetative growth when applied using overhead irrigation.

Similar results were achieved when drip irrigation was used. This comparable response suggests that deficit irrigation can be applied without precision drip irrigation. Scheduling two very different crops with respect to their water use and uptake potential, however, highlighted challenges in the application of appropriate deficits for very different crops under one system. Responses to deficit irrigation are more consistent where nursery management allows for scheduling of crops with very different architecture and water use under different regimes.

6. Conclusions and Future Prospective

The drought tolerance of ornamental plants widely varies with genotypes, environmental conditions, and soil or substrate characteristics. Landscape plants have similar mechanisms of drought tolerance to agricultural crops, but assessment of drought tolerance for these plants should be based primarily on aesthetic value rather than growth effects. Because of the wide number of plant species potentially available for ornamental purpose, it should be possible to choose genotypes suitable for drought environments.

Problems in research that occur are linked to: (i) the necessity to experimentally analyze a wide range of plant species to find those most suitable for specific sites; (ii) identifying parameters with simple measurements to discriminate tolerance to drought stress, and (iii) tailoring irrigation methods or plant management strategies to enable the chosen species to cope with water stress.

The study of the mechanism of plant response to drought stress and particularly of signal transduction and development of drought tolerance allow for the identification of suitable plants and management strategies for the cultivation or utilization of ornamental plants in drought-prone environments.

Author Contributions: D.R. and A.F. projected the design of the review. S.T. and D.R. wrote the introduction and the information related with growth and morpho-anatomical modification under drought stress; S.T. wrote all the information related with physiological parameters and oxidative stress; A.F. wrote all the information related with mechanism of signal transduction and development of drought tolerance. D.R. and S.T. wrote all the information related with effect of drought stress on ornamental value, the tool use for mitigating the drought. S.T. and D.R. made the Figure 1 and Table 1. AF made the Figure 2. D.R. and S.T. ordered all the references. All authors wrote the conclusions and revised and approved the manuscript.

References

1. Rundel, P.W.; Arroyo, M.T.; Cowling, R.M.; Keeley, J.E.; Lamont, B.B.; Pausas, J.G.; Vargas, P. Fire and plant diversification in mediterranean-climate regions. *Front. Plant Sci.* **2018**, *9*, 851. [CrossRef] [PubMed]
2. Paz, S.; Negev, M.; Clermont, A.; Green, M.S. Health aspects of climate change in cities with mediterranean climate, and local adaptation plans. *Int. J. Environ. Res. Public Health* **2016**, *13*, 438. [CrossRef] [PubMed]
3. Cowling, R.M.; Rundel, P.W.; Lamont, B.B.; Arroyo, M.K.; Arianoutsou, M. Plant diversity in Mediterranean-climate regions. *Trends Ecol. Evol.* **1996**, *11*, 362–366. [CrossRef]
4. Medrano, H.; Flexas, J.; Galmés, J. Variability in water use efficiency at the leaf level among Mediterranean plants with different growth forms. *Plant Soil* **2009**, *317*, 17–29. [CrossRef]
5. WWAP (World Water Assessment Programme). *The United Nations World Water Development Report 2014: Water and Energy*; UNESCO: Paris, France, 2014.
6. Ouzounidou, G.; Vekiari, S.; Asfi, M.; Gork, M.G.; Sakcali, M.S.; Ozturk, M. Photosynthetic characteristics of carob tree (*Ceratonia siliqua* L.) and chemical composition of its fruit on diurnal and seasonal basis. *Pak. J. Bot.* **2012**, *44*, 1689–1695.
7. Thayer, R.L. Visual ecology: Revitalizing the esthetics of landscape architecture. *Landscape* **1976**, *20*, 37–43.
8. Savé, R. What is stress and how to deal with it in ornamental plants? *Acta Hort.* **2009**, *813*, 241–254. [CrossRef]
9. Kjelgren, R.; Rupp, L.; Kilgren, D. Water conservation in urban landscapes. *HortScience* **2000**, *35*, 1037–1040.
10. Cameron, R.W.F.; Wilkinson, S.; Davies, W.J.; Harrison Murray, R.S.; Dunstan, D.; Burgess, C. Regulation of plant growth in container-grown ornamentals through the use of controlled irrigation. *Acta Hortic.* **2004**, *630*, 305–312. [CrossRef]

11. Niu, G.; Rodriguez, D.S.; Aguiniga, L.; Mackay, W. Salinity tolerance of *Lupinus havardii* and *Lupinus texenis*. *HortScience* **2007**, *42*, 526–528.
12. Álvarez, S.; Rodríguez, P.; Broetto, F.; Sánchez-Blanco, M.J. Long term responses and adaptive strategies of *Pistacia lentiscus* under moderate and severe deficit irrigation and salinity: Osmotic and elastic adjustment, growth, ion uptake and photosynthetic activity. *Agric. Water Manag.* **2018**, *202*, 253–262. [CrossRef]
13. Sánchez-Blanco, M.J.; Rodriguez, P.; Morales, M.A.; Torrecillas, A. Comparative growth and water relations of *Cistus albidus* and *Cistus monspeliensis* plants during water deficit conditions and recovery. *Plant Sci.* **2002**, *162*, 107–113. [CrossRef]
14. Torrecillas, A.; Rodriguez, P.; Sánchez-Blanco, M.J. Comparison of growth, leaf water relations and gas exchange of *Cistus albidus* and *C. monspeliensis* plants irrigated with water of different NaCl salinity levels. *Sci. Hortic. (Amsterdam)* **2003**, *97*, 353–368. [CrossRef]
15. Farieri, E.; Toscano, S.; Ferrante, A.; Romano, D. Identification of ornamental shrubs tolerant to saline aerosol for coastal urban and peri-urban greening. *Urban For. Urban Green.* **2016**, *18*, 9–18. [CrossRef]
16. Hansen, C.W.; Petersen, K.K. Reduced nutrient and water availability to *Hibiscus rosa-sinensis* 'Cairo Red' as a method to regulate growth and improve post-production quality. *Eur. J. Hort. Sci.* **2004**, *69*, 159–166. [CrossRef]
17. Silber, A.; Levi, M.; Cohen, M.; David, N.; Shtaynmetz, Y.; Assouline, S. Response of *Leucadendron* 'Safari Sunset' to regulated deficit irrigation: Effects of stress timing on growth and yield quality. *Agric. Water Manag.* **2007**, *87*, 162–170. [CrossRef]
18. Bernal, M.; Estiarte, M.; Peñuelas, J. Drought advances spring growth phenology of the Mediterranean shrub *Erica multiflora*. *Plant Biol.* **2011**, *13*, 252–257. [CrossRef]
19. Elansary, H.O.; Salem, M.Z.M. Morphological and physiological responses and drought resistance enhancement of ornamental shrubs by trinexapac-ethyl application. *Sci. Hortic.* **2015**, *189*, 1–11. [CrossRef]
20. Cirillo, C.; De Micco, V.; Rouphael, Y.; Balzano, A.; Caputo, R.; De Pascale, S. Morpho-anatomical and physiological traits of two *Bougainvillea* genotypes trained to two shapes under deficit irrigation. *Trees Struct. Funct.* **2017**, *31*, 173–187. [CrossRef]
21. Álvarez, S.; Sánchez-Blanco, M.J. Comparison of individual and combined effects of salinity and deficit irrigation on physiological, nutritional and ornamental aspects of tolerance in *Callistemon laevis* plants. *J. Plant Physiol.* **2015**, *185*, 65–74. [CrossRef]
22. Toscano, S.; Scuderi, D.; Giuffrida, F.; Romano, D. Responses of Mediterranean ornamental shrubs to drought stress and recovery. *Sci. Hortic.* **2014**, *178*, 145–153. [CrossRef]
23. Toscano, S.; Ferrante, A.; Tribulato, A.; Romano, D. Leaf physiological and anatomical responses of *Lantana* and *Ligustrum* species under different water availability. *Plant Physiol. Biochem.* **2018**, *127*, 380–392. [CrossRef]
24. Álvarez, S.; Sánchez-Blanco, M.J. Changes in growth rate, root morphology and water use efficiency of potted *Callistemon citrinus* plants in response to different levels of water deficit. *Sci. Hortic.* **2013**, *156*, 54–62. [CrossRef]
25. Bacelar, E.A.; Santos, D.L.; Moutinho-Pereira, J.M.; Lopes, J.I.; Gonçalves, B.C.; Ferreira, T.C.; Correia, C.M. Physiological behaviour, oxidative damage and antioxidative protection of olive trees grown under different irrigation regimes. *Plant Soil* **2007**, *292*, 1–12. [CrossRef]
26. Chyliňsku, W.K.; Łukaszewska, A.J.; Kutnik, K. Drought response of two bedding plants. *Acta Physiol. Planta* **2007**, *29*, 399–406. [CrossRef]
27. Bacelar, E.A.; Correia, C.M.; Moutinho-Pereira, J.M.; Gonçalves, B.C.; Lopes, J.I.; Torres-Pereira, J.M. Sclerophylly and leaf anatomical traits of five field-grown olive cultivars growing under drought conditions. *Tree Physiol.* **2004**, *24*, 233–239. [CrossRef]
28. Rafi, Z.N.; Kazemi, F.; Tehranifar, A. Morpho-physiological and biochemical responses of four ornamental herbaceous species to water stress. *Acta Physiol. Planta* **2019**, *41*, 7. [CrossRef]
29. Dghim, F.; Abdellaoui, R.; Boukhris, M.; Neffati, M.; Chaieb, M. Physiological and biochemical changes in *Periploca angustifolia* plants under withholding irrigation and rewatering conditions. *S. Afr. J. Bot.* **2018**, *114*, 241–249. [CrossRef]
30. Kebbas, S.; Lutts, S.; Aid, F. Effect of drought stress on the photosynthesis of *Acacia tortilis* subsp. *raddiana at the young seedling stage. Photosynthetica* **2015**, *53*, 288–298. [CrossRef]

31. Ugolini, F.; Bussotti, F.; Raschi, A.; Tognetti, R.; Ennos, A.R. Physiological performance and biomass production of two ornamental shrub species under deficit irrigation. *Trees Struct. Funct.* **2015**, *29*, 407–422. [CrossRef]
32. Ugolini, F.; Tognetti, R.; Bussotti, F.; Raschi, A.; Ennos, A.R. Wood hydraulic and mechanical properties induced by low water availability on two ornamental species *Photinia×fraseri* var. Red Robin and *Viburnum opulus* L. *Urban For. Urban Green.* **2014**, *13*, 158–165. [CrossRef]
33. Cirillo, C.; Rouphael, Y.; Caputo, R.; Raimondi, G.; De Pascale, S. The influence of deficit irrigation on growth, ornamental quality, and water use efficiency of three potted *Bougainvillea* genotypes grown in two shapes. *HortScience* **2014**, *49*, 1284–1291.
34. Kumar, D.; Al Hassan, M.; Naranjo, M.A.; Agrawal, V.; Boscaiu, M.; Vicente, O. Effects of salinity and drought on growth, ionic relations, compatible solutes and activation of antioxidant systems in oleander (*Nerium oleander*). *PLoS ONE* **2017**, *12*, e0185017. [CrossRef]
35. Álvarez, S.; Bañón, S.; Sánchez-Blanco, M.J. Regulated deficit irrigation in different phenological stages of potted geranium plants: Water consumption, water relations and ornamental quality. *Acta Physiol. Plant.* **2013**, *35*, 1257–1267. [CrossRef]
36. Toscano, S.; Farieri, E.; Ferrante, A.; Romano, D. Physiological and biochemical responses in two ornamental shrubs to drought stress. *Front. Plant Sci.* **2016**, *7*, 645. [CrossRef]
37. Navarro, A.; Álvarez, S.; Castillo, M.; Bañón, S.; Sánchez-Blanco, M.J. Changes in tissue-water relations, photosynthetic activity, and growth of *Myrtus communis* plants in response to different conditions of water availability. *J. Hortic. Sci. Biotechnol.* **2009**, *84*, 541–547. [CrossRef]
38. Sánchez-Blanco, M.J.; Álvarez, S.; Navarro, A.; Bañón, S. Changes in leaf water relations, gas exchange, growth and flowering quality in potted geranium plants irrigated with different water regimes. *J. Plant Physiol.* **2009**, *166*, 467–476. [CrossRef]
39. Souza, P.U.; Lima, L.K.S.; Soares, T.L.; de Jesus, O.N.; Filho, M.A.C.; Girardi, E.A. Biometric, physiological and anatomical responses of *Passiflora* spp. to controlled water deficit. *Sci. Hortic.* **2018**, *229*, 77–90. [CrossRef]
40. Wu, F.; Bao, W.; Li, F.; Wu, N. Effects of drought stress and N supply on the growth, biomass partitioning and water-use efficiency of *Sophora davidii* seedlings. *Environ. Exp. Bot.* **2008**, *63*, 248–255. [CrossRef]
41. Smirnoff, N. Plant resistance to environmental stress. *Curr. Opin. Biotechnol.* **1998**, *9*, 214–219. [CrossRef]
42. Fraser, L.H.; Greenall, A.; Carlyle, C.; Turkington, R.; Ross Friedman, C. Adaptive phenotypic plasticity of *Pseudoroegneria spicata*: Response of stomatal density, leaf area and biomass to changes in water supply and increased temperature. *Ann. Bot.* **2009**, *103*, 769–775. [CrossRef] [PubMed]
43. Liu, F.; Stützel, H. Biomass partitioning, specific leaf area, and water use efficiency of vegetable amaranth (*Amaranthus* spp.) in response to drought stress. *Sci. Hortic.* **2004**, *102*, 15–27. [CrossRef]
44. Chaves, M.M.; Maroco, J.P.; Pereira, J.S. Understanding plant responses to drought from genes to the whole plant. *Funct. Plant Biol.* **2003**, *30*, 239–264. [CrossRef]
45. Campbell, D.R.; Wu, C.A.; Travers, S.E. Photosynthetic and growth responses of reciprocal hybrids to variation in water and nitrogen availability. *Am. J. Bot.* **2010**, *97*, 925–933. [CrossRef] [PubMed]
46. Hu, X.; Liu, R.; Li, Y.; Wang, W.; Tai, F.; Xue, R.; Li, C. Heat shock protein 70 regulates the abscisic acid-induced antioxidant response of maize to combined drought and heat stress. *J. Plant Growth Regul.* **2010**, *60*, 225–235. [CrossRef]
47. Ahmadi, U.; Baker, D.A. The effect of water stress on grain filling processes in wheat. *J. Agric. Sci.* **2001**, *136*, 257–269. [CrossRef]
48. Del Blanco, I.A.; Rajaram, S.; Kronstad, W.E.; Reynolds, M.P. Physiological performance of synthetic hexaploid wheat–derived populations. *Crop Sci.* **2000**, *40*, 1257–1263. [CrossRef]
49. Samarah, N.H.; Alqudah, A.M.; Amayreh, J.A.; McAndrews, G.M. The effect of late-terminal drought stress on yield components of four barley cultivars. *J. Agron. Crop Sci.* **2009**, *195*, 427–441. [CrossRef]
50. Anjum, S.; Xie, X.Y.; Wang, L.C.; Saleem, M.F.; Man, C.; Wang, L. Morphological, physiological and biochemical responses of plants to drought stress. *Afr. J. Agric. Res.* **2011**, *6*, 2026–2032. [CrossRef]
51. Xiong, L.; Zhu, J. Molecular and genetic aspects of plant responses to osmotic stress. *Plant Cell Environ.* **2002**, *25*, 131–139. [CrossRef] [PubMed]
52. Ludlow, M.M. Stress physiology of tropical pasture plants. *Trop. Grassl.* **1980**, *14*, 136–145.

53. Nilsen, E.; Orcutt, D. *The Physiology of Plants under Deficit. Abiotic Factors*; Willey: New York, NY, USA, 1996; Volume 689, ISBN-13: 978-0471031529.
54. Guerrier, G. Fluxes of Na^+, K^+ and Cl^-, and osmotic adjustment in *Lycopersicon pimpinellifolium* and *L. esculentum* during short- and long-term exposures to NaCl. *Physiol. Plant.* **1996**, *97*, 583–591. [CrossRef]
55. Munns, R. Comparative physiology of salt and water stress. *Plant Cell Environ.* **2002**, *25*, 239–250. [CrossRef] [PubMed]
56. Inan, G.; Zhang, Q.; Li, P.H.; Wang, Z.L.; Cao, Z.Y.; Zhang, H.; Zhang, C.Q.; Quist, T.M.; Goodwin, S.M.; Zhu, J.; et al. Salt cress: A halophyte and cryophyte Arabidopsis relative model system and its applicability to molecular genetic analyses of growth and development of extremophiles. *Plant Physiol.* **2004**, *135*, 1718–1737. [CrossRef] [PubMed]
57. Riaz, A.; Younis, A.; Taj, A.R.; Riaz, S. Effect of drought stress on growth and flowering of marigold (*Tagetes erecta* L.). *Pak. J. Bot.* **2013**, *45*, 123–131.
58. Galmés, J.; Medrano, H.; Flexas, J. Photosynthesis and photoinhibition in response to drought in a pubescent (var. minor) and a glabrous (var. palaui) variety of *Digitalis minor*. *Environ. Exp. Bot.* **2007**, *60*, 105–111. [CrossRef]
59. Lichtenthaler, H.K.; Rinderle, U. The role of chlorophyll fluorescence in the detection of stress conditions in plants. *Crit. Rev. Anal. Chem.* **1988**, *19*, S29–S85. [CrossRef]
60. Reddy, A.R.; Chiatanya, K.V.; Vivekanandan, M. Drought induced responses of photosynthesis and antioxidant metabolism in higher plants. *J. Plant Physiol.* **2004**, *161*, 1189–1202. [CrossRef]
61. Maxwell, K.; Johnson, G.N. Chlorophyll fluorescence—A practical guide. *J. Exp. Bot.* **2000**, *51*, 659–668. [CrossRef]
62. Demmig, B.; Björkman, O. Comparison of the effect of excessive light on chlorophyll fluorescence (77K) and photon yield of O_2 evolution in leaves of higher plants. *Planta* **1987**, *171*, 171–184. [CrossRef]
63. Álvarez, S.; Navarro, A.; Nicolás, E.; Sánchez-Blanco, M.J. Transpiration, photosynthetic responses, tissue water relations and dry mass partitioning in *Callistemon* plants during drought conditions. *Sci. Hortic.* **2011**, *129*, 306–312. [CrossRef]
64. Flexas, J.; Medrano, H. Energy dissipation in C3 plants under drought. *Funct. Plant Boil.* **2002**, *29*, 1209–1215. [CrossRef]
65. Demmig-Adams, B.; Adams, W.W. Photoprotection in an ecological context: The remarkable complexity of thermal energy dissipation. *New Phytol.* **2006**, *172*, 11–21. [CrossRef] [PubMed]
66. Munné-Bosch, S.; Alegre, L. Drought-induced changes in the redox state of α-tocopherol, ascorbate, and the diterpene carnosic acid in chloroplasts of Labiatae species differing in carnosic acid contents. *Plant Physiol.* **2003**, *131*, 1816–1825. [CrossRef]
67. Smirnoff, N. The role of active oxygen in the response of plants to water deficit and dessication. *New Phytol.* **1993**, *125*, 27–58. [CrossRef]
68. Schwanz, P.; Picon, C.; Vivin, P.; Dreyer, E.; Guehi, J.M.; Polle, A. Responses of antioxidative system to drought stress in penduncs... oak and maritime pine as modulated by elevated CO2. *Plant Physiol.* **1996**, *110*, 393–402. [CrossRef] [PubMed]
69. Noctor, G.; Foyer, C.H. Ascorbate glutathione: Keeping active oxygen under control. *Annu. Rev. Plant Physiol. Plant Mol. Biol.* **1998**, *49*, 249–279. [CrossRef]
70. Mittler, R. Oxidative stress, antioxidants and stress tolerance. *Trends Plant Sci.* **2002**, *7*, 405–410. [CrossRef]
71. Foyer, C.H.; Noctor, G. Oxidant and antioxidant signalling in plants: A re-evaluation of the concept of oxidative stress in a physiological context. *Plant Cell Environ.* **2005**, *8*, 1056–1071. [CrossRef]
72. Impa, S.M.; Nadaradjan, S.; Jagadish, S.V.K. Drought stress induced reactive oxygen species and anti-oxidants in plants. In *Abiotic Stress Responses in Plants*; Springer: New York, NY, USA, 2012; pp. 131–147.
73. Foyer, C.H.; Descourvieres, O.; Kunert, K.J. Protection against oxygen radicals: An important defence mechanism studied in transgenic plants. *Plant Cell Environ.* **1994**, *17*, 507–523. [CrossRef]
74. Garratt, L.C.; Janagoudar, B.S.; Lowe, K.C.; Anthony, P.; Power, J.B.; Davey, M.R. Salinity tolerance and antioxidant status in cotton cultures. *Free Radic. Biol. Med.* **2002**, *33*, 502–511. [CrossRef]
75. Cruz de Carvalho, R.; Catala, M.; Silva, J.M.D.; Branquinho, C.; Barreno, E. The impact of dehydration rate on the production and cellular location of reactive oxygen species in an aquatic moss. *Ann. Bot.* **2012**, *110*, 1007–1016. [CrossRef] [PubMed]

76. Foyer, C.H.; Lopez-Delgado, H.; Dat, J.F.; Scott, I.M. Hydrogen peroxide and glutathione-associated mechanisms of acclamatory stress tolerance and signaling. *Physiol. Plant.* **1997**, *100*, 241–254. [CrossRef]
77. Matamorous, M.A.; Dalton, D.A.; Ramos, J.; Clemente, M.R.; Rubio, M.C.; Becana, M. Biochemistry and molecular biology of antioxidants in the rhizobia-legume symbiosis. *Plant Physiol.* **2003**, *133*, 499–509. [CrossRef] [PubMed]
78. Bhattacharjee, S. Reactive oxygen species and oxidative burst: Roles in stress, senescence and signal transduction in plant. *Curr. Sci.* **2005**, *89*, 1113–1121.
79. Lawlor, D.W.; Tezara, W. Causes of decreased photosynthetic rate and metabolic capacity in water-deficient leaf cells: A critical evaluation of mechanisms and integration of processes. *Ann. Bot.* **2009**, *103*, 561–579. [CrossRef]
80. Sankar, B.; Jaleel, C.A.; Manivannan, P.; Kishorekumar, A.; Somasundaram, R.; Panneerselvam, R. Effect of paclobutrazol on water stress amelioration through antioxidants and free radical scavenging enzymes in *Arachis hypogaea* L. *Colloids Surf B Biointerfaces* **2007**, *60*, 229–235. [CrossRef] [PubMed]
81. Jaleel, C.A.; Sankar, B.; Murali, P.V.; Gomathinayagam, M.; Lakshmanan, G.M.A.; Panneerselvam, R. Water deficit stress effects on reactive oxygen metabolism in *Catharanthus roseus*: Impacts on ajmalicine accumulation. *Colloids Surf. B Biointerfaces* **2008**, *62*, 105–111. [CrossRef] [PubMed]
82. Manivannan, P.; Jaleel, C.A.; Somasundaram, R.; Panneerselvam, R. Osmoregulation and antioxidant metabolism in drought stressed *Helianthus annuus* under triadimefon drenching. *C. R. Biol.* **2008**, *331*, 418–425. [CrossRef]
83. Jaleel, C.A.; Gopi, R.; Manivannan, P.; Gomathinayagam, M.; Murali, P.V.; Panneerselvam, R. Soil applied propiconazole alleviates the impact of salinity on *Catharanthus roseus* by improving antioxidant status. *Pestic. Biochem. Phys.* **2008**, *90*, 135–139. [CrossRef]
84. Manivannan, P.; Jaleel, C.A.; Kishorekumar, A.; Sankar, B.; Somasundaram, R.; Panneerselvam, R. Protection of *Vigna unguiculata* (L.) Walp. plants from salt stress by paclobutrazol. *Colloids Surf. B Biointerfaces* **2008**, *61*, 315–318. [CrossRef] [PubMed]
85. Puckette, M.C.; Weng, H.; Mahalingam, R. Physiological and biochemical responses to acute ozone-induced oxidative stress in *Medicago truncatula*. *Plant Physiol. Biochem.* **2007**, *45*, 70–79. [CrossRef] [PubMed]
86. Chen, C.; Dickman, M.B. Proline suppresses apoptosis in the fungal pathogen *Colletotrichum trifolii*. *Proc. Natl. Acad. Sci. USA* **2005**, *102*, 3459–3464. [CrossRef] [PubMed]
87. Jaleel, C.A.; Riadh, K.; Gopi, R.; Manivannan, P.; Ines, J.; Al-Juburi, H.J.; Chang-Xing, Z.; Hong-Bo, S.; Panneerselvam, R. Antioxidant defense responses: Physiological plasticity in higher plants under abiotic constraints. *Acta Physiol. Plant.* **2009**, *31*, 427–436. [CrossRef]
88. Gong, H.; Zhu, X.; Chen, K.; Wang, S.; Zhang, C. Silicon alleviates oxidative damage of wheat plant in pot under drought. *Plant Sci.* **2005**, *169*, 313–321. [CrossRef]
89. Yin, Y.; Li, S.; Liao, W.; Lu, Q.; Wen, X.; Lu, C. Photosystem II photochemistry, photoinhibition, and the xanthophylls cycle in heat-stressed rice leaves. *J. Plant Physiol.* **2010**, *167*, 959–966. [CrossRef] [PubMed]
90. Gill, S.S.; Tuteja, N. Reactive oxygen species and antioxidant machinery in abiotic stress tolerance in crop plants. *Plant Physiol. Biochem.* **2010**, *48*, 909–930. [CrossRef]
91. Ahmad, P.; Jaleel, C.A.; Salem, M.A.; Nabi, G.; Sharma, S. Roles of enzymatic and non-enzymatic antioxidants in plants during abiotic stress. *Crit. Rev. Biotechnol.* **2010**, *30*, 161–175. [CrossRef]
92. Lima, L.H.C.; Návia, D.; Inglis, P.W.; De Oliveira, M.R.V. Survey of *Bemisia tabaci* (Gennadius) (Hemiptera: Aleyrodidae) biotypes in Brazil using RAPD markers. *Genet. Mol. Res.* **2000**, *23*, 781–785. [CrossRef]
93. Alscher, R.G.; Erturk, N.; Heath, L.S. Role of superoxide dismutases (SODs) in controlling oxidative stress in plants. *J. Exp. Biol.* **2002**, *53*, 1331–1341.
94. Matsui, A.; Ishida, J.; Morosawa, T.; Mochizuki, Y.; Kaminuma, E.; Endo, T.A.; Satou, M. Arabidopsis transcriptome analysis under drought, cold, high-salinity and ABA treatment conditions using a tiling array. *Plant Cell Physiol.* **2008**, *49*, 1135–1149. [CrossRef] [PubMed]
95. Wilkins, O.; Bräutigam, K.; Campbell, M.M. Time of day shapes Arabidopsis drought transcriptomes. *Plant J.* **2010**, *63*, 715–727. [CrossRef]
96. Tommasini, L.; Svensson, J.T.; Rodriguez, E.M.; Wahid, A.; Malatrasi, M.; Kato, K.; Wanamaker, S.; Resnik, J.; Close, T.J. Dehydrin gene expression provides an indicator of low temperature and drought stress: Transcriptome-based analysis of barley (*Hordeum vulgare* L.). *Funct. Integr. Genom.* **2008**, *8*, 387–405. [CrossRef] [PubMed]

97. Uno, Y.; Furihata, T.; Abe, H.; Yoshida, R.; Shinozaki, K.; Yamaguchi-Shinozaki, K. Arabidopsis basic leucine zipper transcription factors involved in an abscisic acid-dependent signal transduction pathway under drought and high-salinity conditions. *Proc. Natl. Acad. Sci. USA* **2000**, *97*, 11632–11637. [CrossRef] [PubMed]
98. Klay, I.; Gouia, S.; Liu, M.; Mila, I.; Khoudi, H.; Bernadac, A.; Bouzayen, M.; Pirrello, J. Ethylene Response Factors (ERF) are differentially regulated by different abiotic stress types in tomato plants. *Plant Sci.* **2018**, *274*, 137–145. [CrossRef] [PubMed]
99. Abe, H.; Urao, T.; Ito, T.; Seki, M.; Shinozaki, K.; Yamaguchi-Shinozaki, K. Arabidopsis AtMYC2 (bHLH) and AtMYB2 (MYB) function as transcriptional activators in abscisic acid signaling. *Plant Cell* **2003**, *15*, 63–78. [CrossRef] [PubMed]
100. Koornneef, M.; Leon-Kloosterziel, K.M.; Schwartz, S.H.; Zeevaart, J.A. The genetic and molecular dissection of abscisic acid biosynthesis and signal transduction in Arabidopsis. *Plant Physiol. Biochem.* **1998**, *36*, 83–89. [CrossRef]
101. Magwanga, R.O.; Lu, P.; Kirungu, J.N.; Lu, H.; Wang, X.; Cai, X.; Zhou, Z.; Zhang, Z.; Salih, H.; Wang, K.; et al. Characterization of the late embryogenesis abundant (LEA) proteins family and their role in drought stress tolerance in upland cotton. *BMC Genet.* **2018**, *19*, 6. [CrossRef]
102. Hundertmark, M.; Hincha, D.K. LEA (late embryogenesis abundant) proteins and their encoding genes in *Arabidopsis thaliana*. *BMC Genom.* **2008**, *9*, 118. [CrossRef]
103. Bray, E.A. Plant responses to water deficit. *Trends Plant Sci.* **1997**, *2*, 48–54. [CrossRef]
104. Yoshiba, Y.; Kiyosue, T.; Nakashima, K.; Yamaguchi-Shinozaki, K.; Shinozaki, K. Regulation of levels of proline as an osmolyte in plants under water stress. *Plant Cell Physiol.* **1997**, *38*, 1095–1102. [CrossRef]
105. Hosseini, M.S.; Samsampour, D.; Ebrahimi, M.; Abadía, J.; Khanahmadi, M. Effect of drought stress on growth parameters, osmolyte contents, antioxidant enzymes and glycyrrhizin synthesis in licorice (*Glycyrrhiza glabra* L.) grown in the field. *Phytochemistry* **2018**, *156*, 124–134. [CrossRef] [PubMed]
106. Zandalinas, S.I.; Mittler, R.; Balfagón, D.; Arbona, V.; Gómez-Cadenas, A. Plant adaptations to the combination of drought and high temperatures. *Physiol. Plant.* **2018**, *162*, 2–12. [CrossRef]
107. Flexas, J.; Carriquí, M.; Nadal, M. Gas exchange and hydraulics during drought in crops: Who drives whom? *J. Exp. Bot.* **2018**, *69*, 3791–3795. [CrossRef] [PubMed]
108. Foyer, C.H.; Shigeoka, S. Understanding oxidative stress and antioxidant functions to enhance photosynthesis. *Plant Physiol.* **2011**, *155*, 93–100. [CrossRef]
109. Mariani, L.; Ferrante, A. Agronomic management for enhancing plant tolerance to abiotic stresses-drought, salinity, hypoxia, and lodging. *Horticulturae* **2017**, *3*, 52. [CrossRef]
110. Niu, G.; Rodriguez, D.S.; Wang, Y.T. Impact of drought and temperature on growth and leaf gas exchange of six bedding plant species under greenhouse conditions. *HortScience* **2006**, *41*, 1408–1411.
111. García-Castro, A.; Volder, A.; Restrepo-Diaz, H.; Starman, T.W.; Lombardini, L. Evaluation of different drought stress regimens on growth, leaf gas exchange properties, and carboxylation activity in purple Passionflower plants. *J. Am. Soc. Hortic. Sci.* **2017**, *142*, 57–64. [CrossRef]
112. Zollinger, N.; Kjelgren, R.; Cerny-Koenig, T.; Kopp, K.; Koenig, R. Drought responses of six ornamental herbaceous perennials. *Sci. Hortic.* **2006**, *109*, 267–274. [CrossRef]
113. Schroeder, J.I.; Kwak, J.M.; Allen, G.J. Guard cell abscisic acid signalling and engineering drought hardiness in plants. *Nature* **2001**, *410*, 327–330. [CrossRef] [PubMed]
114. Reid, M.S. Ethylene and abscission. *HortScience* **1985**, *20*, 45–50.
115. Prerostova, S.; Dobrev, P.I.; Gaudinova, A.; Knirsch, V.; Körber, N.; Pieruschka, R.; Fiorani, F.; Brzobohatý, B.; Cerný, M.; Spichal, L.; et al. Cytokinins: Their impact on molecular and growth responses to drought stress and recovery in Arabidopsis. *Front. Plant Sci.* **2018**, *9*, 655. [CrossRef]
116. Werner, T.; Nehnevajova, E.; Köllmer, I.; Novák, O.; Strnad, M.; Krämer, U.; Schmülling, T. Root-specific reduction of cytokinin causes enhanced root growth, drought tolerance, and leaf mineral enrichment in Arabidopsis and tobacco. *Plant Cell* **2010**, *22*, 3905–3920. [CrossRef]
117. Liu, F.; Xing, S.; Ma, H.; Du, Z.; Ma, B. Cytokinin-producing, plant growth-promoting rhizobacteria that confer resistance to drought stress in Platycladus orientalis container seedlings. *Appl. Microbiol. Biotechnol.* **2013**, *97*, 9155–9164. [CrossRef] [PubMed]
118. Peleg, Z.; Blumwald, E. Hormone balance and abiotic stress tolerance in crop plants. *Curr. Opin. Plant Biol.* **2011**, *14*, 290–295. [CrossRef] [PubMed]

119. Yong, J.W.; Letham, D.S.; Wong, S.C.; Farquhar, G.D. Effects of root restriction on growth and associated cytokinin levels in cotton (*Gossypium hirsutum*). *Funct. Plant Biol.* **2010**, *37*, 974–984. [CrossRef]
120. Franco, J.A.; Martinéz-Sanchéz, J.J.; Fernández, J.A.; Bañón, S. Selection and nursery production of ornamental plants for landscaping and xerogardening in semi-arid and environments. *J. Hortic. Sci. Biotechnol.* **2006**, *81*, 3–17. [CrossRef]
121. Lenzi, A.; Pittas, L.; Martinelli, T.; Lombardi, P.; Tesi, R. Response to water stress of some oleander cultivars suitable for pot plant production. *Sci. Hortic.* **2009**, *122*, 426–431. [CrossRef]
122. Rafi, Z.N.; Kazemi, F.; Tehranifar, A. Effects of various irrigation regimes on water use efficiency and visual quality of some ornamental herbaceous plants in the field. *Agric. Water Manag.* **2019**, *212*, 78–87. [CrossRef]
123. Iles, J.K. The science and practice of stress reduction in managed landscapes. *Acta Hortic.* **2003**, *618*, 117–124. [CrossRef]
124. Cho, K.; Toler, H.; Lee, J.; Ownley, B.; Stutz, J.C.; Moore, J.L.; Augé, R.M. Mycorrhizal symbiosis and response of sorghum plants to combined drought and salinity stresses. *J. Plant Physiol.* **2006**, *163*, 517–528. [CrossRef] [PubMed]
125. Barea, J.M.; Azcón-Aguilar, C. Production of plant growth-regulating substances by the vesicular-arbuscular mycorrhizal fungus Glomus mosseae. *Appl. Environ. Microbiol.* **1982**, *43*, 810–813.
126. Wong, W.S.; Tan, S.N.; Ge, L.; Chen, X.; Yong, J.W.H. The importance of phytohormones and microbes in biofertilizers. In *Bacterial Metabolites in Sustainable Agroecosystem*; Springer: Cham, Switzerland, 2015; pp. 105–158. [CrossRef]
127. Glick, B.R. Bacteria with ACC deaminase can promote plant growth and help to feed the world. *Microbiol. Res.* **2014**, *169*, 30–39. [CrossRef]
128. du Jardin, P. Plant biostimulants: Definition, concept, main categories and regulation. *Sci. Hortic.* **2015**, *196*, 3–14. [CrossRef]
129. Yakhin, O.I.; Lubyanov, A.A.; Yakhin, I.A.; Brown, P.H. Biostimulants in plant science: A global perspective. *Front. Plant Sci.* **2017**, *7*, 2049. [CrossRef]
130. Toscano, S.; Romano, D.; Massa, D.; Bulgari, R.; Franzoni, G.; Ferrante, A. Biostimulant applications in low input horticultural cultivation systems. *Italus Hortus* **2018**, *25*, 27–36. [CrossRef]
131. Calvo, P.; Nelson, L.; Kloepper, J.W. Agricultural uses of plant biostimulants. *Plant Soil* **2014**, *383*, 3–41. [CrossRef]
132. Massa, D.; Lenzi, A.; Montoneri, E.; Ginepro, M.; Prisa, D.; Burchi, G. Plant response to biowaste soluble hydrolysates in hibiscus grown under limiting nutrient availability. *J. Plant Nutr.* **2018**, *41*, 396–409. [CrossRef]
133. Battacharyya, D.; Babgohari, M.Z.; Rathor, P.; Prithiviraj, B. Seaweed extracts as biostimulants in horticulture. *Sci. Hortic.* **2015**, *196*, 39–48. [CrossRef]
134. Vernieri, P.; Ferrante, A.; Borghesi, E.; Mugnai, S. I biostimolanti: Uno strumento per migliorare la qualità delle produzioni. *Fertil. Agrorum* **2006**, *1*, 17–22.
135. Davies, M.J.; Harrison-Murray, R.; Atkinson, C.J.; Grant, O.M. Application of deficit irrigation to container-grown hardy ornamental nursery stock via overhead irrigation, compared to drip irrigation. *Agric. Water Manag.* **2016**, *163*, 244–254. [CrossRef]

3

Growth Responses and Root Characteristics of Lettuce Grown in Aeroponics, Hydroponics and Substrate Culture

Qiansheng Li [1], Xiaoqiang Li [2], Bin Tang [3] and Mengmeng Gu [1,*]

1 Department of Horticultural Sciences, Texas A & M AgriLife Extension Service, Texas A & M University, College Station, TX 77843, USA; qianshengli@tamu.edu
2 Shanghai Jieyou Agriculture Sci & Tech Co., Ltd., Shanghai 201210, China; xql992@126.com
3 Spraying Systems (Shanghai) Co., Shanghai 201611, China; billtb@spray.com.cn
* Correspondence: mgu@tamu.edu.

Abstract: Aeroponics is a relatively new soilless culture technology which may produce food in space-limited cities or on non-arable land with high water-use efficiency. The shoot and root growth, root characteristics, and mineral content of two lettuce cultivars were measured in aeroponics, and compared with hydroponics and substrate culture. The results showed that aeroponics remarkably improved root growth with a significantly greater root biomass, root/shoot ratio, and greater total root length, root area, and root volume. However, the greater root growth did not lead to greater shoot growth compared with hydroponics, due to the limited availability of nutrients and water. It was concluded that aeroponics systems may be better for high value true root crop production. Further research is necessary to determine the suitable pressure, droplet size, and misting interval in order to improve the continuous availability of nutrients and water in aeroponics, if it is to be used to grow crops such as lettuce for harvesting above-ground parts.

Keywords: soilless culture; root growth; root/shoot ratio

---

1. Introduction

Soilless culture, including aeroponics, aquaponics, and hydroponics, is considered one of the more innovative agricultural strategies to produce more from less, in order to feed the estimated 11 billion people in the world by 2100 [1]. Aeroponics is a promising technology that grows plants with their root systems exposed to a nutrient mist in a closed chamber [2]. Plants grow well in aeroponics, primarily because of the highly aerobic environment it creates. It is even possible to control the root-zone atmosphere when it is combined with a gas delivery system [3]. Integrated vertical aeroponic farming systems with manipulation of temperature and CO_2 in the root-zone environment can achieve more efficient use of land area to secure a vegetable supply in space-limited cities [4]. Aeroponics is also an excellent option for space mission life support systems that require optimum control of growth parameters [5].

Aeroponics has been widely used in plant physiology research, but is not as commonly used as hydroponic methods on a commercial scale [6]. However, aeroponics has been increasingly used for growing numerous vegetable crops such as lettuce, cucumber, melon, tomato, herbs, potato, and floral crops, and especially for those crops where roots are harvested as the end product. Seed potato production may be the most successful application of aeroponics on a commercial scale, done mostly in China, Korea, South America, and African countries in recent years [7–10]. Aeroponics is able to produce large numbers of minitubers in one generation that can be harvested sequentially, eliminating the need for field production, thereby reducing costs and saving time [7]. This technique was applied to effectively produce minitubers of yam (*Dioscorea rotundata* and *D. alata*) in Nigeria and Ghana [11,12]. Aeroponics could be an alternative production system for other high-value root and rhizome crops, such as great burdock (*Arctium lappa*) [13], ginger (*Zingiber officinale*) [14], medicinal crops, such as *Urtica dioica* and *Anemopsis californica* [15] and saffron (*Crocus sativus*) [16]. Essential oil production of herbs like valerian (*Valeriana officinalis*) grown using aeroponics has also been reported [17]. Aeroponics has also been reported as an economic method for rapid root induction and clonal propagation of three endangered and medicinally important plants [18]. Aeroponics was used to produce tree saplings (*Acacia mangium*) with arbuscular mycorrhiza (AM) fungi inoculation [19]. The well-aerated root environment of aeroponics was beneficial for root initiation and subsequent root growth in woody (*Ficus*) and herbaceous (*Chrysanthemum*) cuttings [20].

Many studies have clearly shown that aeroponics promotes plant growth rates through optimization of root aeration because the plant is totally suspended in air, giving the plant stem and root systems access to 100% of the available oxygen in the air [7]. Droplet size and frequency of exposure of the roots to the nutrient solution are the critical factors which may affect oxygen availability [2]. Large droplets lead to less oxygen being available to the root system, while fine droplets produce excessive root hair without developing a lateral root system for sustained growth [10]. Three broad categories are generally used to classify droplet forming systems and droplet size; regular spray nozzles with droplet size >100 μm (spray), compressed gas atomizers with droplet size between 1 to 100 μm (fog), and ultrasonic systems with droplet size 1 to 35 μm (mist) [21]. The most common type is when the nutrient solution is compressed through nozzles by a high pressure pump, forming a fine mist in the growth chambers [7]. An ultrasonic misting system was adopted in a sterile aeroponics culture system for in vitro propagation [22].

In this study, air atomizing nozzles (1/4J Series) were employed for the aeroponics system. The air atomizing nozzles require a single air source for atomizing the air and to provide independent control of liquid, atomizing air, and fan air pressure for fine tuning of the flow rate, droplet size, spray distribution, and coverage. These air atomizing nozzles were equipped with clean-out needles to eliminate clogging and ensure optimum performance. The objectives of the present study were to compare shoot and root growth, root characteristics, and mineral contents of two lettuce cultivars grown in aeroponics, hydroponics (nutrient film technique, NFT) and substrate culture.

2. Materials and Methods

2.1. Cultivation Systems

The experiment was carried out in a 12.8 × 24 m experimental Venlo type glasshouse, which was equipped with outside and inside shade nets, fans and pad, and misting system. The aeroponics units were built in an A-frame shape, 1.4 m wide, 1.4 m high, and 6 m long; both sides were covered with multiple Styrofoam panels at 60° angles. The planting density was 25 plants per m^2 at a spacing of 20 × 20 cm. Six nozzles (AEROJSUMAX-6SS, Spraying Systems (Shanghai) Co., Shanghai, China) were placed horizontally at the end and middle of the A-frame. The nozzles can be operated under an air pressure from 0.7×10^5 to 4×10^5 Pa, with liquid capacity from 7.6 to 63 L/h. AutoJet® Spray System (Spraying Systems (Shanghai) Co.) and was installed to monitor and automatically adjust the spray pattern, flow rate, droplet size, liquid pressure, and atomizing air pressure. Misting lasted 20 s with a 30 s interval before misting again. The droplet size was adjusted to 50 μm and the nutrient solution was recycled.

The NFT hydroponics system consisted of a PVC trough on a slope of 1 percent. The trough was 10 cm wide, 5 cm deep, and 6 m long. The cascade troughs were suspended one above the other, up to 7 levels. The nutrient solution entered the high end of the slightly sloping top trough, exited at the low end of that trough into the high end of the next one, and so forth, and back to the reservoir from where it was pumped. The flow rate was set at 5 L/min. Plants were set 20 cm apart in each trough.

The substrate culture system was conducted using square PVC plastic pots, which were 46 cm long, 40 cm wide, and 18 cm high. Six plants were planted in each pot at a 20 × 20 cm spacing. A mix of 50% peat and 50% perlite was used as the substrate. The substrate depth was approximately 18 cm. The nutrient solution of 1.2 L/day was supplied through 3 drip lines in each pot twice a day.

2.2. Planting and Experimental Arrangement

Lettuce seeds of cultivars 'Nenglv naiyou' and 'Dasusheng' (Institute of vegetable and flower, CAAS, Beijing, China) were sown in 72-cell polystyrene trays. Each cell was filled with a hydroponic planting basket with a sponge for support. At the two true leaves stage, all plants were watered with a half strength Hoagland's nutrient solution [23] until the seedlings were ready for transplanting. Four weeks after sowing, lettuce seedlings were transplanted to the aeroponics, hydroponics, and substrate culture systems. Plants were supplied with full strength Hoagland's nutrient solution (containing N 210 mg/L, K 235 mg/L, Ca 200 mg/L, P 31 mg/L, S 64 mg/L, Mg 48 mg/L, B 0.5 mg/L, Fe 1 to 5 mg/L, Mn 0.5 mg/L, Zn 0.05 mg/L, Cu 0.02 mg/L, Mo 0.01 mg/L). Three A-shape aeroponics units, 12 hydroponic troughs, and 24 substrate culture pots were planted for the comparison experiment.

2.3. Harvesting and Measurement

2.3.1. Biomass and Root/Shoot Ratio

Nine plants from each cultivation system were harvested and washed with tap water. Substrates in the roots of the plants from the substrate cultivation treatment were gently washed off. The fresh weight of shoots and roots was recorded immediately after removing the free surface moisture with soft paper towels. Shoot and root samples were then oven dried at 85 °C for 48 h, and weighed for dry weight on a scale accurate to 0.0001 g. The root/shoot ratio was calculated as the root dry weight/shoot dry weight.

2.3.2. Root Characteristics

Five plants from each cultivation treatment were randomly sampled for measurement of root characteristics. Washed roots were immersed and spread out in a 40 × 25 × 10 cm square blue plastic container which was filled with tap water to a depth of 3 cm. The entire root system was photographed from above with a digital camera (Nikon D90, Nikon Corporation, Tokyo, Japan) and saved using the jpeg format (Figure 1A). The photographs were re-cropped, scaled (Figure 1B), and processed with GiA Roots software (Georgia Tech Research Corporation and Duke University, USA) to obtain a threshold image (Figure 1C) for measuring the characteristics of all the roots. The measured root characteristics included average root diameter (width), root length (network length), root area (network surface area), root volume (network volume), maximum number of roots, median number of roots, and network perimeter [24].

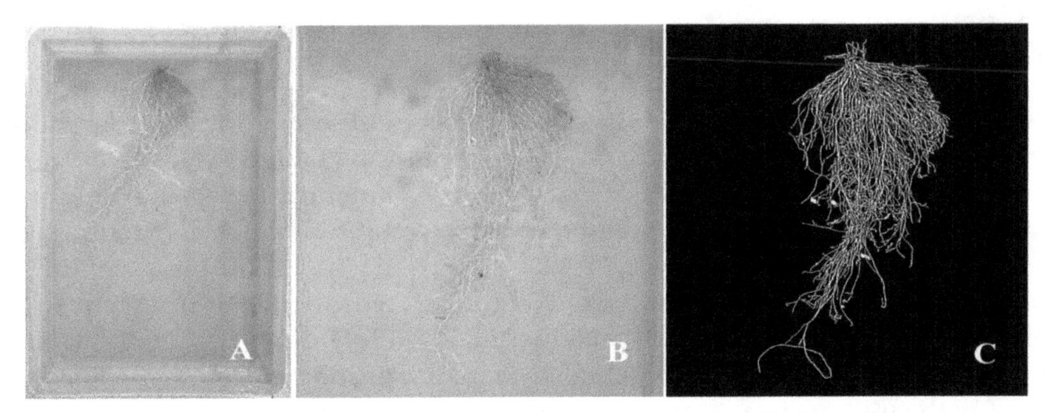

Figure 1. The entire root was immersed and spread out in a 40 × 25 × 10 cm square plastic container filled with tap water to a depth of 3 cm (**A**), the image was re-cropped after being scaled (**B**), and changed to a threshold image (**C**) with GiA Roots software, for root characteristics measurements.

2.3.3. Plant Leaf Nitrogen, Phosphorus, and Potassium Content

Three plants from each cultivation treatment were randomly sampled for mineral nutrient analysis and oven dried as above. The dried shoots from the different treatments were milled to passed through a 1 mm screen. The ground dry material (~0.2000 g) was wet digested using a H_2SO_4-H_2O_2 solution. Nitrogen content was determined using the Kjeldahl method [25]. Phosphorus was determined by the ascorbic acid molybdenum blue method [26]. Potassium was determined by flame emission spectrophotometry [27].

2.4. Statistical Analysis

All results were subjected to a two-way analysis of variance (ANOVA) using SPSS Statistics 19.0; the effects of the cultivation system, genotypes (cultivars), and their interaction were analyzed. Within each cultivar, means were separated using Duncan's multiple range test at $P = 0.05$. The results were expressed as means $\pm$ SE.

3. Results

3.1. Plant Growth and Biomass

The cultivation systems significantly influenced the growth of both lettuce cultivars. Lettuce grown in hydroponics had larger above-ground parts, while the aeroponic lettuce had greater root dry weight and root/shoot ratio, and plants from substrate cultivation had the smallest size (Figure 2, Table 1).

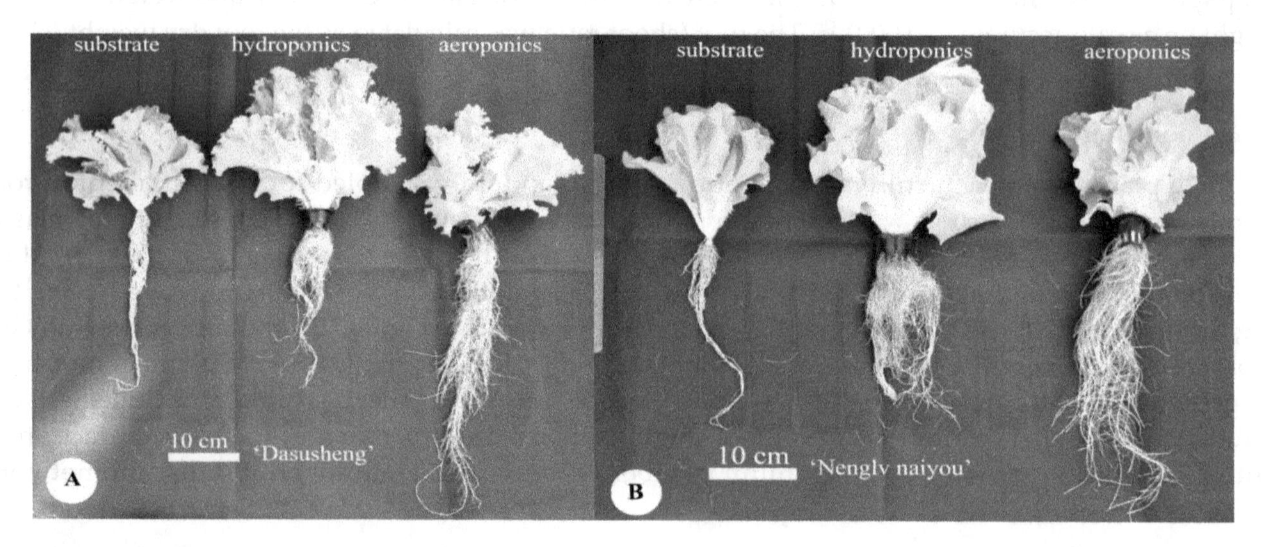

Figure 2. Whole plants of lettuce cultivars 'Dasushen' (**A**) and 'Nenglv naiyou' (**B**) 45 days after transplanting in aeroponic, hydroponic, and substrate cultivation systems.

The two-way ANOVA showed significant effects of cultivation system on shoot and root fresh weight, shoot and root dry weight, and root/shoot ratio (Table 1). However, genotype only showed significant effects on shoot and root dry weight. An interaction between cultivation system and genotype was significant on fresh weight and dry weight of shoot and root, but not on the root/shoot ratio.

In both cultivars, the shoot fresh and dry weights of hydroponic lettuce were approximately twice that of aeroponic and substrate cultivated lettuce (Table 1). Root fresh weights of aeroponics and hydroponics lettuce were significantly higher than that of substrate cultivated lettuce. The root dry weights of both cultivars in aeroponics were significantly higher than that of hydroponics and substrate cultivation. The most remarkable difference between the three growing methods was the root/shoot ratio. In both cultivars, the root/shoot ratio of aeroponics lettuce was almost three times that of the hydroponics lettuce, and was also significantly higher than that of the substrate culture (Table 1).

Table 1. Shoot and root fresh weight (FW), dry weight (DW), and root to shoot ratio of two lettuce cultivars grown in three cultivation systems.

	Shoot FW (g)	Root FW (g)	Shoot DW (g)	Root DW (g)	Root/Shoot Ratio
			Lactuca sativa 'Dasusheng'		
aeroponics	37.8 ± 2.67 [bz]	8.67 ± 1.20 [a]	2.40 ± 0.17 [b]	0.80 ± 0.15 [a]	0.32 ± 0.04 [a]
hydroponics	88.8 ± 9.47 [a]	8.78 ± 1.24 [a]	4.86 ± 0.54 [a]	0.59 ± 0.06 [b]	0.12 ± 0.01 [c]
substrate	49.2 ± 2.34 [b]	6.92 ± 0.43 [b]	3.23 ± 0.13 [b]	0.69 ± 0.03 [b]	0.22 ± 0.02 [b]
			Lactuca sativa 'Nenglv Naiyou'		
aeroponics	50.9 ± 2.60 [b]	10.3 ± 0.46 [a]	2.58 ± 0.11 [b]	0.77 ± 0.07 [a]	0.30 ± 0.03 [a]
hydroponics	96.1 ± 4.23 [a]	11.5 ± 1.07 [a]	4.80 ± 0.16 [a]	0.54 ± 0.03 [b]	0.11 ± 0.01 [b]
substrate	39.4 ± 1.72 [c]	3.9 ± 0.35 [b]	2.03 ± 0.06 [c]	0.26 ± 0.03 [c]	0.13 ± 0.01 [b]
			Significance		
Cultivation system (CS)	*** [y]	***	***	**	***
Genotype (G)	Ns	Ns	*	*	ns
CS × G	*	**	*	*	ns

[z] Values are mean ± SE (n = 9). In the same cultivar, values followed by the same superscript letter are not significantly different ($P \leq 0.05$). [y] *** = $P < 0.001$; ** = $P < 0.01$; * = $P < 0.05$; ns, not significant at $P \geq 0.05$.

3.2. Root Characteristics

Analysis of the root characteristics by GiA Roots software revealed details of the influence of growing methods on root growth. The root length, area, volume, and network perimeter of aeroponic lettuce (both cultivars) were significantly greater than that of hydroponic and substrate cultivated lettuce (Table 2). In particular, for the cultivar 'Dasusheng', the root length, root area, root volume, and perimeter in aeroponic cultivation were four to five times that of the hydroponic and substrate cultivation. However, the average root diameter did not significantly differ among treatments in the cultivar 'Dasusheng'. Average root diameter of hydroponically-grown 'Nenglv naiyou' was significantly greater than that from substrate cultivation. The maximum and median numbers of roots of aeroponic 'Dasusheng' lettuce were two to three times higher than that from hydroponic and substrate cultivation; such differences in maximum root number were not found for the cultivar 'Nenglv naiyou' where only median root number for aeroponic cultivation was greater than hydroponic but not substrate cultivation.

Table 2. Root characteristics of lettuce grown in aeroponics, hydroponics, and substrate culture systems.

	Average Root Diameter (mm)	Root Length (cm)	Root Area (cm^2)	Root Volume (cm^3)	Maximum No. of Roots	Median No. of Roots	Network Perimeter (cm)
			Lactuca sativa 'Dasusheng'				
aeroponics	0.501 ± 0.017 [az]	3043 ± 231 [a]	479 ± 42 [a]	7.24 ± 0.77 [a]	75.0 ± 5.8 [a]	39.0 ± 4.1 [a]	6019 ± 473 [a]
hydroponics	0.551 ± 0.025 [a]	581 ± 113 [b]	100 ± 20 [b]	1.64 ± 0.35 [b]	24.6 ± 2.9 [b]	13.8 ± 1.6 [b]	1164 ± 221 [b]
substrate	0.501 ± 0.007 [a]	724 ± 126 [b]	114 ± 21 [b]	1.67 ± 0.35 [b]	32.6 ± 5.3 [b]	17.4 ± 2.2 [b]	1437 ± 245 [b]
			Lactuca sativa 'Nenglv Naiyou'				
aeroponics	0.511 ± 0.0023 [ab]	2634 ± 260 [a]	424 ± 46 [a]	6.63 ± 0.84 [a]	63.0 ± 7.4 [a]	34.4 ± 3.1 [a]	5197 ± 557 [a]
hydroponics	0.554 ± 0.0009 [a]	1688 ± 239 [b]	296 ± 46 [b]	4.91 ± 0.85 [ab]	54.4 ± 3.0 [a]	19.0 ± 4.8 [b]	3379 ± 473 [b]
substrate	0.487 ± 0.0006 [b]	1378 ± 58 [b]	211 ± 10 [b]	2.99 ± 0.15 [b]	51.2 ± 1.6 [a]	37.0 ± 2.6 [a]	2755 ± 113 [b]
			Significance				
Cultivation system (CS)	ns [y]	***	**	*	***	***	*
Genotype (G)	Ns	***	***	***	***	**	***
CS × G	*	*	ns	Ns	ns	*	*

[z] Values are mean ± SE (n = 9). In the same cultivar, values followed by the same superscript letter are not significantly different ($P \leq 0.05$). [y] *** = $P < 0.001$; ** = $P < 0.01$; * = $P < 0.05$; ns, not significant at $P \geq 0.05$.

The two-way ANOVA indicated that both the cultivation system and genotype significantly affected all the root characteristics except the average root diameter, i.e., average root diameter, root length, median number of roots, and root perimeter (Table 2).

3.3. *Leaf N, P, and K Contents*

In both cultivars, the leaf N content of hydroponic lettuce was significantly higher than that of aeroponic and substrate cultivated lettuce, but there was no difference between aeroponic and substrate cultivation (Figure 3A). The leaf P content of hydroponic lettuce was significantly higher than that of the aeroponic and substrate cultivated lettuce (Figure 3B). Leaf K content of both aeroponic and hydroponic lettuce was significantly higher than that of substrate cultivated lettuce, but there was no difference between the aeroponic and hydroponic lettuce (Figure 3C).

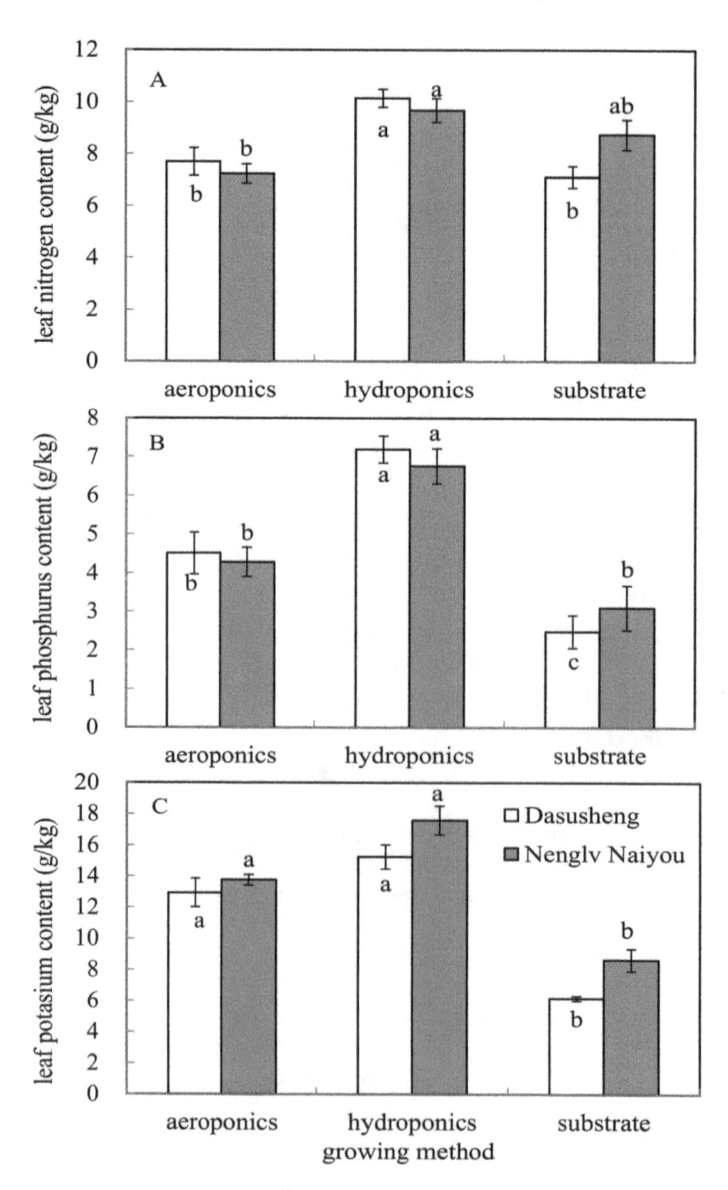

Figure 3. Leaf nitrogen (**A**), phosphorus (**B**), and potassium (**C**) content of two lettuce cultivars grown on aeroponics, hydroponics, and substrate cultivation systems. Values are mean ± SE ($n = 3$). In the same cultivar, values followed by the same letter are not significantly different ($P \leq 0.05$).

The two-way ANOVA indicated that the cultivation system significantly affected leaf N, P, and K contents (Table 3), while genotype only had a significant effect on the leaf K content, and there was no significant interaction between cultivation system and genotype.

Table 3. *F*-values from a two-way analysis of variance (ANOVA) of leaf nitrogen, phosphorus, and potassium content as affected by three cultivation systems and two genotypes.

	N	P	K
Cultivation system (CS)	11.6 *** z	50.6 ***	84.6 **
Genotype (G)	0.28 ns	0.01 ns	10.7 **
CS × G	2.51 ns	0.88 ns	0.84 ns

z *** = $P < 0.001$; ** = $P < 0.01$; ns, not significant, $P \geq 0.05$.

4. Discussion

The most impressive result of this study was the significant improvement of root growth of lettuce in the aeroponic system. The cultivation systems also affected the fresh weight, dry weight, and root/shoot ratio, while the genotypes only had significant effects on dry weight, and the interaction between them also significantly affected the biomass but not the root/shoot ratio. In both cultivars, the root dry weight of aeroponic lettuce was significantly higher than that of hydroponic and substrate cultivation, and the root/shoot ratio of aeroponic lettuce was two to three times that of the other two systems. The two-way ANOVA results indicated that the root characteristics were more dependent on genotype; however, the cultivation systems also had significant effects on the root characteristics except on root diameter. The greater total root length, root area, and root volume further proved that aeroponics was beneficial to root growth. However, the greater root system of aeroponics did not lead to more shoot biomass (yield) than hydroponics. Instead, shoot biomass of aeroponic lettuce was significantly less than that of hydroponics. This may be due to sufficient nutrient and water supply when the root system was submerged continuously in nutrient solution. The cultivation systems had significant influences on leaf N, P, and K content, while genotype only showed significant effects on K content, and there was no cultivation system by genotype interaction on the mineral contents.

In aeroponics, the nutrient solution was only sprayed as fine droplets at intervals, which may limit shoot growth and improve root growth, as the plant's response may be to adapt to the relative deficit of water and nutrients during the intervals. In valerian (*Valeriana officinalis*) cultivation trials, it was also found that both leaf area and biomass production in an aeroponic system were lower than in floating raft hydroponic and substrate cultivation systems; it was concluded that this may be caused by the higher proliferation of roots inside the frame reducing the performance of nozzles [17]. The root number of saffron (*Crocus sativus*) plants was also significantly greater in aeroponics than that in hydroponics and soil culture, but no significant difference in shoot growth was found [16]. The larger distance between misting sprayers and roots restricted root access to the water micro droplets, resulted in decreasing nutrient availability and absorbance. In this case, plants were forced to compensate by increasing root surface area and weight [28]. Thus, the droplet size and the misting interval will have a great effect on plant growth in aeroponic culture.

Good aeration of the root environment is the most important advantage of aeroponics. Aeroponics significantly improved adventitious root formation in rapid root induction and clonal propagation of three endangered and medicinally important plants over soil grown stem cuttings [18]. Aeroponics showed higher yield and better size distribution in potato minituber production, but growth was influenced by such factors as the genotype, the availability of nutrients, the stretching of the cycle, and the culture density [8]. Higher root vitality of plants was observed in aeroponics and aerohydroponics than that of deep water culture [29]. If the roots of plants in aeroponics can absorb nutrients and water readily, better growth of the above-ground part will result. However, the major disadvantage of aeroponics is the possibility of irreversible damage or complete loss, since there is no substrate at all (neither solid nor water) that could enable the plants to survive in the event of a technical or power failure [7,9]. During our experiment, there was a one-day mechanical breakdown of the aeroponic system, which could also have affected growth.

5. Conclusions

From this study, it can be concluded that aeroponics is beneficial to root growth, with significantly greater root/shoot ratio, root length, root area, and root volume. Thus, an aeroponic system may be superior for producing high value, true root crops, particularly for medicinal plants as suggested by Hayden [13,15]. When growing root crops in aeroponics, clean products may be harvested sequentially. To grow crops like lettuce in aeroponics for harvesting above-ground parts, further research is necessary to determine suitable pressure, droplet size, and misting interval in order to improve the continuous availability of nutrients and water so that growth of above-ground parts of plants can be optimized.

Author Contributions: Conceptualization, Q.L.; data curation, Q.L.; formal analysis, Q.L.; investigation, X.L.; methodology, Q.L.; resources, B.T.; supervision, M.G.; writing—original draft, Q.L.; writing—review and editing, M.G.

Acknowledgments: We thank Xue Zhang and Minmin Wan for their help with measurements.

References

1. Lal, R. Feeding 11 billion on 0.5 billion hectare of area under cereal crops. *Food Energy Secur.* **2016**, *5*, 239–251. [CrossRef]
2. Jones, J.B., Jr. *Complete Guide for Growing Plants Hydroponically*; CRC Press: Boca Raton, FL, USA, 2014.
3. Kratsch, H.A.; Graves, W.R.; Gladon, R.J. Aeroponic system for control of root-zone atmosphere. *Environ. Exp. Bot.* **2006**, *55*, 70–76. [CrossRef]
4. He, J. Farming of Vegetables in Space-Limited Environments. *Cosmos* **2015**, *11*, 21–36. [CrossRef]
5. Clawson, J.; Hoehn, A.; Stodieck, L.; Todd, P.; Stoner, R. *Re-Examining Aeroponics for Spaceflight Plant Growth*; SAE Technical Paper 0148-7191; SAE International: Warrendale, PA, USA, 2000.
6. Resh, H.M. *Hydroponic Food Production: A Definitive Guidebook for the Advanced Home Gardener and the Commercial Hydroponic Grower*; CRC Press: Boca Raton, FL, USA, 2012.
7. Buckseth, T.; Sharma, A.K.; Pandey, K.K.; Singh, B.P.; Muthuraj, R. Methods of pre-basic seed potato production with special reference to aeroponics—A review. *Sci. Hortic.* **2016**, *204*, 79–87. [CrossRef]
8. Tierno, R.; Carrasco, A.; Ritter, E.; de Galarreta, J.I.R. Differential Growth Response and Minituber Production of Three Potato Cultivars Under Aeroponics and Greenhouse Bed Culture. *Am. J. Potato Res.* **2013**, *91*, 346–353. [CrossRef]
9. Mateus-Rodriguez, J.R.; de Haan, S.; Andrade-Piedra, J.L.; Maldonado, L.; Hareau, G.; Barker, I.; Chuquillanqui, C.; Otazú, V.; Frisancho, R.; Bastos, C.; et al. Technical and Economic Analysis of Aeroponics and other Systems for Potato Mini-Tuber Production in Latin America. *Am. J. Potato Res.* **2013**, *90*, 357–368. [CrossRef]
10. Margaret, C. Potential of aeroponics system in the production of quality potato (*Solanum tuberosum* L.) seed in developing countries. *Afr. J. Biotechnol.* **2012**, *11*, 3993–3999.
11. Oteng-Darko, P.; Kyei-Baffour, N.; Otoo, E.; Agyare, W.A. Growing Seed Yams in the Air: The Agronomic Performance of Two Aeroponics Systems Developed in Ghana. *Sustain. Agric. Res.* **2017**, *6*, 106–116. [CrossRef]
12. Maroya, N.; Balogun, M.; Asiedu, R.; Aighewi, B.; Kumar, P.L.; Augusto, J. Yam propagation using aeroponics technology. *Annu. Res. Rev. Biol.* **2014**, *4*, 3894–3903. [CrossRef]
13. Hayden, A.; Yokelsen, T.; Giacomelli, G.; Hoffmann, J. Aeroponics: An alternative production system for high-value root crops. *Acta Hortic.* **2004**, *629*, 207–213. [CrossRef]
14. Hayden, A.L.; Brigham, L.A.; Giacomelli, G.A. Aeroponic cultivation of ginger (*Zingiber officinale*) rhizomes. *Acta Hortic.* **2004**, *659*, 397–402. [CrossRef]
15. Hayden, A.L. Aeroponic and hydroponic systems for medicinal herb, rhizome, and root crops. *Hortscience* **2006**, *41*, 536–538.
16. Souret, F.F.; Weathers, P.J. The Growth of Saffron (*Crocus sativus* L.) in Aeroponics and Hydroponics. *J. Herbs Spices Med. Plants* **2000**, *7*, 25–35. [CrossRef]
17. Tabatabaei, S.J. Effects of Cultivation Systems on the Growth, and Essential Oil Content and Composition of Valerian. *J. Herbs Spices Med. Plants* **2008**, *14*, 54–67. [CrossRef]

18. Mehandru, P.; Shekhawat, N.S.; Rai, M.K.; Kataria, V.; Gehlot, H.S. Evaluation of aeroponics for clonal propagation of Caralluma edulis, Leptadenia reticulata and Tylophora indica—Three threatened medicinal Asclepiads. *Physiol. Mol. Biol. Plants* **2014**, *20*, 365–373. [CrossRef] [PubMed]
19. Martin-Laurent, F.; Lee, S.-K.; Tham, F.-Y.; Jie, H.; Diem, H.G. Aeroponic production of Acacia mangium saplings inoculated with AM fungi for reforestation in the tropics. *For. Ecol. Manag.* **1999**, *122*, 199–207. [CrossRef]
20. Soffer, H.; Burger, D.W. Effects of dissolved oxygen concentrations in aero-hydroponics on the formation and growth of adventitious roots. *J. Am. Soc. Hortic. Sci.* **1988**, *113*, 218–221.
21. Weathers, P.; Liu, C.; Towler, M.; Wyslouzil, B. Mist reactors: Principles, comparison of various systems, and case studies. *Electron. J. Integr. Biosci.* **2008**, *3*, 29–37.
22. Tisserat, B.; Jones, D.; Galletta, P.D. Construction and use of an inexpensive in vitro ultrasonic misting system. *HortTechnology* **1993**, *3*, 75–78.
23. Hoagland, D.R.; Arnon, D.I. *The Water-Culture Method for Growing Plants without Soil*, 2nd ed.; Circular 347; California Agricultural Experiment Station: Berkeley, CA, USA, 1950.
24. Galkovskyi, T.; Mileyko, Y.; Bucksch, A.; Moore, B.; Symonova, O.; Price, C.A.; Topp, C.N.; Iyer-Pascuzzi, A.S.; Zurek, P.R.; Fang, S. GiA Roots: Software for the high throughput analysis of plant root system architecture. *BMC Plant Biol.* **2012**, *12*, 116. [CrossRef] [PubMed]
25. Horneck, D.A.; Miller, R.O. Determination of total nitrogen in plant tissue. In *Handbook of Reference Methods for Plant Analysis*; Kalra, Y.P., Ed.; CRC Press LLC: Boca Raton, FL, USA, 1998; pp. 75–83.
26. Jones, J.B., Jr. *Laboratory Guide for Conducting Soil Tests and Plant Analysis*; CRC Press: Boca Raton, FL, USA, 2001.
27. Horneck, D.; Hanson, D. Determination of potassium and sodium by flame emission spectrophotometry. In *Handbook of Reference Methods for Plant Analysis*; Kalra, Y.P., Ed.; CRC Press LLC: Boca Raton, FL, USA, 1998; pp. 153–155.
28. Salachas, G.; Savvas, D.; Argyropoulou, K.; Tarantillis, P.; Kapotis, G. Yield and nutritional quality of aeroponically cultivated basil as affected by the available root-zone volume. *Emir. J. Food Agric.* **2015**, *27*, 911–918. [CrossRef]
29. Chang, D.C.; Park, C.S.; Kim, S.Y.; Lee, Y.B. Growth and Tuberization of Hydroponically Grown Potatoes. *Potato Res.* **2012**, *55*, 69–81. [CrossRef]

4

Timing of a Short-Term Reduction in Temperature and Irradiance Affects Growth and Flowering of Four Annual Bedding Plants

Jennifer K. Boldt * and James E. Altland

United States Department of Agriculture, Agricultural Research Service, Application Technology Research Unit, Wooster, OH 44691, USA; James.Altland@ars.usda.gov

* Correspondence: Jennifer.Boldt@ars.usda.gov.

Abstract: Heating and supplemental lighting are often provided during spring greenhouse production of bedding plants, but energy inputs are a major production cost. Different energy-savings strategies can be utilized, but effects on plant growth and flowering must be considered. We evaluated the impact and timing of a two-week low-energy (reduced temperature and irradiance) interval on flowering and growth of impatiens (*Impatiens walleriana* Hook.f. 'Accent Orange'), pansy (*Viola* × *wittrockiana* Gams. 'Delta Premium Blue Blotch'), petunia (*Petunia* × *hybrida* Hort. Vilm.-Andr. 'Dreams Pink'), and snapdragon (*Antirrhinum majus* L. 'Montego Violet'). Flowering was delayed 7 to 10 days when the low-energy exposure occurred before flowering. Flower number was reduced 40–61% in impatiens, 33–35% in petunia (low-energy weeks 5–6 and weeks 7–8, respectively), and 35% in pansy (weeks 5–6). Petunia and impatiens dry mass gradually decreased as the low-energy exposure occurred later in production; petunias were 26% (weeks 5–6) and 33% (weeks 7–8) smaller, and impatiens were 20% to 31% smaller than ambient plants. Estimated energy savings were 14% to 16% for the eight-week period, but only up to 7% from transplant to flowering. Growers can consider including a two-week reduction in temperature and irradiance to reduce energy, provided an additional week of production is scheduled.

Keywords: temperature; irradiance; ornamental plants; greenhouse production

---

1. Introduction

Greenhouse production of annual bedding plants for spring markets occurs in late winter and early spring. Heating and supplemental lighting are often provided to offset low outdoor temperatures and augment low solar irradiation intensity and duration. Consequently, energy inputs are a major production cost for greenhouse-grown plants. In the United States, energy is the third largest expense after labor and plant material, and it accounted for 9% of total production costs in the 2014 Census of Horticultural Specialties [1]. Approximately 65% to 85% of total greenhouse energy consumption is for heating [2].

Irradiance drives photosynthesis and primarily influences crop growth and dry weight gain. Temperature primarily influences crop development, including rates of leaf unfolding, flower initiation, and flower development. Together, light and temperature impact crop timing and quality. The ratio of radiant energy (light) to thermal energy (temperature), or RRT, is one way to describe this [3]. A higher RRT increases crop quality; for example, plants grown at lower temperatures and higher irradiance (high RRT) will be of higher quality than plants grown at higher temperatures and lower irradiance (low RRT). This ratio is an indicator of plant carbon balance, which becomes depleted under prolonged exposure to high temperature and low irradiance [4]. For example, starch levels in rose (*Rosa* × *hybrida*

'Red Berlin') were similar when plants were exposed to high temperature and high irradiance or low temperature and low irradiance but were diminished under high temperature and low irradiance [4].

Crops need to meet both a target market date and minimum quality standards. Lighting and temperature set points can be adjusted during production to address economic and environmental concerns. Growers are continually looking for strategies to reduce energy consumption and costs without sacrificing plant growth, quality, and/or finished time. Therefore, it is important to evaluate different production strategies and determine when they may be feasible to implement. Surveys have found 55% to 58% of responding U.S. growers have implemented conservation or energy efficient practices [5,6]. Installing an energy or thermal screen is one strategy for reducing energy use [2], but only 12% of respondents in one of the surveys had them [5]. Other viable production strategies for reducing energy use include growing in unheated greenhouses or high tunnels [7], using root zone heating [8], or using a reduced temperature to finish (RTF) [9].

Many growers maintain static air temperature set points, independent of ambient weather conditions outside [10]. Allowing the greenhouse temperature to rise above the desired mean daily temperature (MDT) when heating demand is low and fall below the desired MDT when heating demand is high, but maintain the same MDT, is another strategy to reduce energy use [11]. These temperature integration strategies have been referred to as dynamic temperature control [2], integrating temperature control [12], multi-day temperature setting [13], or dynamic photosynthetic optimization [14]. While MDT remains the same, a wider range of acceptable temperature fluctuations is allowed. This can be accomplished by increasing the daytime ventilation set point and decreasing the nighttime heating set point, using a computer algorithm to maintain a rolling MDT and adjust temperatures based on predicted weather patterns [15], or using a computer algorithm to adjust greenhouse conditions based on photosynthetic optimization [14,16] or plant assimilate balance [13]. Temperatures need to remain within the linear range of plant development rate, between the base temperature (T_{base}; development rate = 0) and the optimum temperature (T_{opt}; development rate is maximal) for each species to minimize delays in development [17].

Dynamic temperature management can be integrated on a 24 h [11,18], multi-day [15], or weekly basis [19,20]. It has been successful for roses [12,21], potted plants [18,22], and vegetables [19] when the amplitude of the bandwidth was $\leq \pm 6$ °C. One drawback, however, is that dynamic heating requires a greenhouse environmental control computer with sophisticated software, and not all growers have or can afford these systems [2]. For greenhouses without environmental control systems, an energy-reduction alternative is to optimize the growing environment on days with a lower heating requirement (i.e. warmer, sunnier, and or less windy) and reduce energy inputs (lower the temperature, turn off supplemental lighting, and close the energy curtain) on days that require more heating. This short-term reduction in temperature and irradiance was successful when implemented 1 to 2 days per week, with reduced energy costs and minimal impact on plant growth and crop timing [23,24]. However, weather patterns often are cyclical on a longer time scale, and growers may contend with days to weeks of continuous cloudy weather during winter and early spring. Condensing the low temperature and irradiance exposure into a continuous time period rather than interspersed throughout production may impact its successful implementation, even though it is for a similar total number of days. Therefore, our objective was to evaluate the timing of a two-week reduction in temperature and irradiance on plant growth and flowering of four popular annual bedding plant crops and estimate potential cost savings using the Virtual Grower software program. We selected cold-tolerant [pansy (*Viola* × *wittrockiana* Gams.) and snapdragon (*Antirrhinum majus* L.)], cold-intermediate [petunia (*Petunia* × *hybrida* Hort. Vilm.-Andr.)], and cold-sensitive [impatiens (*Impatiens walleriana* Hook.f.)] species, as categorized by their T_{base} [25,26].

2. Materials and Methods

Seeds of impatiens 'Accent Orange', pansy 'Delta Premium Blue Blotch', petunia 'Dreams Pink', and snapdragon 'Montego Violet' were sown on 15 December 2014 (replication 1) and 6 January

2015 (replication 2) into 288-cell plug trays filled with a peat-based soilless substrate (LC-1; Sun Gro Horticulture, Bellevue, WA). Trays were placed in a growth chamber (GR48; Environmental Growth Chambers, Chagrin Falls, OH) set to provide 25 °C constant air temperature, 300 $\mu mol \cdot m^{-2} \cdot s^{-1}$ photosynthetic photon flux density (PPFD) from high-pressure sodium (HPS) lamps, and an 8 h photoperiod. They were watered as needed, and provided 75 $mg \cdot L^{-1}$ N constant liquid feed of 20N–4.4P–16.6K (Jack's 20-10-20; JR Peters, Inc., Allentown, PA, USA) at each irrigation once true leaves emerged.

Two greenhouse environments were set up in identical compartments located within a glass-glazed greenhouse (Toledo, OH, USA). The ambient compartment represented typical greenhouse conditions during winter and early spring. Temperature set points were 22 °C day/18 °C night. High-pressure sodium light fixtures (Sunlight Supply, Inc., Vancouver, WA, USA) provided approximately 75 $\mu mol \cdot m^{-2} \cdot s^{-1}$ of supplemental irradiance from 1000-W bulbs (Osram Sylvania Products, Inc., Manchester, NH, USA) when ambient PPFD at the benchtop was less than 300 $\mu mol \cdot m^{-2} \cdot s^{-1}$, and a constant 14 h photoperiod (0600-2000 HR) was maintained. The cool compartment represented a cool, low light environment. Temperature set points were 13 °C day/10 °C night. A spun-woven energy curtain was continuously closed, which reduced ambient irradiance by approximately 50%, relative to ambient conditions. Day-extension lighting with HPS lamps was provided when ambient irradiance was less than 10 $\mu mol \cdot m^{-2} \cdot s^{-1}$ PPFD to achieve a constant 14 h photoperiod. Dataloggers (HOBO®Pro v2; Onset Applications, Bourne, MA, USA) in each environment measured air temperature. Quantum sensors (Model QSO-S; Apogee Instruments, Logan, UT, USA) were connected to a data logger (CR10X; Campbell Scientific, Logan, UT, USA) and mean PPFD was recorded every 15 min.

On 20 January and 9 February 2015, plants were transplanted into 11.5 cm diameter round pots filled with LC-1. They were irrigated as needed during the experiment. Plants were watered once weekly with reverse-osmosis water and fertilized with 20N–4.4P–16.6K at an N concentration of 150 $mg \cdot L^{-1}$ at all other irrigation events. Electrical conductivity (EC) and pH of the substrate solution was monitored every two weeks on three additional plants of each species grown continuously in the ambient and cool, low light environments, using the Pour-Through technique, to ensure values remained within the recommended ranges for all species [27].

Plants were moved from ambient conditions to the cool, low light environment for a two-week interval during the eight-week duration of the experiment (i.e., weeks 1–2, weeks 3–4, weeks 5–6, or weeks 7–8 in cool, low light, with the other 6 weeks in ambient conditions). Two additional treatments included a continuous ambient control and a continuous cool, low light control. There were five plants per species per treatment. Mean air temperatures and daily light integrals (DLIs) are provided in Table 1.

Table 1. Mean air temperature (°C) and daily light integral (DLI, $mol \cdot m^{-2} \cdot d^{-1}$) for each treatment. Plants were grown in ambient conditions [22/18 °C air temperature, 14 h photoperiod, and ambient irradiance +75 $\mu mol \cdot m^{-2} \cdot s^{-1}$ supplemental lighting from high-pressure sodium (HPS) lamps when photosynthetic photon flux density (PPFD) was less than 300 $\mu mol \cdot m^{-2} \cdot s^{-1}$] and transferred to cool, low light conditions (13/10 °C air temperature, ambient irradiance, and a 14 h photoperiod achieved by providing 75 $\mu mol \cdot m^{-2} \cdot s^{-1}$ from HPS lamps when PPFD was less than 10 $\mu mol \cdot m^{-2} \cdot s^{-1}$) at 2-week intervals during an 8-week production cycle.

Treatment	Target Mean Temperature (°C)	Mean Temperature (°C)		DLI ($mol \cdot m^{-2} \cdot d^{-1}$)	
		Replication 1	Replication 2	Replication 1	Replication 2
Ambient	20.3	20.3 ± 1.7	20.3 ± 1.7	10.7 ± 3.0	12.8 ± 3.4
Weeks 1–2 cool	18.2	18.3	18.7	10.2	11.6
Weeks 3–4 cool	18.2	18.5	18.3	9.7	11.3
Weeks 5–6 cool	18.2	18.5	18.6	9.4	10.9
Weeks 7–8 cool	18.2	18.5	18.4	8.9	10.7
Cool	11.8	12.9 ± 1.1	13.2 ± 1.4	6.1 ± 1.5	6.3 ± 1.9

Flowering was checked daily, and date of first flower was recorded. Eight weeks after transplant, flower number was counted. Relative chlorophyll content (CCM-200; Apogee Instruments, Inc., Logan, UT) was measured on three recently mature leaves per plant, and the mean value was used for statistical analysis. Plant height was measured from the substrate surface to the apex. Plant width was measured at the widest point and perpendicular to the widest point, then the two measurements were averaged. Above-ground plant tissue was removed, washed with 0.1 N HCl, rinsed with ultra-purified (18 MΩ) water, dried in a forced-air oven at 60 °C for 3 days, and weighed for dry mass.

A virtual greenhouse was constructed to estimate daily heating and supplemental lighting costs for each environment, using the USDA-ARS software program Virtual Grower 3.0.9 (USDA-ARS, Toledo, OH, USA). Greenhouse dimensions, materials, and components were as described previously [24]. Total energy costs for each treatment were calculated (1) as the sum of daily energy costs for the eight-week production duration, and (2) as the sum of daily energy costs from the start of the experiment to mean date of the first flower, based on the temperature and supplemental lighting schedules.

Data were analyzed as a randomized complete block design, with six treatments, five single-plant replications per treatment, and repeated twice in time. A separate analysis was conducted for each species. Data were analyzed in SAS (SAS 9.3; SAS Institute, Inc., Cary, NC, USA) using the GLM procedure (PROC GLM) and mean separation was conducted with Tukey's HSD at $\alpha = 0.05$ for significant treatment effects ($P \leq 0.05$).

3. Results and Discussion

Length of production is a critical benchmark for greenhouse growers, as increased production time reduces the number of crop cycles per season and increases the fixed costs allocated to each crop. Flowering is important for quick sell-through at retail, and therefore, time to flower is an important scheduling metric. Compared to control plants grown at ambient conditions, flowering was delayed when plants were provided with a two-week low-energy interval early in production. This occurred in all crops grown in the cool, low light environment in weeks 1–2 or weeks 3–4, and additionally in petunia in weeks 5–6 (Table 2). The delay in flowering was 7 to 10 days. Ambient control plants began flowering during week 5 of production (29 to 35 days after transplant); mean days to flower was 30, 32, 34, and 34 days in pansy, impatiens, snapdragon, and petunia, respectively. As such, the absence of delayed flowering on plants exposed to the low-energy exposure in weeks 5–6 or weeks 7–8 occurred because plants were budded or flowering before the start of the cool, low light exposure.

Continuous exposure to low-energy conditions delayed flowering by more than 3 weeks, compared to ambient controls (Table 2). Incomplete flowering occurred in impatiens and snapdragon by the end of the experiment, 8 weeks after transplant, and none of the petunias had flowered. Minimum temperature (T_{min}) values for flower development of 2.0 to 4.0 °C have been reported for snapdragon, 2.8 to 5.5 °C for petunia, 4.1 °C for viola (*Viola cornuta* L. 'Sorbet Plum Velvet'), and 7.2 °C for impatiens 'Blitz 3000 Deep Orange' [25,26]. Although we used different cultivars in our study, the 10 °C nighttime set point in the low-energy treatment was greater than the T_{min} for all species. Therefore, flowering would have eventually occurred, but the delay and variability of flowering does not make continuous low-energy conditions a viable production strategy for most species.

Table 2. Plant growth and flowering of impatiens (*Impatiens walleriana* 'Accent Orange'), pansy (*Viola* × *wittrockiana* 'Delta Premium Blue Blotch'), petunia (*Petunia* × *hybrida* 'Dreams Pink'), and snapdragon (*Antirrhinum majus* 'Montego Violet') grown in ambient conditions (ambient), grown in ambient conditions and transferred to a low-energy environment at two-week intervals during an eight-week production cycle, or grown continuously in the low-energy environment (continuous). Ambient conditions were 22/18 °C air temperature, 14 h photoperiod, and ambient irradiance + 75 $\mu mol \cdot m^{-2} \cdot s^{-1}$ supplemental lighting from high-pressure sodium (HPS) lamps when photosynthetic photon flux density (PPFD) was less than 300 $\mu mol \cdot m^{-2} \cdot s^{-1}$. Low-energy conditions were 13/10 °C air temperature, ambient irradiance, and a 14 h photoperiod achieved by providing 75 $\mu mol \cdot m^{-2} \cdot s^{-1}$ from HPS lamps when PPFD was less than 10 $\mu mol \cdot m^{-2} \cdot s^{-1}$.

Crop	Treatment	CCI (4 Weeks after Transplant)	CCI (8 Weeks after Transplant)	Height (cm)	Width (cm)	Dry Mass (g)	Flower Number	Days to Flower
Impatiens	Ambient	55.5 ± 3.1	56.7 ± 3.4	19.9 ± 0.7	42.1 ± 1.4	14.9 ± 0.6	58.2 ± 5.9	32 ± 2
	Weeks 1–2	37.9 ± 1.5	57.6 ± 2.9	18.4 ± 0.9	38.4 ± 0.9	11.9 ± 0.5	34.8 ± 2.6	43 ± 2
	Weeks 3–4	30.2 ± 1.5	51.4 ± 2.7	16.8 ± 0.8	35.8 ± 1.4	10.9 ± 0.6	22.9 ± 2.2	42 ± 2
	Weeks 5–6	57.4 ± 3.1	59.5 ± 3.6	18.0 ± 1.0	36.2 ± 1.5	10.9 ± 0.3	28.3 ± 2.6	37 ± 3
	Weeks 7–8	57.1 ± 3.3	53.5 ± 4.0	16.0 ± 0.9	34.3 ± 1.4	10.3 ± 0.6	26.8 ± 4.6	38 ± 2
	Continuous	13.1 ± 1.0	34.7 ± 1.4	9.9 ± 1.0	15.1 ± 0.8	1.1 ± 0.1	0.3 ± 0.2	54 ± 2
	ANOVA [z]	<0.0001	<0.0001	<0.0001	<0.0001	<0.0001	<0.0001	<0.0001
	$HSD_{0.05}$ [y]	10.3	13.3	2.8	4.0	2.2	14.8	8
Pansy	Ambient	75.4 ± 1.3	60.8 ± 4.1	12.3 ± 0.6	16.6 ± 1.1	3.3 ± 0.4	11.1 ± 0.9	30 ± 1
	Weeks 1–2	61.5 ± 2.8	63.5 ± 2.3	13.0 ± 0.6	17.7 ± 1.2	3.6 ± 0.6	8.9 ± 0.7	38 ± 2
	Weeks 3–4	54.1 ± 2.5	65.3 ± 4.3	12.1 ± 0.7	17.5 ± 0.9	3.7 ± 0.3	8.8 ± 1.3	38 ± 2
	Weeks 5–6	71.2 ± 3.2	74.7 ± 3.3	11.5 ± 0.6	17.2 ± 0.9	3.9 ± 0.5	7.2 ± 1.0	37 ± 3
	Weeks 7–8	65.7 ± 4.2	60.6 ± 3.0	14.1 ± 0.4	18.8 ± 0.9	3.4 ± 0.3	8.0 ± 0.8	33 ± 2
	Continuous	43.1 ± 2.3	67.3 ± 3.1	11.2 ± 0.8	15.3 ± 0.4	1.8 ± 0.1	0.9 ± 0.2	54 ± 1
	ANOVA	<0.0001	0.0505	0.0051	0.0211	<0.0001	<0.0001	<0.0001
	$HSD_{0.05}$	12.1	-	2.3	2.9	1.1	3.6	7
Petunia	Ambient	27.8 ± 1.2	38.9 ± 1.7	21.8 ± 0.7	42.7 ± 1.0	15.3 ± 0.9	41.7 ± 2.4	35 ± 1
	Weeks 1–2	20.1 ± 1.0	34.3 ± 1.5	20.7 ± 0.6	42.9 ± 1.4	13.5 ± 0.4	37.6 ± 2.8	42 ± 1
	Weeks 3–4	16.0 ± 0.4	35.8 ± 1.2	20.6 ± 1.0	40.6 ± 0.7	14.1 ± 0.5	34.1 ± 3.0	42 ± 1
	Weeks 5–6	25.7 ± 1.2	34.1 ± 1.4	22.1 ± 1.0	37.5 ± 1.2	11.4 ± 0.8	28.0 ± 2.8	42 ± 1
	Weeks 7–8	26.4 ± 1.3	30.7 ± 2.4	20.7 ± 0.9	37.4 ± 1.3	10.2 ± 0.7	27.2 ± 1.5	33 ± 1
	Continuous	18.8 ± 1.2	23.1 ± 1.1	13.7 ± 0.8	23.1 ± 0.7	3.3 ± 0.2	0.0 ± 0.0	-
	ANOVA	<0.0001	<0.0001	<0.0001	<0.0001	<0.0001	<0.0001	<0.0001
	$HSD_{0.05}$	4.6	6.6	3.5	3.8	2.1	8.1	5
Snapdragon	Ambient	65.2 ± 4.9	60.2 ± 3.1	18.6 ± 0.7	22.5 ± 0.6	7.2 ± 0.7	61.2 ± 8.3	34 ± 1
	Weeks 1–2	66.0 ± 2.0	62.4 ± 3.7	19.5 ± 0.8	23.8 ± 0.6	7.1 ± 0.6	48.3 ± 5.6	41 ± 2
	Weeks 3–4	62.8 ± 3.3	62.2 ± 3.4	20.4 ± 0.4	23.1 ± 0.7	6.2 ± 0.5	57.5 ± 3.5	40 ± 2
	Weeks 5–6	62.7 ± 5.2	73.3 ± 4.0	21.9 ± 1.0	23.4 ± 0.7	6.4 ± 0.6	53.8 ± 5.0	38 ± 2
	Weeks 7–8	63.9 ± 4.6	63.5 ± 3.0	18.8 ± 0.8	22.4 ± 0.6	6.0 ± 0.4	53.3 ± 4.2	34 ± 1
	Continuous	51.2 ± 2.2	51.9 ± 2.5	22.8 ± 1.0	24.1 ± 0.5	4.3 ± 0.3	0.1 ± 0.1	60 ± 2
	ANOVA	0.0199	0.0003	0.0004	0.1822	<0.0001	<0.0001	<0.0001
	$HSD_{0.05}$	13.2	11.9	3.0	-	1.5	16.7	4

[z] Analysis of variance; [y] Tukey's honest significant difference (α = 0.05).

Delayed flowering with the low-energy exposure was primarily due to the lower overall temperature. Increased days to flower in response to decreased temperature have been reported for impatiens 'Accent Red', pansy 'Delta Yellow Blotch', petunia 'Easy Wave Coral Pink' and 'Wave Purple', and snapdragon 'Chimes White', but not for impatiens 'Super Elfin White' [28–32]. For example, a 23 days delay in flowering occurred in snapdragon 'Chimes White' when temperature decreased from 20 to 10 °C [30]. In previous studies, time to flower for petunia 'Easy Wave Coral Pink' and petunia 'Wave Purple' increased as MDT decreased from 26 to 14 °C, and time to flower for pansy 'Delta Yellow Blotch' increased linearly as MDT decreased from 25.7 to 16.3 °C [28,29]. Finally, a 3 to 4 days delay in flowering of impatiens 'Super Elfin Lipstick', petunia 'Avalanche Pink', and pansy 'Colossus Yellow Blotch' was documented for each 1 °C reduction in temperature [33], which is consistent with the 7 to 10 day delay in flowering we observed in our study when providing a two-week low-energy exposure (overall mean temperature was 1.6 to 2.0 °C lower than the ambient treatment; Table 1).

Decreased irradiance in the low-energy conditions may have also contributed to delayed flowering, which has been reported for impatiens 'Super Elfin White', snapdragon 'Rocket Rose', pansy 'Delta Yellow Blotch', and petunia 'Snow Cloud' [29,32,34]. Snapdragon 'Rocket Rose' flowering was delayed 21 days, and impatiens 'Super Elfin White' flowering was delayed 4 days, but only at high temperatures, when DLI decreased from 21.8 to 10.5 $mol \cdot m^{-2} \cdot d^{-1}$ [32]. The influence of DLI on flowering is often attributed to meristem heating by the increased irradiance intensity, and, therefore, is a temperature effect as well [35].

The two-week low-energy exposure also decreased flower number and plant growth relative to those grown at ambient conditions. Flower number in impatiens, pansy, and petunia, but not snapdragon, decreased compared to ambient controls (Table 2). Impatiens was most sensitive, and a two-week low-energy exposure at any point during production reduced flower number 40% to 61%. Pansy flower number was lower only in the weeks 5–6 exposure (35%), and petunia flower number was lower only in the weeks 5–6 and weeks 7–8 exposures (33% and 35%, respectively). Comparing across the four two-week low-energy treatments, timing did not influence impatiens, pansy, or snapdragon flower number, although petunias in low-energy weeks 1–2 had more flowers than the other low-energy intervals. This suggests implementation of a two-week low-energy exposure would not affect flower number regardless of when it was applied during production. Snapdragon inflorescence number was similar across the ambient control and four low-energy treatments (12 to 13 inflorescences), and higher than in the continuous low-energy treatment (<1 inflorescence; data not shown).

Reduced flower number was likely due to both a delay in flowering and reduced net photosynthesis in the low-energy environment. Light is a primary driver of photosynthesis, and temperature influences the rates of enzymatic activity and carbon loss via photorespiration. A decrease in petunia flower development rate as mean DLI decreased from 14 to 4 $mol \cdot m^{-2} \cdot d^{-1}$ has been reported [28]. Likewise, petunia 'Snow Cloud' grown at temperatures ranging from 10 to 30 °C flowered 3 to 23 days later at a given temperature when provided a DLI of 6 $mol \cdot m^{-2} \cdot d^{-1}$ rather than 13 $mol \cdot m^{-2} \cdot d^{-1}$ [34]. A reduction in flower number was observed in petunia 'Supertunia Vista Bubblegum' and 'Supertunia Mini Strawberry Pink Veined' grown 2 days or more per week in low-energy conditions and in pansy 'Matrix Blue Blotch' grown 4 days per week or continuously in low-energy conditions, compared to plants grown continuously in ambient conditions [24].

Relative chlorophyll content index (CCI) four weeks after transplant was generally lower in plants exposed to low-energy conditions, i.e., the weeks 1–2, weeks 3–4, and continuous treatments (Table 2). After eight weeks, relative CCI was not affected by the timing of the low-energy exposure, when compared to ambient conditions. The continuous low-energy treatment had lower CCI values for impatiens and petunia but not pansy or snapdragon (Table 2). This may be related to their cold tolerance, as impatiens and petunia are more cold-sensitive than pansy and snapdragon [25,26]. The reduction in relative CCI is likely a response to the lower temperature. In cotton (*Gossypium hirsutum*

L. var. Delta Pine 61), chlorophyll concentration decreased as temperature decreased [36]. Chlorophyll concentration in tomato (*Solanum lycopersicon* L. cv. M-19) and pepper (*Capsicum annuum* L. cv. M-71) decreased by almost half after a 12 days chilling (5 °C) treatment [37]. Although relative chlorophyll content rather than total chlorophyll concentration was measured in this study, it is an accepted proxy [38].

Plant height was unaffected by the timing of the low-energy exposure in impatiens and petunia. Except for the weeks 7–8 low-energy exposure, which was similar to the ambient control, snapdragon plant height increased with a two-week low-energy exposure (5% to 18% increase; Table 2). We observed an increase in shoot height previously in dianthus 'Telstar Pink' as the number of days per week in low-energy conditions increased [24]. Additionally, shoot height of some *Kalanchoe* species increased as DLI decreased from 17.2 to 4.3 to mol·m^{-2}·d^{-1} [39]. It appears a period of lower DLI during the vegetative phase elicited a shade avoidance response in snapdragon. The timing of the weeks 7–8 low-energy treatment occurred after flowering had begun, and therefore, did not affect internode length nor cause elongation of the inflorescence peduncle.

Plant diameter generally was unaffected or smaller when the timing of the two-week low-energy exposure occurred later in production. Snapdragon plant diameter was similar across all treatments and pansy was smaller only when grown in continuous low-energy conditions (Table 2). Impatiens and petunia plant diameters were smaller than the ambient controls when the two-week low-energy interval occurred later in production, in weeks 3–4 or later or in weeks 5–6 or later, respectively.

Similar to plant height and width, pansy and snapdragon dry mass were smaller only in plants grown continuously in the low-energy conditions (46% and 40%, respectively; Table 2). Petunia and impatiens dry mass gradually decreased as the two-week low-energy exposure occurred later in production. Petunias grown in weeks 5–6 or weeks 7–8 were 26% and 33% smaller, respectively, than ambient controls. Any two-week low-energy exposure in impatiens, regardless of the timing, reduced final dry mass (20% to 31%). The sensitivity of each species to dry mass accumulation corresponds to their cold tolerance; pansy and snapdragon are considered cold-tolerant annuals, petunia is a cold-intermediate species, and impatiens is a cold-sensitive species. The smaller plant size and dry mass is likely the result of reduced photosynthesis at lower irradiance and temperature, which reduced carbohydrate availability for growth. Previously, a decrease in plant dry mass of eight annual bedding plants was observed as DLI decreased, including impatiens 'Cajun Red' and petunia 'Apple Blossom' [40]. Additionally, pansy 'Delta Yellow Blotch' plant mass decreased in response to decreased irradiance, and pansy 'Universal Violet' dry mass decreased as temperature decreased [29,41].

Estimated energy costs were calculated for two production durations. First, energy costs were calculated for the total eight-week production period and relativized to energy costs in the ambient environment. Energy costs in our greenhouse were 84% to 86% of ambient when the two-week low-energy exposure was provided (Table 3). Energy costs increased slightly as the timing of the low-energy period occurred later in production due to naturally increasing temperatures and irradiance from late winter to early spring. The continuous low-energy environment had the lowest energy costs, at 44% of ambient. The second energy calculation estimated energy costs from transplant to mean date of the first flower for each species. This accounted for the lower per day cost in the low-energy conditions but the extended time of production due to delayed flowering. More modest reductions in energy costs occurred, ranging from a 7% reduction to a 17% increase in energy costs compared to the ambient environment (Table 3). Energy costs for the continuous low-energy treatment were not calculated due to the lack of 100% flowering after eight weeks. The estimated energy savings of 4% to 6% when pansy, petunia, and snapdragon were grown in cool, low light conditions during weeks 1–2 of production are slightly lower than the 8% to 18% energy savings reported for other dynamic temperate integration strategies [11,12,16], but much less than the 19% to 46% savings reported [14]. Additionally, snapdragons grown for two weeks in cool, low light conditions generally exhibited the lowest relative energy costs of the four species evaluated, compared to plants grown in ambient conditions, while impatiens generally had the highest relative energy costs. This corresponds to the plant growth and flowering data and is reflective of their classification as cold-tolerant and cold-sensitive species, respectively.

Table 3. Percent energy costs, relative to ambient conditions, calculated in Virtual Grower 3.0.9 using parameters specified [24]. Ambient conditions were 22/18 °C air temperature, 14 h photoperiod, and ambient irradiance + 75 $\mu mol \cdot m^{-2} \cdot s^{-1}$ supplemental lighting from high-pressure sodium (HPS) lamps when photosynthetic photon flux density (PPFD) was less than 300 $\mu mol \cdot m^{-2} \cdot s^{-1}$. Low-energy conditions were 13/10 °C air temperature, ambient irradiance, and a 14 h photoperiod achieved by providing 75 $\mu mol \cdot m^{-2} \cdot s^{-1}$ from HPS lamps when PPFD was less than 10 $\mu mol \cdot m^{-2} \cdot s^{-1}$. Impatiens (*Impatiens walleriana* 'Accent Orange'), pansy (*Viola* × *wittrockiana* 'Delta Premium Blue Blotch'), petunia (*Petunia* × *hybrida* 'Dreams Pink'), and snapdragon (*Antirrhinum majus* 'Montego Violet') were grown in ambient conditions (ambient), grown in ambient conditions and transferred to a low-energy environment at two-week intervals during an eight-week production cycle, or grown continuously in the low-energy environment (continuous).

Treatment	Relative Cumulative Energy Cost [z]	Relative Cumulative Energy Cost at Flowering [y]			
		Impatiens	Pansy	Petunia	Snapdragon
Ambient	100%	100%	100%	100%	100%
Weeks 1–2 cool	84%	103%	96%	94%	95%
Weeks 3–4 cool	86%	101%	98%	98%	93%
Weeks 5–6 cool	86%	99%	102%	99%	97%
Weeks 7–8 cool	88%	117%	106%	94%	100%
Continuous	44%	- [x]	-	-	-

[z] Total energy costs after 8 weeks, reported as a relative percent compared to plants grown continuously in ambient conditions; [y] Energy costs were calculated from transplant to flowering and reported as a relative percent compared to plants grown continuously in ambient conditions; [x] Not all plants had flowered after 8 weeks, and therefore energy costs were not calculated.

One of our objectives was to evaluate the effectiveness of a two-week low-energy interval as a strategy to reduce energy costs for greenhouse production of spring bedding plants. In all species, flowering was the most sensitive and was delayed in all species if the low-energy exposure was applied before flowering. The timing of the two-week low energy exposure influenced the severity of the impact on plant growth. It minimally affected pansy and snapdragon, regardless of when the two-week interval was applied. Petunia was more impacted when the interval was applied in the second half of production, and impatiens dry mass was affected regardless of the week. Therefore, depending on species and target plant size, the timing of the low-energy exposure could be shifted earlier in production to have a minimal impact on plant size and dry mass, or later in production to have a more pronounced effect on final plant size. For example, shifting the exposure later in production may provide a non-chemical strategy for growth regulation of species and cultivars with a vigorous plant habit.

Additionally, we wanted to evaluate the possibility of applying the intermittent short-term reduction in temperature and irradiance as a continuous two-week exposure rather than for 1 to 2 days per week, as reported previously [24], to better align with weather patterns and energy demands. Overall mean temperature and DLI for a crop will be similar when the cumulative duration of low-energy conditions is the same, whether provided continuously for a period of time or intermittently throughout production. Therefore, it stands to reason flowering time would be similar if crops are able to integrate temperature throughout the period of flower initiation and development, as long as temperatures remain between T_{min} and T_{opt}. In our study, pansy, petunia, and snapdragon grown in low-energy conditions for 2 days per week flowered within 2 days of plants exposed to low-energy conditions for a two-week period in weeks 1–2, weeks 3–4, or weeks 4–6 (data not shown). A comparison was not made with impatiens due to insufficient seedling number for the intermittent low-energy treatments.

In conclusion, flowering and plant growth were negatively impacted by the addition of a two-week low-energy exposure, resulting in delayed flowering, fewer flowers, and reduced plant size, mass, and relative chlorophyll content. Cold-sensitive crops, like impatiens, will be most severely impacted. Results from this study indicate the inclusion of a two-week low-energy exposure would decrease energy costs over an eight-week period but have a nominal reduction or even increase in overall energy consumption from transplant to flowering due to delays in flowering. Therefore, it could be worthwhile for growers to consider a short-term intermittent reduction in temperature and DLI to reduce heating costs, but only for cold-tolerant species like snapdragon and if an extra week of production can be built into the production schedule.

Author Contributions: Conceptualization, J.K.B.; Investigation, J.K.B.; Methodology, J.K.B. and J.E.A.; Project administration, J.K.B.; Writing – original draft, J.K.B.; Writing – review and editing, J.K.B. and J.E.A.

Disclaimer: Mention of trade names or commercial products in this publication is solely for the purpose of providing specific information and does not imply recommendation or endorsement by the U.S. Department of Agriculture. USDA is an equal opportunity provider and employer.

References

1. United States Department of Agriculture. 2012 Census of Agriculture: Census of Horticultural Specialties (2014). AC–12–SS–3. 2015. Available online: https://www.nass.usda.gov/Publications/AgCensus/2012/Online_Resources/Census_of_Horticulture_Specialties/HORTIC.pdf (accessed on 19 November 2018).
2. Runkle, E.; Both, A.J. *Greenhouse Energy Conservation Strategies*; Extension Bulletin E–3160; Michigan State University: East Lansing, MI, USA, 2011.
3. Liu, B.; Heins, R.D. Is plant quality related to the ratio of radiant energy to thermal energy? *Acta Hortic.* **1997**, *435*, 171–182. [CrossRef]
4. Dieleman, J.A.; Meinen, E. Interacting effects of temperature integration and light intensity on growth and development of single-stemmed cut rose plants. *Sci. Hortic.* **2007**, *113*, 182–187. [CrossRef]
5. Dennis, J.H.; Lopez, R.G.; Behe, B.K.; Hall, C.R.; Yue, C.; Campbell, B.J. Sustainable production practices by greenhouse and nursery plant growers. *HortScience* **2010**, *45*, 1232–1237. [CrossRef]
6. Hall, T.J.; Dennis, J.H.; Lopez, R.G.; Marshall, M.I. Factors affecting growers' willingness to adopt sustainable floriculture practices. *HortScience* **2009**, *44*, 1346–1351. [CrossRef]
7. Currey, C.J.; Lopez, R.G.; Mattson, N.S. Finishing bedding plants: A comparison of an unheated high tunnel versus a heated greenhouse in two geographic locations. *HortTechnology* **2014**, *24*, 527–534. [CrossRef]
8. Olberg, M.W.; Lopez, R.G. Growth and development of poinsettia (*Euphorbia pulcherrima*) finished under reduced air temperature and bench-top root-zone heating. *Sci. Hortic.* **2016**, *210*, 197–204. [CrossRef]
9. Camberato, D.M.; Lopez, R.G.; Krug, B.A. Development of *Euphorbia pulcherrima* under reduced finish temperatures. *HortScience* **2012**, *47*, 745–750. [CrossRef]
10. Dieleman, J.A.; Meinen, E.; Marcelis, L.F.M.; de Zwart, H.F.; van Henten, E.J. Optimisation of CO_2 and temperature in terms of crop growth and energy use. *Acta Hortic.* **2005**, *691*, 149–154. [CrossRef]
11. Rijsdijk, A.A.; Vogelezang, J.V.M. Temperature integration on a 24-hour base: A more efficient climate control strategy. *Acta Hortic.* **2000**, *519*, 163–169. [CrossRef]
12. Buwalda, F.; Rijsdijk, A.A.; Vogelezang, J.V.M.; Hattendorf, A.; Batta, L.G.G. An energy efficient heating strategy for cut rose production based on crop tolerance to temperature fluctuations. *Acta Hortic.* **1999**, *507*, 117–125. [CrossRef]
13. Elings, A.; de Zwart, H.F.; Janse, J.; Marcelis, L.F.M.; Buwalda, F. Multiple-day temperature settings on the basis of the assimilate balance: A simulation study. *Acta Hortic.* **2006**, *718*, 219–226. [CrossRef]
14. Lund, J.B.; Andreassen, A.; Ottosen, C.-O.; Aaslyng, J.M. Effect of a dynamic climate on energy consumption and production of *Hibiscus rosa-sinensis* L. in greenhouses. *HortScience* **2006**, *41*, 384–388. [CrossRef]
15. Ottosen, C.-O.; Rosenqvist, E.; Aaslyng, J.M.; Jakobsen, L. Dynamic climate control in combination with average temperature control saves energy in ornamentals. *Acta Hortic* **2005**, *691*, 133–140. [CrossRef]
16. Aaslyng, J.M.; Ehler, N.; Karlsen, P.; Rosenqvist, E. IntelliGrow: A component-based climate control system for decreasing greenhouse energy consumption. *Acta Hortic.* **1999**, *507*, 35–41. [CrossRef]

17. Heins, R.D.; Liu, B.; Runkle, E.S. Regulation of crop growth and development based on environmental factors. *Acta Hortic.* **2000**, *511*, 15–24. [CrossRef]
18. Fink, M. Effects of short-term temperature fluctuations on plant growth and conclusions for short-term temperature optimization in greenhouses. *Acta Hortic.* **1993**, *328*, 147–154. [CrossRef]
19. De Koning, A.N.M. Long-term temperature integration of tomato. Growth and development under alternating temperature regimes. *Scientia Hortic.* **1990**, *45*, 117–127. [CrossRef]
20. Liebig, H.-P. Temperature integration by kohlrabi growth. *Acta Hortic.* **1988**, *230*, 371–380. [CrossRef]
21. Dieleman, J.A.; Meinen, E.; Dueck, T.A. Effects of temperature integration on growth and development of roses. *Acta Hortic.* **2005**, *691*, 51–58. [CrossRef]
22. Körner, O.; Challa, H. Temperature integration and process-based humidity control in chrysanthemum. *Comput. Electron. Agric.* **2004**, *43*, 1–21. [CrossRef]
23. Boldt, J.K.; Gesick, E.Y.; Meyer, M.H.; Erwin, J.E. Alternative periodic energy-efficient light and temperature strategies for herbaceous ornamental production. *HortScience* **2011**, *46*, S354.
24. Boldt, J.K. Short-term reductions in irradiance and temperature minimally affect growth and development of five floriculture species. *HortScience* **2018**, *53*, 33–37. [CrossRef]
25. Blanchard, M.G.; Runkle, E.S. Quantifying the thermal flowering rates of eighteen species of annual bedding plants. *Scientia Hortic.* **2011**, *128*, 30–37. [CrossRef]
26. Vaid, T.M.; Runkle, E.S. Developing flowering rate models in response to mean temperature for common annual ornamental crops. *Scientia Hortic.* **2013**, *161*, 15–23. [CrossRef]
27. LeBude, A.V.; Bilderback, T.E. *The Pour-Through Extraction Procedure: A Nutrient Management Tool for Nursery Crops*; AG–717–W; NC State University Coop Extension Publication: Raleigh, NC, USA, 2009.
28. Blanchard, M.G.; Runkle, E.S.; Fisher, P.R. Modeling plant morphology and development of petunia in response to temperature and photosynthetic daily light integral. *Scientia Hortic.* **2011**, *129*, 313–320. [CrossRef]
29. Niu, G.; Heins, R.D.; Cameron, A.C.; Carlson, W.H. Day and night temperatures, daily light integral, and CO2 enrichment affect growth and flower development of pansy (*Viola* × *wittrockiana*). *J. Am. Soc. Hortic. Sci.* **2000**, *125*, 436–441. [CrossRef]
30. Munir, M.; Jamil, M.; Baloch, J.; Khattak, K.R. Growth and flowering of *Antirrhinum majus* L. under varying temperatures. *Int. J. Agric. Biol.* **2004**, *6*, 173–178.
31. Pramuk, L.A.; Runkle, E.S. Modeling growth and development of celosia and impatiens in response to temperature and photosynthetic daily light integral. *J. Ame. Soc. Hortic. Sci.* **2005**, *130*, 813–818. [CrossRef]
32. Warner, R.M.; Erwin, J.E. Prolonged high temperature exposure and daily light integral impact growth and flowering of five herbaceous ornamental species. *J. Am. Soc. Hortic. Sci.* **2005**, *130*, 319–325. [CrossRef]
33. Mattson, N.S.; Erwin, J.E. Temperature affects flower initiation and development rate of *Impatiens*, *Petunia*, and *Viola*. *Acta Hortic.* **2003**, *624*, 191–197. [CrossRef]
34. Kaczperski, M.P.; Carlson, W.H.; Karlsson, M.G. Growth and development of *Petunia* × *hybrida* as a function of temperature and irradiance. *J. Am. Soc. Hortic. Sci.* **1991**, *116*, 232–237. [CrossRef]
35. Mattson, N.S.; Erwin, J.E. The impact of photoperiod and irradiance on flowering of several herbaceous ornamentals. *Sci. Hortic.* **2005**, *104*, 275–292. [CrossRef]
36. Winter, K.; Königer, M. Dry matter production and photosynthetic capacity in *Gossypium hirsutum* L. under conditions of slightly suboptimum leaf temperatures and high levels of irradiance. *Oecologia* **1991**, *87*, 190–197. [CrossRef] [PubMed]
37. Javanmardi, J.; Rahemi, M.; Nasirzadeh, M. Physiological and reproductive responses of tomato and pepper transplants to low-temperature conditioning. *Int. J. Veg. Sci.* **2013**, *19*, 294–310. [CrossRef]
38. Ruiz-Espinoza, F.H.; Murillo-Amador, B.; García-Hernández, J.L.; Fenech-Larios, L.; Rueda-Puente, E.O.; Troyo-Diéguez, E.; Kaya, C.; Beltrán-Morales, A. Field evaluation of the relationship between chlorophyll content in basil leaves and a portable chlorophyll meter (SPAD-502) readings. *J. Plant Nutr.* **2010**, *33*, 423–438. [CrossRef]
39. Currey, C.J.; Erwin, J.E. Photosynthetic daily light integral impacts growth and flowering of several kalanchoe species. *HortTechnology* **2011**, *21*, 98–102. [CrossRef]

40. Faust, J.E.; Holcombe, V.; Rajapakse, N.C.; Layne, D.R. The effect of daily light integral on bedding plant growth and flowering. *HortScience* **2005**, *40*, 645–649.
41. Adams, S.R.; Pearson, S.; Hadley, P. An analysis of the effects of temperature and light integral on the vegetative growth of pansy cv. Universal Violet (*Viola* × *wittrockiana* Gams.). *Ann. Bot.* **1997**, *79*, 219–225. [CrossRef]

5

Irrigation of Greenhouse Crops

Georgios Nikolaou [1,*], Damianos Neocleous [2], Nikolaos Katsoulas [1] and Constantinos Kittas [1]

1 Department of Agriculture Crop Production and Rural Environment, School of Agricultural Sciences, University of Thessaly, Fytokou Str., 38446 Volos, Greece; nkatsoul@gmail.com (N.K.); ckittas@uth.gr (C.K.)

2 Department of Natural Resources and Environment, Agricultural Research Institute, 1516 Nicosia, Cyprus; d.neocleous@ari.gov.cy

* Correspondence: gnicolaounic@gmail.com.

Abstract: Precision agricultural greenhouse systems indicate considerable scope for improvement of irrigation management practices, since growers typically irrigate crops based on their personal experience. Soil-based greenhouse crop irrigation management requires estimation on a daily basis, whereas soilless systems must be estimated on an hourly or even shorter interval schedule. Historically, irrigation scheduling methods have been based on soil or substrate monitoring, dependent on climate or time with each having both strengths and weaknesses. Recently, plant-based monitoring or plant reflectance-derived indices have been developed, yet their potential is limited for estimating the irrigation rate in order to apply proper irrigation scheduling. Optimization of irrigation practices imposes different irrigation approaches, based on prevailing greenhouse environments, considering plant-water-soil relationships. This article presents a comprehensive review of the literature, which deals with irrigation scheduling approaches applied for soil and soilless greenhouse production systems. Irrigation decisions are categorized according to whether or not an automatic irrigation control has the ability to support a feedback irrigation decision system. The need for further development of neural networks systems is required.

Keywords: accumulated radiation method; feedback irrigation system; fuzzy control system; irrigation dose; precision irrigation; phyto-sensing; soilless culture; transpiration; water use efficiency

---

1. Introduction

The concept of "precision agriculture" is used to define technologies that support customized agricultural practices aimed at higher efficiency and a lower impact on the environment [1]. Greenhouse production systems decrease crop water requirements by as much as 20% to 40% compared to open field cultivation; however, growers routinely apply more irrigation water than the estimated water consumption [2–4]. Irrigation practices are generally based on the personal perspective of the grower; i.e., irrigation without monitoring the soil or plant water status [5]. Considering the number of different plant species grown in prevailing greenhouse environments, the types of substrate and container sizes, field and soil characteristics, and the different irrigation systems, it becomes obvious why irrigation scheduling becomes complex if it is to be achieved with any level of precision [6–8]. Therefore, an accurate short term estimation of crop water requirements in protected cultivation are a prerequisite for optimal irrigation scheduling; as evapotranspiration (ET_C) could occur so rapidly that water loss can cause plant damage before wilting symptoms become visible [9,10]. Irrigation management is typically expected to achieve maximum water supply for plant growth and production, with soil or substrate water content being maintained close to field capacity [11].

Even in soilless cultivation systems, irrigation represents a very large and potentially important loss of nutrients and a source of environmental pollution (i.e., drain to waste hydroponics systems) as a surplus of 20% to 50% of the plant's water uptake in each irrigation cycle is often recommended [12–16].

Indeed, annual use of irrigation water ranges from 150 to 200 mm (e.g., leafy vegetable) in soil-based greenhouse crops to 1000 to 1500 mm in soilless-grown (e.g., Solanaceae, cucurbits) [11]. For container nursery production, as cited by Fulcher et al. [17], those values could be as high as 2900 mm.

Considering the scarcity of water resources combined with the operational energy irrigation costs, maintaining the sustainable use of water is a major water-climate policy challenge since excessive irrigation results in low water use efficiency, increases in runoff and contributes to higher CO_2 emissions [18–20]. Several institutions have worked to improve water use in irrigation, developed various models of water efficiency, reducing the environmental problems associated with irrigation in order to mitigate severe structural water deficits, yet these models are not commercialized [4,21].

This paper presents a review of the literature dealing with irrigation of greenhouse crops. The necessity, the advantages and the limitations of each irrigation approach used are discussed in relation to different greenhouse types and the ability of an irrigation controller unit to support a feedback irrigation decision system.

2. Background

The exact time and volume of irrigation are probably the most important factors for efficient irrigation management and saving water, and these in turn also improve the productivity and quality of crops grown in the greenhouse [22,23]. This is especially true as the high potential efficiency of fertigation (i.e., irrigation combined with fertilization) has become a routine cultural practice, therefore the terms "irrigation" and "fertigation" are often used interchangeably [24–27]. Yet, irrigation management of substrate-based greenhouse crops still requires much more accurate control than for the same crop grown in soil, taking into account that substrates have very little nutrient buffering capacity [28].

Soilless growth systems in readily made artificial media commonly use organic (i.e., coconut coir, peat moss, pine bark) or inert substrates (i.e., perlite, rockwool, vermiculite); with substrate volume at approximately 10 to 40 L m^{-2} as is the case of rockwool or perlite slabs [29,30]. Horticultural production has historically been increasingly based on those ready-made substrates produced on an industrial scale with unique characteristics such as a limited cation exchange and low buffering capacity, good water permeability and adequate aeration [31]. Compared to soil cultivation systems, soilless growth systems are superior for plant growth as less energy is required by plants to extract water at field capacity, therefore experiencing a lower risk of oxygen deficiency [32,33]. In the same manner, all containerized production systems can be considered as hydroponic (i.e., soilless growth system) since they consist of an artificial root zone aimed at optimizing water and nutrient availability [34]. However, the restricted root volume may negatively affect the supply of nutrients to the plants as the water in the substrate may be rapidly decreased [32,35–37]. In addition, changes are induced in air and water retention characteristics of organic and inorganic substrates when they are used for longer periods than one growing season [38,39]. Therefore, according to Deepagoda et al. [29], a porous media should preferably be inert to prevent chemical and biological interactions.

Irrespective of the type of greenhouse cultivation system used (i.e., soil or soilless), irrigation scheduling should be managed (I) to supply plants with the volume of water equal to the volume of transpired water for maintaining crop productivity, (II) to overcome the differences in water discharge achieving high water uniformity (III) to move excessive salts towards the rooting system, avoiding soil salination [30,40]. Even in the latter case, for greenhouse cultivation systems there is always a risk of erroneous choices in the matching irrigation supply to crop evapotranspiration, as it may be affected by sudden changes in outside weather conditions or the use of climate control systems such as heating and ventilation [41]. That is another reason why for open hydroponic systems the main irrigation strategy is to supply nutrient solutions, with a surplus of 30% to 50% of the water uptake by the plants [14].

The leaching requirement in greenhouse soil-grown cultivation can be estimated based on irrigation water salinity and crop salt tolerance following FAO [42] as below:

$$LR = \frac{EC_{iw}}{5EC_e - EC_{iw}}, \quad (1)$$

where LR is the minimum leaching requirement needed to control salts within the tolerance EC_e of the crop; EC_{iw} is the electrical conductivity of the irrigation water applied (dS m^{-1}); and, EC_e the average soil salinity tolerated by the crop as measured on a soil saturation extract (dS m^{-1}).

However, as cited by Ben-Gal et al. [43], traditional guidelines for the calculation of the crop-specific leaching requirement is imprecise due to failure to consider soil type, climate, or salinity-induced reduction in plant transpiration. Such omissions could possibly result in underestimating actual leaching and over-estimation of leaching requirements.

Micro-irrigation is often promoted as a technology that can increase the application efficiency of water, and improve crop production and quality. The sub irrigation system also applies for the production of many ornamental hydroponic crops. However, the tendency for salts to build up in the upper portion of the root zone represents a drawback [44,45]. Harmanto et al. [46] working with soil-based greenhouse tomatoes (*Solanum lycopersicum*) in a tropical environment indicated that by applying drip irrigation, the water savings inside the greenhouse could be as much as 20% to 25% higher compared to an open field drip irrigated farming system.

For scheduling irrigation in soil or soilless greenhouses, it is essential to estimate the crop evapotranspiration and, according to the soil or substrate, the irrigation dose. In addition, as cited in Incrocci et al. [16], the irrigation dose of container growing medium could be estimated based on water potential or volumetric water content, with the use of soil moisture sensors. In the meantime, the adoption of soil moisture monitoring in vegetables has been restricted by means of sensor accuracy and price as well as labor required for installation, removal, and collection of readings [7]. A recent review by Bianchi et al. [47] summarized the four macro-groups of soil water potential devices and their operational characteristics.

According to Cahn and Johnson [7], an advantage of tension thresholds is the lesser influence by soil texture in comparison to volumetric moisture thresholds. Even so, as cited by Nikolaou et al. [48], sensors that estimate dielectric capacitance or dielectric permittivity of substrates (e.g., time domain reflectometry, frequency domain) have a propensity to be more reliable for soilless culture systems, as opposed to sensors measuring water availability through the matric potential such as the tensiometers.

According to Baille [49], in the short-term, decision level irrigation can be triggered based either on greenhouse microclimate or on soil/substrate moisture status. Irrigation scheduling based on direct or indirect measurement of plant water status and plant physiological responses to drought by using plant-based methods was comprehensively reviewed by several authors [50,51]. The different methods of irrigation scheduling in greenhouses is summarized below (Table 1).

Table 1. Greenhouse irrigation methods for soil and soilless greenhouse cultivation systems.

Scheduling Irrigation	Based on	Method/Device Use	Decisions Made	Reference
Time clock based	Time	Irrigation controllers	Irrigation frequency	[52,53]
Climate monitoring	Evapotranspiration	Lysimeters	Determine evapotranspiration (ET_C)	[54–56]
		Class A Pan	Determine reference evapotranspiration (ET_O)	[57,58]
		Reduce Class A Pan	Determine reference evapotranspiration (ET_O)	[2,59]
		Atmometer	Determine reference evapotranspiration (ET_O)	[15]
		Evapotranspiration models	Crop water used	[9,41]
	Solar radiation	Pyranometer	Irrigation frequency	[60,61]
Soil or substrate monitoring	Water potential	Tensiometer	Irrigation frequency/dose mainly for soil cultivations	[62]
		Electrical resistance sensor (e.g., gypsum blocks)	Irrigation frequency for soil	[62]
	Volumetric water content	Dielectric sensor (e.g., time domain reflectometry, frequency domain)	Irrigation frequency for soilless and soil cultivations	[62–64]
	Electrical conductivity	Electrical conductivity sensor	Irrigation frequency for soilless cultivation	[65–67]
	Physical properties	Mathematic formula	Irrigation dose/frequency for soilless and soil cultivations	[23,51,65, 68]
	Percentage of drainage	Mathematic formula, weighting devices	Irrigation volume and frequency based on trial and error for soilless	[69,70]
Phyto-sensing	Leaf water potential	Pressure chamber	Irrigation timing	[33]
	Stomata resistance	Diffusion porometer	Irrigation timing	[33]
	Canopy temperature	Infrared thermometry	Irrigation timing	[33,71,72]
	Flow on water in the stem	Heat balance sap flow sensor	Irrigation timing/detect water shortages	[33,73,74]
	Changes in stem diameter	Dentrometer	Irrigation timing	[33]
	Crop reflectance	Sensing system equipment and plant reflectance indices (e.g., photochemical reflectance index, normalized difference vegetation index)	Detect water stress	[51,75]

2.1. Monitoring Irrigation in Greenhouse Crops

Irrigation scheduling may have an impact on crop water productivity, affecting fruit yield and quality as well [76–78]. However, the targeted performance of a crop is largely situational; as irrigation might also be used as a tool for increasing water use efficiency, for maximizing yield or economic return [79].

For soil-based greenhouse cucumber (*Cucumis sativus*), Alomran [77] indicated that applying deficit irrigation at specific crop stages with 80% ET_C (i.e., decrease irrigation water up to 40% ET_C) is the most appropriate irrigation strategy for high crop water productivity and yield. For greenhouse tomatoes, partial root drying resulted in a water savings of 50%, but negatively affected the total fruit and total dry mass. However, the considerable savings of water could make partial root drying feasible in areas where water is scarce and expensive [80].

For soilless greenhouse cucumber, between transplanting and flowering, irrigation should be scheduled so as to induce slight water stress and increase root growth, while tomatoes should be stressed for a longer period (i.e., about three weeks) in order to set the first and second trusses [81]. In addition, several authors [53,60,82–84] indicated that increasing the irrigation intervals in soilless culture with the same daily amount of water applied positively influenced crop growth and production and minimized the outflow of water and nutrients from the greenhouse into the environment. However, that is not always the case, because results are often crop and substrate specific, and are also dependent on the experimental conditions and the limiting growth factor(s) [85].

A more rational approach for optimizing irrigation is through automatic irrigation controllers. Therefore, irrigation management approaches may be categorized according to the ability of a controller unit to support a feedback system [86]. Irrigation operations are often automated by using timers, specialized controllers, or computer control [87]. In the simplest form of automation in an "open loop irrigation control system", no measurements of the system outputs are used to modify the inputs and irrigation is based on preset time intervals (i.e., time clock scheduling) [86,88,89]. In a "feedback based irrigation closed-loop control system", the system provides growers with output data in real time (i.e., percentage of drainage, plant water status) which are evaluated in order to reschedule or perform irrigation. In a "feed forward irrigation control system" water uptake is predicted by using growth and transpiration models [14,71]. In addition, computerized-controlled irrigation systems can utilize a range of data to achieve accurate delivery of water according to crop requirements [71]. These systems are often mentioned as a fuzzy-logic control system, artificial intelligent system or multicriterion decision-making system. They are gaining importance because of their inherent ability to judge alternative scenarios for the selection of the best alternative which may be further analyzed before implementation [90].

2.2. The Soil/Substrate Physical Properties and the Irrigation Dose

Evapotranspiration rates depend on greenhouse environmental conditions, and are also affected by the water supply to the roots [91]. For scheduling irrigation, hydraulic properties and water content dependence on substrate suction must be known as they influence the water movement and retention in the substrate [91–93]. Water retention curves or moisture characteristic curves relate the water content in a specific substrate to the matric potential at a given tension or height [94]. Different kinds of substrates, as Fields et al. [95] indicated, have different water retention curves (Figure 1).

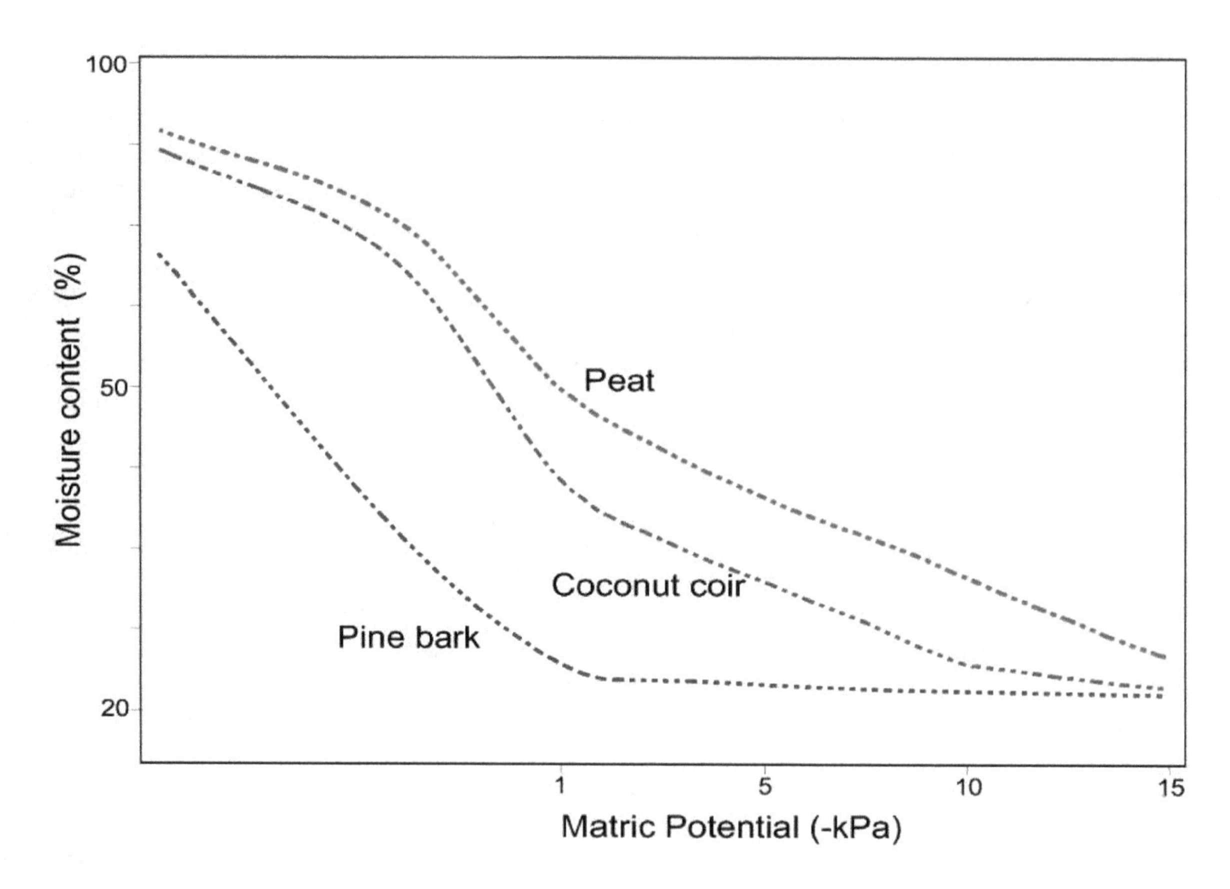

Figure 1. Moisture retention curves of peat, coconut coir and aged pine bark substrate components. Data adapted from [95].

From Figure 1, we can observe that the easily available water content (i.e., water released between 1 and 5 kPa) in coir is higher in comparison to pine bark; therefore, the crop may absorb more available water, reducing the need for applying a high frequency irrigation program. In addition, a taller container proportionately holds less water, as a percentage of water content by volume. Irrigation should take into account indices related to substrate availability of water, to container geometry and to specific substrate characteristics [64]. In general, water held by tensions higher than 10 kPa are considered unavailable to the crop, and water held between 5 to 10 kPa tensions are referred to as the substrate water buffer capacity. The available water in the container can be estimated according to Baudoin as follows [96]:

$$AWcont. = +0.64AW + 0.30P - 67h + 4.1, \quad (2)$$

where AW is the water that is available in a specific substrate as obtained from the water release curve (%); P represents the substrate porosity; h is the height of the container (m).

However, according to Raviv [97], water and nutrient availability to plants depends on the actual moisture flux from the medium to the plant roots rather than on the water volume in the container. By measuring water contents at different pressure heads, the soil water retention function can be determined. However, the unsaturated hydraulic conductivity measurement is often difficult as it may require expensive equipment [91,94,98]. The reason is because substrates containing predominantly organic components decompose during crop production cycles resulted in changes in air to water ratios. Additionally, shrinkage and compaction of substrates generates problems with watering, hydration, and generally leads to worsening the air to water relationship [99,100]. In the same way, hydraulic conductivity of rockwool and similar substrates is high when well-watered, but declines drastically as it dries out and plants experience a water deficit [101].

Mavrogianopoulos [65] proposed a simple equation for the estimation of the irrigation dose based on substrate characteristics as below:

$$Q = \frac{Y \times W_W \times N}{(1 - dr)}, \quad (3)$$

where Q is the irrigation dose (L $slab^{-1}$); Y is the water holding capacity of the substrate inside the slab (L $slab^{-1}$); W_w is the percentage of the water holding capacity that is easily available water (%); N represents a critical percentage of the easily available water that when reached, irrigation should start (i.e., typical values between 5% to 35%); dr is the percentage of drainage (%).

According to the same author, the substrate water holding capacity could be estimated by weighing it when dry, then filling it up with water for up to 24 h to complete the wetting process, draining up to 12 h, and reweighing it. The difference in weight is the water holding capacity, in kg $slab^{-1}$ or L $slab^{-1}$; however, this procedure should be repeated at different stages throughout plant growth.

In addition, another equation for estimating the irrigation dose was proposed by Katsoulas [82], taking into account the crop transpiration rate plus an extra amount of water for leaching purposes. This method simply reflects the substrate influence on crop water uptake, but requires frequent measurements of crop's transpiration rates.

$$E = \frac{T_r}{(1 - D)}, \quad (4)$$

$$Tr = \zeta R G_o, \quad (5)$$

$$\zeta = \frac{K_c \tau \alpha}{\lambda}, \quad (6)$$

where E is the amount of water applied in (Kg m^{-2}); Tr is the crop transpiration (kg m^{-2}); K_c is the species-specific crop coefficient; τ is the greenhouse radiation transmission coefficient; RG_o is the solar radiation measured outside the greenhouse (Kj m^{-2}); a is the evaporation coefficient; λ is the latent heat of vapourization of water (kJ kg^{-1}); and D is the drainage rate equal to 50% of irrigation water apply.

For greenhouse soil-based cultivation the amount of water which is "available" for root water uptake is defined as the amount of soil water between field capacity (i.e., soil matric pressure −10 or −33 kPa) and the permanent wilting point (i.e., −1500 kPa) expressed in m of water per m of soil depth. However, besides its use in irrigation management, field capacity is not an adequate soil physical quantity to assess soil water availability to crops, as a considerable (10% to 50%) fraction of transpired water is acquired from the soil at water contents above field capacity [102].

The most common values for typical soil texture classes are cited by Snider [103] in Table 2.

Table 2. Average values of available water holding capacity of the main soil texture groups (cm of water per cm of soil). Data adapted from [103].

Common Name	Field Capacity	Wilting Point	Available Water
Sandy soils	0.06–0.20	0.02–0.08	0.04–0.12
Loamy soils	0.23–0.27	0.10–0.12	0.13–0.15
Clayey soils	0.28–0.40	0.13–0.25	0.15–0.18

When the available soil moisture within the rooting zone has attained a predefined level of available water (i.e., the management allowable deficit—MAD), irrigation is triggered. The estimation of MAD is difficult because it depends on plant species and the evaporative conditions [104]. In general, the MAD can be calculated as a percentage of the available water, usually 30% to 50% in soil and 10% in soilless cropping systems. Then, the irrigation dose can be calculated by multiplying the MAD with a coefficient with typical values from 1.15 to 2, to account for water application uniformity and salinity. Typically, the frequency of irrigation can be estimated when the accumulated daily ET_C for the periods between irrigations approaches the MAD [96]. In line with this, the irrigation frequency of greenhouse soil cultivations can be estimated by dividing the readily available water with crop evapotranspiration [68].

Zeng et al. [23], working with drip-irrigated greenhouse soil cultivation of muskmelon, defined the irrigation dose by measuring the soil water content daily. When the water content was reduced to the irrigation start point, then the irrigation amount was decided:

$$Ir = \frac{\gamma \times h \times \theta_{fc} \times (g1 - g2)}{IE} \times 10, \quad (7)$$

$$SWC = \frac{FW - DW}{DW} \times 100, \quad (8)$$

where Ir represents the water amount by the drip irrigation system (mm); γ (gamma) is the soil bulk density (1.36 g·cm^{-3}); h is the depth of the soil which is irrigated in accordance with the vegetative stage; θ_{fc} is the water field capacity (32.9%); g_1 is the irrigation application rate (%); g_2 is the irrigation start point (60%); IE is the irrigation efficiency; SWC is the soil water content (%); FW is the fresh weight; and, DW is the dry weight.

3. Open and Feed Forward Irrigation Control System

3.1. Time Clock Scheduling and the Accumulated Radiation Method

In greenhouses with no feedback system control (i.e., open loop system), irrigation scheduling is determined according to the grower's perspective. Usually growers use a standard irrigation dose and change the frequency of irrigation; thus, they automate irrigation only on the basis of time [8,88]. For soil-based greenhouse crops the irrigation frequency is usually on a daily basis under warm and sunny conditions, and every 3–4 days under cooler and cloudy conditions [28]. In soilless systems, irrigation usually starts one hour after sunrise and stops one hour before sunset, with hourly or even shorter irrigation intervals during a day under high radiation conditions. For some substrates (i.e., rockwool) night irrigation is also recommended, avoiding drying to account for crop transpiration [105,106]. In line with this, according to Schröder and Lieth [81], irrigation at night is advised if the moisture content of the substrate has fallen below 8% to 10% from the previous morning. However, those rules-of-thumb obviously do not apply when the weather conditions are changing quickly from day to day [52,107]. Lizarraga et al. [52] evaluated the efficiency of timed scheduling, and concluded that this method does not actually meet the irrigation requirements of hydroponic tomatoes resulting in over and under irrigation during the morning and in the afternoon, respectively. Similarly, Incrocci et al. [16], working with several species of ornamentals in container nursery crops, reported an increase of the water use with timed irrigation scheduling by 20% to 40% and nutrient emissions of 39% to 74% in comparison with model-based irrigation.

A more rational approach for irrigation scheduling is the accumulated radiation method, allowing more closely matched water supply to the ET rate, which is primarily a day time phenomenon depending strongly on solar radiation [14,69,89,108,109]. However, Shin et al. [110] indicated that the transpiration rate of soilless paprika (*Capsicum annuum*) plants did not proportionally increase with an increase in light intensity, especially in high light conditions.

For estimating how much energy the crop has received, a light sensor (e.g., pyranometer) is used to measure incident solar radiation. Once this has been intercepted, a threshold value of light energy, an irrigation event, is triggered. Schröder and Lieth [81] suggested light sums inside greenhouses between 0.4 and 0.6 MJ m^{-2} in closed and 1.4 and 1.8 MJ m^{-2} in open hydroponic cultures with drainage volume factors of 30% and 15%, respectively. However, for rockwool substrate in a free drainage system, Lee [111] recommended accumulated values of 0.8 MJ m^{-2} with a minimum rest time set (i.e., not irrigated if the target value is reached) between 20 min in bright and 50 min on dark days according to the crop growth phase. Additionally, for bell pepper in container growth with peat mix, perlite, and pine bark media, Jovicich et al. [61] indicated that the first-quality fruit weight was enhanced at solar radiation integral levels of 0.34 MJ m^{-2}, while Lizarraga et al. [52] proposed indoor values of 0.81 MJ m^{-2} for tomatoes grown in perlite in bags of 40 L. In addition, Nikolaou et al. [60], working with cucumber in rockwool, indicated a 9% lower drainage amount between high and low irrigation frequency treatments (i.e., accumulated radiation outside greenhouse 1.3 MJ m^{-2} as opposed to 3 MJ m^{-2}), with no negative impact on production.

Despite the fact that the frequency of irrigation can be calculated based on the accumulated radiation method, the threshold values of light energy requires frequent evaluation, as affected by changes of crop coefficient values and cultivation techniques (e.g., defoliation). In any case, the solar radiation method was used only in soilless systems.

3.2. *Crop Evapotranspiration and the Water Balance Method*

Crop evapotranspiration assessment is necessary to correctly quantify crop irrigation water needs, playing a crucial role in cooling greenhouse crop canopies [112]. In order to evaluate crop evapotranspiration (ET_C), environmental conditions and physical, morphological and physiological features of soil-plant systems have to be considered [113].

A lot of research has been conducted in the field of estimating crop water irrigation needs, in real time, similar to the initial Penman–Monteith evapotranspiration models, which were initially developed for open field cultivations. However, the majority of these studies indicated a drastic effect of different greenhouse types to the transpiration rate and the need for model recalibration in prevailing greenhouse environments [72,114]. A brief summary of the most common evapotranspiration models in different greenhouse types, from literature cited by Fazlil Ilahi [115].

Class A evaporation pans (Figure 2) are considered one of the most widely used systems for climatic measurements in the estimation of the evapotranspiration rate for open field and protected cultivation, because of their simplicity and low cost [2,48]. The pan has proven its practical value and has been used successfully to estimate reference evapotranspiration by observing the evaporation loss from a water surface and applying empirical coefficients to relate pan evaporation to reference evapotranspiration (ET_O) [116].

Figure 2. Class A evaporation pan.

The estimation of crop evapotranspiration with the use of a class A evaporation pan can be calculated according to Allen et al. [117].

$$ET_C = E_P \times K_P \times K_C, \quad (9)$$

where ET_C is the maximum daily crop evapotranspiration measured (mm); E_P is the daily evaporation from class A Pan (mm); K_P is the pan coefficient;and, K_C is the crop coefficient.

However, there is difficulty in obtaining accurate field measurements with the use of pan A for herbaceous plants, because the crop coefficient (K_C value) is constantly changing throughout the growth, pruning and harvesting phases [48]. Abdel-Razzak et al. [118], working with cherry tomatoes, verified crop coefficients between 0.4 and 1.1, depending on the growth stage, while Gallardo et al. [28] indicated higher K_C values for supporter melon (*Cucumis* spp.) crops in relation to non supporter types. Regarding the pan A coefficient, a constant value of 0.79 provides a good estimation of reference evapotranspiration (ET_O) rates in plastic greenhouses under Mediterranean conditions [119], while Çakir et al. [57], working with cucumber in a solar greenhouse covered with netting material, indicated a plant-pan coefficient between 1.25 to 1.50. In line with this, for several crops in Cyprus, the evapotranspiration rate was calculated from reference evapotranspiration based on pan evaporation data, following the methodology proposed by Allen et al. [118] as indicated in Table 3 derived from Markou and Papadavid and Christou et al. [120,121]:

Table 3. Monthly and yearly estimated evapotranspiration requirements for several crops in protected cultivation (mm). Data adapted from [122,123].

Crop	J	F	M	A	M	J	J	A	S	O	N	D	Total
Tomato	42	60	85	120	180	168				12	40	36	743
Cucumber	42	48	72	120	208						40	36	566
French bean	42	48	84	140	70						24	28	436
Aubergines	12	24	40	60	76	100	68						380
Pepper	12	24	40	60	76	100	112						424
Watermelon	10	20	32	48	84	28							222
Sweet melon	10	20	32	48	84	28							222
Zucchini	12	24	50	78	136	88							388

The estimation of reference evapotranspiration is common in China and in Japan, the use of a reduced-size 20 cm diameter pan, eliminating the disadvantage of the large area occupied by a class A pan, (i.e., 121 cm diameter) [2,109,122]. Zeng et al. [23] working in soil-based greenhouse cultivation, indicated that K_{CP} values of cucumber equal to one can be recommended for the most appropriate irrigation scheduling from a standard 0.2 m diameter pan.

Commercially available atmometers can be used as an alternative approach to estimate evapotranspiration rate [7]. The estimated evapotranspiration values using a Piche atmometer (evaporated surface of paper disc), a modified atmometer (evaporated surface of a porous-porcelain plate) and a reduced evaporation pan were compared with the Penman–Monteith evapotranspiration method. As results indicated, atmometers had the best performance for estimating crop evapotranspiration in a greenhouse and could be used advantageously in relation to the evaporation pans [123].

According to several authors, devices that measure actual plant–soil evapotranspiration confined within containers (i.e., lysimeters) provide the only direct measure of water flux from a vegetative surface (i.e., can detect losses as small as 0.01 mm of water) and as such, they provide a standard against which other methods can be tested and calibrated [54,117]. Weighing lysimeters, measuring ET_C directly through changes in mass, while drainage lysimeters calculate ET_C through water budgets, where excess water removed by drainage or vacuum is subtracted from a known water volume applied to the soil surface [54]. In addition, Shin and Son [69] used load cells for the direct estimation of irrigation and drainage water amounts in soilless systems. Measurement practices, as cited by Sabeh [124], have ranged from weighting lysimeters measuring output every 10 min to calculating a 60 min average of 1 min measurements. However, the expense of these lysimeters limits their use to research and plants grown in containers (i.e., soilless culture systems; Figure 3) [7].

Figure 3. Mounting a lysimeter in a greenhouse structure (left), weighing device S-Beam load cell (centered) and plants grown in a lysimeter (right).

In soil-based systems simple portable scales must be replaced by expensive lysimeters. In this case sampling and representativeness problems become serious [125]. In order to eliminate this problem for soil-based greenhouse systems, it is preferable to use the water balance method, although is not very accurate, by monitoring all additions to and losses from a field's water [103]. In low volume/high frequency irrigation systems, the method has generally been sufficiently robust under a wide range of conditions [50]. Çakir et al. [57] and Mao et al. [126] estimated cucumber greenhouse crop evapotranspiration as follows:

$$ET = I \pm \Delta\Sigma - D - R, \quad (10)$$

where ET is crop evapotranspiration (mm); I is the total irrigation amount applied (mm); $\Delta\Sigma$ is change in soil water storage (mm); R and D are run-off and water loss, respectively, through deep percolation (mm) which is assumed to be zero since the amount of irrigation water is controlled and the precipitation or discharge rate of the irrigation system is below the soil infiltration rate.

4. Feedback Irrigation System

4.1. Soil/Substrate Monitoring

The frequency of irrigation could be monitored in soilless systems by measuring the change in salt concentration inside the cultivation slab. In this case, irrigation starts when the substrate electrical conductivity increased in relation to the nutrient solution electrical conductivity to a certain limit (e.g., 0.3–1 m·S·dm^{-1}) [65]. Using sensors for monitoring the EC, the pH and the amount of drainage could also be used as a tool for evaluating irrigation scheduling, taking into account seasonal transpiration differences [74]. As cited by Lizarraga et al. [52], the EC of the drainage solution should not be higher than 1 m·S·cm^{-1} compared to the EC of the irrigation solution. In addition, the percentage of the drainage amount could be tuned for irrigation control in greenhouses using a trial-end-error approach (e.g., the percentage of drainage should not be higher than 30% of the irrigation applied). Although the irrigation control system considers drainage amount as a single variable, it could not calculate the exact water amount used by the plant [69].

Greenhouse soil cultivation thresholds of soil potential have been used by many authors as a tool for irrigation management; even though soil matric potential values have been used, they appear to be based on experience [127]. For example, the irrigation of tomatoes and cucumbers growing in clay soils, with water potential set-points of −40 and −30 kPa, resulted in water savings of 35% and 46%, respectively, compared with irrigation set points at −10 kPa [128]. In line with this, for zucchini grown in artificial sand-mulched soil, a threshold soil matric potential of −25 kPa favored production and water savings in comparison with irrigation at −10 or −40 kPa [129]. On the other hand, for soilless crops, Depardieu et al. [130] indicated that plant growth and fruit production for strawberries (*Fragaria X ananassa*) grown in organic substrate (i.e., peat-sawdust mixture, aged bark, coconut fiber) were enhanced if irrigation started at −1.0 to −1.5 kPa, instead of −1.5 to −2.5 kPa.

4.2. Plant Monitoring

Plant phyto-sensing (e.g., leaf water potential, canopy temperature, crop reflectance) has been developed for an early, quantitative detection of plant responses to actual soil water availability, in order to define in real time, irrigation strategies to maximize plant growth [131]. However, a significant limitation is the fact that they do not provide a direct measure of the irrigation volume required. Hence, plant based sensing (Figure 4) is commonly used in conjunction with other irrigation techniques such as soil moisture measurement and the water balance approach [68].

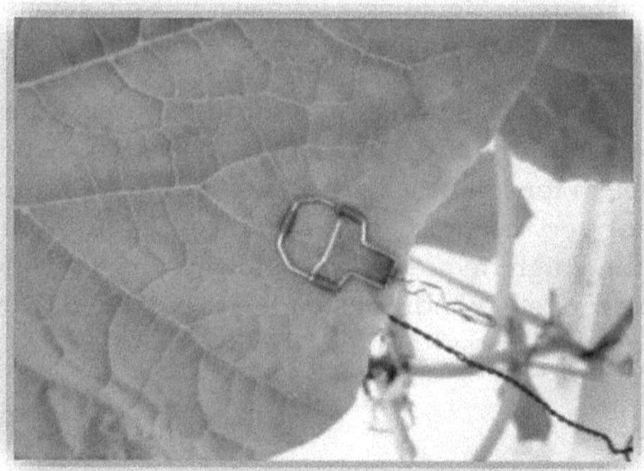

Figure 4. Stem micro-variation sensor (left) and a contact leaf temperature sensor (right).

In general, the use of plant phyto-sensing indicators as a tool for irrigation scheduling requires the estimation of reference or threshold values [50]. For example, Seelig et al. [132] used leaf thickness as an input parameter for automated irrigation control of container soil greenhouse cowpea (*Vigna unguiculata*) plants; indicating that 25% to 45% of irrigation water could be conserved compared with a typical timed irrigation schedule. Similarly, Nikolaou et al. [72] indicated a good correlation between leaf temperature of soilless cucumber with transpiration, and established a relationship between transpiration and leaf temperature by modified the simplified Penman−Monteith equation.

Although remote plant phyto-sensing and crop reflectance indices have been applied with great success in open fields, in greenhouses it has not yet been fully tested, as there are problems associated with greenhouse cover and structure shading [133].

The sensors and approaches used for crop reflectance measurements, and the indices used for crop water and nutrient status detection in greenhouse crops, have been reviewed by Katsoulas et al. [51]

5. Artificial Neural Networks and Fuzzy-Logic Control Systems

Artificial neural networks are analogue computer systems, which are made up of a large number of highly interconnected processing units which encompass computer algorithms that can solve several types of problems, based on different input units [134]. The use of artificial neural networks in agricultural systems is supported, as the plants' responses to their environment can be considered chaotic [135]. Until now, those systems have been applied mainly for open field cultivation, in the estimation of soil moisture content based on various soil and environmental parameters and for irrigation planning [90,136,137]. Pérez-Castro et al. [138] indicated that the water requirements within a greenhouse (i.e., evapotranspiration) can be calculated based on virtual sensors by monitoring external greenhouse climatic conditions. In line with this, virtual water sensors for soilless greenhouse tomato based on crop growth, substrate water and crop transpiration rate were also used by Sánchez-Molina et al. [139]. On the other hand Ben Ali et al. [140], developed a fuzzy logic control system in order to promote a suitable microclimate by activating the appropriate actuators installed inside the greenhouse with the appropriate rate.

Agriculture in developed countries seems to be in a transition, with increasing use of ICT (Information and Communications Technology) within the agricultural ecosystems [141]. Additionally, virtual plants have already been used to develop a case study for the irrigation processes of a greenhouse [142].

6. Concluding Remarks

This paper presents a review of irrigation management in soil and soilless crop production in greenhouses where irrigation scheduling should match the diurnal course of evapotranspiration as affected by the prevailing greenhouse environment through soil/substrate and crop characteristics. The majority of irrigation methods used in the past implement a feed forward or a feedback irrigation decision support system, and, in addition, water stress indices were developed based on plant-sensing. However, a gap in commercialized solutions exists despite the significant research work in the field of precision irrigation of greenhouse crops. It is important that a large margin of progress in greenhouse water and fertilizer use efficiency is managed by farmers [3]. The information presented reveals a need for the development of a commercial irrigation controller unit, in order to model and monitor the soil-plant-atmosphere utilizing artificial intelligence analyses.

Author Contributions: G.N. and D.N. conducted the literature review and produced final drafts of the manuscript. Commentary and review of manuscript drafts were conducted by N.K. and C.K.

References

1. Kittas, C.; Elvanidi, A.; Katsoulas, N.; Ferentinos, K.P.; Bartzanas, T. Reflectance indices for the detection of water stress in greenhouse tomato (*Solanum lycopersicum*). *Acta Hortic.* **2016**, *1112*, 63–70. [CrossRef]
2. Fernandes, C.; Corá, J.; Araújo, J. Reference evapotranspiration estimation inside greenhouses. *Sci. Agric.* **2003**, *60*, 591–594. [CrossRef]
3. Kitta, E.; Bartzanas, T.; Katsoulas, N.; Kittas, C. Benchmark irrigated under cover agriculture crops. *Agric. Agric. Sci. Procedia* **2015**, *4*, 348–355. [CrossRef]
4. Levidow, L.; Zaccaria, D.; Maia, R.; Vivas, E.; Todorovic, M.; Scardigno, A. Improving water-efficient irrigation: Prospects and difficulties of innovative practices. *Agric. Water Manag.* **2014**, *146*, 84–94. [CrossRef]
5. Bonachela, S.; González, A.M.; Fernández, M.D. Irrigation scheduling of plastic greenhouse vegetable crops based on historical weather data. *Irrig. Sci.* **2006**, *25*, 53–62. [CrossRef]
6. Lea-Cox, J.D.; Ross, D.S.; Teffeau, K.M. A Water and Nutrient Management Planning Process for Container Nursery and Greenhouse Production Systems in Maryland. *J. Environ. Hortic.* **2001**, *19*, 230–236.
7. Cahn, M.D.; Johnson, L.F. New Approaches to Irrigation Scheduling of Vegetables. *Horticulturae* **2017**, *3*, 1–20. [CrossRef]
8. Belayneh, B.E.; Lea-Cox, J.D.; Lichtenberg, E. Costs and benefits of implementing sensor-controlled irrigation in a commercial pot-in-pot container nursery. *Horttechnology* **2013**, *23*, 760–769. [CrossRef]
9. Qiu, R.; Kang, S.; Du, T.; Tong, L.; Hao, X.; Chen, R.; Chen, J.; Li, F. Effect of convection on the Penman–Monteith model estimates of transpiration of hot pepper grown in solar greenhouse. *Sci. Hortic.* **2013**, *160*, 163–171. [CrossRef]

10. Zimmermann, D.; Reus, R.; Westhoff, M.; Gessner, P.; Bauer, W.; Bamberg, E.; Bentrup, F.W.; Zimmermann, U. A novel, non-invasive, online-monitoring, versatile and easy plant-based probe for measuring leaf water status. *J. Exp. Bot.* **2008**, *59*, 3157–3167. [CrossRef]
11. Bacci, L.; Battista, P.; Cardarelli, M.; Carmassi, G.; Rouphael, Y.; Incrocci, L.; Malorgio, F.; Pardossi, A.; Rapi, B.; Colla, G. Modelling Evapotranspiration of Container Crops for Irrigation Scheduling. In *Evapotranspiration—From Measurements to Agricultural and Environmental Applications*; Gerosa, G., Ed.; IntechOpen Limited: London, UK, 2011; pp. 263–282. ISBN 978-953-307-512-9.
12. Van der Linden, A.M.A.; Hoogsteen, M.J.J.; Boesten, J.J.T.I.; Van Os, E.A.; Wipfler, E.L. *Fate of Plant Protection Products in Soilless Cultivations after Drip Irrigation: Measured vs. Modelled Concentrations*; National Institute for Public Health and the Environment: Bilthoven, The Netherlands, 2016; pp. 2–61.
13. Llorach-Massan, P.; Muñoz, P.; Riera, M.R.; Gabarrell, X.; Rieradevall, J.; Montero, J.I.; Villalba, G. N_2O emissions from protected soilless crops for more precise food and urban agriculture life cycle assessments. *J. Clean. Prod.* **2017**, *149*, 1118–1126. [CrossRef]
14. Kläring, H.K. Strategies to control water and nutrient supplies to greenhouse crops. A review. *Agronomie* **2001**, *21*, 311–321. [CrossRef]
15. Schiattone, M.I.; Viggiani, R.; Di Venereb, D.; Sergiob, L.; Cantore, V.; Todorovic, M.; Perniola, M.; Candido, V. Impact of irrigation regime and nitrogen rate on yield, quality and water use efficiency of wild rocket under greenhouse conditions. *Sci. Hortic.* **2018**, *229*, 182–192. [CrossRef]
16. Incrocci, L.; Marzialetti, P.; Incrocci, G.; Di Vita, A.; Balendonck, J.; Bibbiani, C.; Spagnol, S.; Pardossi, A. Substrate water status and evapotranspiration irrigation scheduling in heterogenous container nursery crops. *Agric. Water Manag.* **2014**, *131*, 30–40. [CrossRef]
17. Fulcher, F.A.; Buxton, J.W.; Geneve, R.L. Developing a physiological-based, on-demand irrigation system for container production. *Sci. Hortic.* **2012**, *138*, 221–226. [CrossRef]
18. Daccache, A.; Ciurana, J.S.; Rodriguez Diaz, J.A.; Knox, J.W. Water and energy footprint of irrigated agriculture in the Mediterranean region. *Environ. Res. Lett.* **2014**, *9*, 1–12. [CrossRef]
19. Egea, G.; Fernández, J.E.; Alcon, F. Financial assessment of adopting irrigation technology for plant-based regulated deficit irrigation scheduling in super high-density olive orchards. *Agric. Water Manag.* **2017**, *187*, 47–56. [CrossRef]
20. Montesano, F.F.; Van Iersel, M.W.; Boari, F.; Cantore, V.; D'Amato, G.; Parente, A. Sensor-based irrigation management of soilless basil using a new smart irrigation system: Effects of set-point on plant physiological responses and crop performance. *Agric. Water Manag.* **2018**, *203*, 20–29. [CrossRef]
21. Pawlowski, A.; Sánchez-Molina, J.A.; Guzmán, J.L.; Rodríguez, F.; Dormido, S. Evaluation of event-based irrigation system control scheme for tomato crops in greenhouses. *Agric. Water Manag.* **2017**, *183*, 16–25. [CrossRef]
22. Sezen, S.M.; Celikel, G.; Yazar, A.; Tekin, S.; Kapur, B. Effect of irrigation management on yield and quality of tomatoes grown in different soilless media in a glasshouse. *Sci. Res. Essays* **2010**, *5*, 41–48.
23. Zeng, C.Z.; Bie, Z.L.; Yuan, B.Z. Determination of optimum irrigation water amount for drip-irrigated muskmelon (*Cucumis melo* L.) in plastic greenhouse. *Agric. Water Manag.* **2009**, *96*, 595–602. [CrossRef]
24. Putra, P.A.; Yuliando, H. Soilless Culture System to Support Water Use Efficiency and Product Quality: A Review. *Agric. Agric. Sci. Procedia* **2015**, *3*, 283–288. [CrossRef]
25. Montesano, F.F.; Serio, F.; Mininni, C.; Signore, A.; Parente, A.; Santamaria, P. Tensiometer-Based Irrigation Management of Subirrigated Soilless Tomato: Effects of Substrate Matric Potential Control on Crop Performance. *Front. Plant Sci.* **2015**, *6*, 1–11. [CrossRef] [PubMed]
26. Van Os, E.A.; Gieling, T.H.; Ruijs, M.N.A. Equipment for hydroponic installations. In *Hydroponic Production of Vegetables and Ornamental*; Savvas, D., Passam, H., Eds.; Embryo Publications: Athens, Greece, 2002; pp. 104–140. ISBN 960-8002-12-5.
27. Breś, W.; Kleiber, T.; Trelka, T. Quality of water used for drip irrigation and fertigation of horticultural plants. *Folia Hortic.* **2010**, *22*, 67–74. [CrossRef]
28. Gallardo, M.; Thompson, R.B.; Fernández, M.D. Water requirements and irrigation management in Mediterranean greenhouses: The case of the southeast coast of Spain. In *Good Agricultural Practices for Greenhouse Vegetable Crops*; Plant Production and Protection Paper 217; FAO: Rome, Italy, 2013; pp. 109–136.

29. Chamindu Deepagoda, T.K.K.; Chen Lopez, J.C.; Møldrup, P.; de Jonge, L.W.; Tuller, M. Integral parameters for characterizing water, energy, and aeration properties of soilless plant growth media. *J. Hydrol.* **2013**, *502*, 120–127. [CrossRef]
30. Pardossi, A.; Carmassi, G.; Diara, C.; Incrocci, L.; Maggini, R.; Massa, D. *Fertigation and Substrate Management in Closed Soilless Culture*; University of Pisa, Dipartimento di Bioologia delle Piante Agrarrie (DBPA): Pisa, Italy, 2011; pp. 1–63.
31. Maślanka, M.; Magdziarz, R. The influence of substrate type and chlormequat on the growth and flowering of marigold (*Tagetes* L.). *Folia Hortic.* **2017**, *29*, 189–198. [CrossRef]
32. Raviv, M.; Lieth, J.M. Significance of Soilless Culture in Agriculture. In *Soilless Culture. Theory and Practice*; Raviv, M., Lieth, J.M., Eds.; Elsevier: Amsterdam, The Netherlands, 2008; pp. 1–10. ISBN 978-0-444-52975-6.
33. Raviv, M.; Blom, T.J. The effect of water availability and quality on photosynthesis and productivity of soilless-grown cut roses. *Sci. Hortic.* **2001**, *88*, 257–276. [CrossRef]
34. Adams, P. Nutritional control in hydroponics. In *Hydroponic Production of Vegetables and Ornamental*; Savvas, D., Passam., H., Eds.; Embryo Publications: Athens, Greece, 2002; pp. 211–261. ISBN 960-8002-12-5.
35. Asaduzzaman, Md.; Saifullah, Md.; Mollick, S.R.; Hossain, M.Md.; Halim, G.M.A.; Asao, T. Influence of Soilless Culture Substrate on Improvement of Yield and Produce Quality of Horticultural Crops. In *Soilless Culture-Use of Substrates for the Quality Horticultural Crops*; Asaduzzaman, Md., Ed.; IntechOpen Limited: London, UK, 2015; pp. 1–31.
36. Martínez-Gutiérrez, G.A.; Morales, I.; Aquino-Bolaños, T.; Escamirosa-Tinoco, C.; Hernández-Tolentino, M. Substrate volume and nursery times for earliness and yield of greenhouse tomato. *Emirates J. Food Agric.* **2016**, *28*, 897–902. [CrossRef]
37. Rouphael, Y.; Cardarelli, M.; Rea, E.; Colla, G. The influence of irrigation system and nutrient solution concentration on potted geranium production under various conditions of radiation and temperature. *Sci. Hortic.* **2008**, *118*, 328–337. [CrossRef]
38. Vox, G.; Teitel, M.; Pardossi, A.; Minuto, A.; Tinivella, F.; Schettini, E. Sustainable greenhouse systems. In *Sustainable Agriculture: Technology, Planning and Management*; Salazar, A., Rios, I., Eds.; Nova Science Publishers, Inc.: New York, NY, USA, 2010; pp. 1–78. ISBN 978-1-60876-269-9.
39. Warren, S.L.; Bilderback, T.E. More plant per gallon: Getting more out of your water. *Horttechnology* **2005**, *15*, 14–18. [CrossRef]
40. Leteya, J.; Hoffmanb, G.J.; Hopmansc, J.W.; Grattanc, S.R.; Suarezd, D.; Corwind, D.L.; Ostera, J.D.; Wua, L.; Amrhein, C. Evaluation of soil salinity leaching requirement guidelines Agricultural Water Management Evaluation of soil salinity leaching requirement guidelines. *Agric. Water Manag.* **2011**, *98*, 502–506. [CrossRef]
41. Baille, M.; Baille, A.; Laury, J.C. A simplified model for predicting evapotranspiration rate of nine ornamental species vs. climate factors and leaf area. *Sci. Hortic.* **1994**, *59*, 217–232. [CrossRef]
42. Ayers, R.S.; Westcot, D.W. *Water Quality for Agriculture*; Irrigation and Drainage Paper 29; FAO: Rome, Italy, 1985; pp. 1–131. ISBN 92-5-102263-1.
43. Ben-Gal, A.; Ityel, E.; Dudley, L.; Cohen, S.; Yermiyahu, U.; Presnov, E.; Zigmond, L.; Shani, U. Effect of irrigation water salinity on transpiration and on leaching requirements: A case study for bell peppers. *Agric. Water Manag.* **2008**, *95*, 587–597. [CrossRef]
44. Skaggs, R.K. Predicting drip irrigation use and adoption in a desert region. *Agric. Water Manag.* **2001**, *51*, 125–142. [CrossRef]
45. Rouphael, Y.; Colla, G. Growth, yield, fruit quality and nutrient uptake of hydroponically cultivated zucchini squash as affected by irrigation systems and growing seasons. *Sci. Hortic.* **2005**, *105*, 177–195. [CrossRef]
46. Harmanto; Salokhe, V.M.; Babel, M.S.; Tantau, H.J. Water requirement of drip irrigated tomatoes grown in greenhouse in tropical environment. *Agric. Water Manag.* **2005**, *71*, 225–242. [CrossRef]
47. Bianchi, A.; Masseroni, D.; Thalheimer, M.; Medici, L.O.; Facchi, A. Field irrigation management through soil water potential measurements: A review. *Ital. J. Agrometeorol.* **2017**, *2*, 25–38.
48. Nikolaou, G.; Neocleous, D.; Katsoulas, N.; Kittas, C. Irrigation management techniques used in soilless cultivation. In *Advances in Hydroponic Research*; Webster, D.J., Ed.; Nova Science Publishers, Inc.: New York, NY, USA, 2017; pp. 1–33. ISBN 978-1-53612-131-5.
49. Baille, A. Water management in soilless cultivation in relation to inside and outside climatic conditions and type of substrate. *Italus Hortus.* **2001**, *8*, 16–22.

50. Jones, H.G. Irrigation scheduling: Advantages and pitfalls of plant-based methods. *J. Exp. Bot.* **2004**, *55*, 2427–2436. [CrossRef]
51. Katsoulas, N.; Elvanidi, A.; Ferentinos, K.P.; Kacira, M.; Bartzanas, T.; Kittas, C. Crop reflectance monitoring as a tool for water stress detection in greenhouses: A review. *Biosyst. Eng.* **2016**, *151*, 374–398. [CrossRef]
52. Lizarraga, A.; Boesveld, H.; Huibers, F.; Robles, C. Evaluating irrigation scheduling of hydroponic tomato in Navarra, Spain. *Irrig. Drain.* **2003**, *52*, 177–188. [CrossRef]
53. Silber, A.; Xu, G.; Levkovitch, I.; Soriano, S.; Bilu, A.; Wallach, R. High irrigation frequency: The effect on plant growth and on uptake of water and nutrients. *Plant Soil* **2003**, *253*, 466–477. [CrossRef]
54. Beeson, R.C., Jr. Weighing lysimeter systems for quantifying water use and studies of controlled water stress for crops grown in low bulk density substrates. *Agric. Water Manag.* **2011**, *98*, 967–976. [CrossRef]
55. Libardi, L.G.P.; de Faria, R.T.; Dalri, A.B.; de Souza Rolim, G.; Palaretti, L.F.; Coelho, A.P.; Martins, I.P. Evapotranspiration and cropcoefficient (Kc) of presprouted sugarcane plantlets for greenhouse irrigation management. *Agric. Water Manag.* **2019**, *212*, 306–316. [CrossRef]
56. Vera-Repulloa, J.A.; Ruiz-Peñalverb, L.; Jiménez-Buendíaa, M.; Rosillob, J.J.; Molina-Martínez, J.M. Software for the automatic control of irrigation using weighing-drainage lysimeters. *Agric. Water Manag.* **2015**, *151*, 4–12. [CrossRef]
57. Çakir, R.; Kanburoglu-Çebi, U.; Altintas, S.; Ozdemir, A. Irrigation scheduling and water use efficiency of cucumber grown as a spring-summer cycle crop in solar greenhouse. *Agric. Water Manag.* **2017**, *180*, 78–87. [CrossRef]
58. Abou-Hadid, A.F.; El-Shinawy, M.Z.; El-Oksh, I.; Gomaa, H.; El-Beltagy, A.S. Studies on Water Consumption of Sweet Pepper Plant Under Plastic Houses. *Acta Hortic.* **1994**, *366*, 365–372. [CrossRef]
59. Zhang, C.; Gao, H.; Deng, X.; Lu, Z.; Lei, Y.; Zhou, H. Design method and theoretical analysis for wheel-hub driving solar tractor. *Emirates J. Food Agric.* **2016**, *28*, 903–911. [CrossRef]
60. Nikolaou, G.; Neocleous, D.; Katsoulas, N.; Kittas, C. Effect of irrigation frequency on growth and production of a cucumber crop under soilless culture. *Emirates J. Food Agric.* **2017**, *29*, 863–871. [CrossRef]
61. Jovicich, E.; Cantliffe, D.J.; Stoffella, P.J.; Haman, D.Z. Bell pepper fruit yield and quality as influenced by solar radiation-based irrigation and container media in a passively ventilated greenhouse. *HortScience* **2007**, *42*, 642–652.
62. Pardossi, A.; Incrocci, L.; Incrocci, G.; Fernando, M.; Bacci, L.; Rapi, B.; Marzialetti, P.; Hemming, J.; Balendonck, J. Root Zone Sensors for Irrigation Management in Intensive Agriculture. *Sensors* **2009**, *9*, 2809–2835. [CrossRef] [PubMed]
63. Nemali, K.S.; Montesano, F.; Dove, S.K.; Van Iersel, M.W. Calibration and performance of moisture sensors in soilless substrates: ECH_2O and Theta probes. *Sci. Hortic.* **2007**, *112*, 227–234. [CrossRef]
64. Murray, J.D.; Lea-Cox, J.D.; Ross, D.S. Time domain reflectometry accurately monitors and controls irrigation water applications in soilless substrates. *Acta Hortic.* **2004**, *633*, 75–82. [CrossRef]
65. Mavrogianopoulos, G.N. Irrigation dose according to substrate characteristics, in hydroponic systems. *Open Agric.* **2015**, *1*, 1–6. [CrossRef]
66. Dorai, M.; Papadopoulos, A.P.; Gosselin, A. Influence of electric conductivity management on greenhouse tomato yield and fruit quality. *Agronomie* **2001**, *4*, 367–383. [CrossRef]
67. Liopa-Tsakalidi, A.; Barouchas, P.; Salahas, G. Response of Zucchini to the Electrical Conductivity of the Nutrient Solution in Hydroponic Cultivation. *Agric. Agric. Sci. Procedia* **2015**, *4*, 459–462. [CrossRef]
68. White, S.; Raine, S.R. *A Grower Guide to Plant Based Sensing for Irrigation Scheduling*; National Centre for Engineering in Agriculture Publication 1001574/6; USQ: Toowoomba, Australia, 2008; pp. 1–52.
69. Shin, J.H.; Son, J.E. Development of a real-time irrigation control system considering transpiration, substrate electrical conductivity, and drainage rate of nutrient solutions in soilless culture of paprika (*Capsicum annuum* L.). *Eur. J. Hortic. Sci.* **2015**, *80*, 271–279. [CrossRef]
70. Nikolaou, G.; Neocleous, D.; Katsoulas, N.; Kittas, C. Dynamic assessment of whitewash shading and evaporative cooling on the greenhouse microclimate and cucumber growth in a Mediterranean climate. *Ital. J. Agrometeorol.* **2018**, *2*, 15–26.
71. Prenger, J.J.; Ling, P.P.; Hansen, R.C.; Keener, H.H. Plant response-based irrigation in a greenhouse: System evaluation. *Trans. ASAE Am. Soc. Agric. Eng.* **2005**, *48*, 1175–1183. [CrossRef]

72. Nikolaou, G.; Neocleous, D.; Katsoulas, N.; Kittas, C. Modelling transpiration of soilless greenhouse cucumber and its relationship with leaf temperature in a mediterranean climate. *Emirates J. Food Agric.* **2017**, *29*, 911–920. [CrossRef]
73. De Swaef, T.; Steppe, K. Linking stem diameter variations to sap flow, turgor and water potential in tomato. *Funct. Plant Biol.* **2010**, *37*, 429–438. [CrossRef]
74. Ehret, D.L.; Lau, A.; Bittman, S.; Lin, W.; Shelford, T. Automated monitoring of greenhouse crops. *Agronomie* **2001**, *21*, 403–414. [CrossRef]
75. Kim, M.; Kim, S.; Kim, Y.; Choi, Y.; Seo, M. Infrared Estimation of Canopy Temperature as Crop Water Stress Indicator. *Korean J. Soil Sci. Fertil.* **2015**, *48*, 499–504. [CrossRef]
76. Nuruddin, M.; Madramootoo, C.A.; Dodds, G.T. Effects of Water Stress at Different Growth Stages on Greenhouse Tomato Yield and Quality. *HortScience* **2003**, *38*, 1389–1393.
77. Alomran, A.M.; Louki, I.I.; Aly, A.A.; Nadeem, M.E. Impact of deficit irrigation on soil salinity and cucumber yield under greenhouse condition in an arid environment. *J. Agric. Sci. Technol.* **2013**, *15*, 1247–1259.
78. Saleh, S.; Liu, G.; Liu, M.; Ji, Y.; He, H.; Gruda, N. Effect of Irrigation on Growth, Yield, and Chemical Composition of Two Green Bean Cultivars. *Horticulturae* **2018**, *4*, 1–10. [CrossRef]
79. Saha, U.K.; Papadopoulos, A.P.; Hao, X.; Khosla, S. Irrigation strategies for greenhouse tomato production on rockwool. *HortScience* **2008**, *43*, 484–493.
80. Zegbe, J.A.; Behboudian, M.H.; Clothier, B.E. Yield and fruit quality in processing tomato under partial rootzone drying. *Eur. J. Hortic. Sci.* **2006**, *71*, 252–258.
81. Schröder, F.G.; Lieth, J.H. Irrigation control in hydroponics. In *Hydroponic Production of Vegetables and Ornamentals*; Savvas, D., Passam, H., Eds.; Embryo Publications: Athens, Greece, 2002; pp. 263–297. ISBN 960-8002-12-5.
82. Katsoulas, N.; Kittas, C.; Dimokas, G.; Lykas, C. Effect of irrigation frequency on rose flower production and quality. *Biosyst. Eng.* **2006**, *93*, 237–244. [CrossRef]
83. Pires, R.C.M.; Furlani, P.R.; Ribeiro, R.V.; Bodine, J.; Décio, S.; Emílio, L.; André, L.; Torre Neto, A. Irrigation frequency and substrate volume effects in the growth and yield of tomato plants under greenhouse conditions. *Sci. Agric.* **2011**, *68*, 400–405. [CrossRef]
84. Rodriguez-Ortega, W.M.; Martinez, V.; Rivero, R.M.; Camara-Zapata, J.M.; Mestre, T.; Garcia-Sanchez, F. Use of a smart irrigation system to study the effects of irrigation management on the agronomic and physiological responses of tomato plants grown under different temperatures regimes. *Agric. Water Manag.* **2016**, *183*, 158–168. [CrossRef]
85. Tsirogiannis, I.; Katsoulas, N.; Kittas, C. Effect of irrigation scheduling on gerbera flower yield and quality. *HortScience* **2010**, *45*, 265–270.
86. Romero, R.; Muriel, J.L.; García, I.; Muñoz de la Peña, D. Research on automatic irrigation control: State of the art and recent results. *Agric. Water Manag.* **2012**, *114*, 59–66. [CrossRef]
87. Challa, H.; Bakker, J.C. Crop growth. In *Greenhouse Climate Control: An Integrated Approach*; Bakker, J.C., Bot, G.P.A., Challa, H., Van de Braak, N.J., Eds.; Wageningen Academic Publishers: Wageningen, The Netherlands, 1995; pp. 15–97. ISBN 978-90-74134-17-0.
88. Lea-Cox, J.D.; Bauerle, W.L.; Van Iersel, M.W.; Kantor, G.F.; Bauerle, T.L.; Lichtenberg, E.; King, D.M.; Crawford, L. Advancing wireless sensor networks for irrigation management of ornamental crops: An overview. *Horttechnology* **2013**, *23*, 717–724. [CrossRef]
89. Shelford, T.J.; Lau, A.K.; Ehret, D.L.; Chieng, S.T. Comparison of a new plant-based irrigation control method with light-based irrigation control for greenhouse tomato production. *Can. Biosyst. Eng.* **2004**, *4*, 1–6.
90. Raju, K.S.; Kumar, D.N. Multicriterion decision making in irrigation planning. *Irrig. Drain.* **2005**, *54*, 455–465. [CrossRef]
91. Londra, P.A. Simultaneous determination of water retention curve and unsaturated hydraulic conductivity of substrates using a steady-state laboratory method. *HortScience* **2010**, *45*, 1106–1112.
92. Bougoul, S.; Boulard, T. Water dynamics in two rockwool slab growing substrates of contrasting densities. *Sci. Hortic.* **2006**, *107*, 399–404. [CrossRef]
93. Schindler, U.; Müller, L.; Eulenstein, F. Hydraulic Performance of Horticultural Substrates-1. Method for Measuring the Hydraulic Quality Indicators. *Horticulturae* **2017**, *3*, 1–7. [CrossRef]
94. Altland, J.E.; Owen, J.S.; Fonteno, W.C. Developing Moisture Characteristic Curves and Their Descriptive Functions at Low Tensions for Soilless Substrates. *J. Amer. Soc. Hortic. Sci.* **2010**, *135*, 563–567.

95. Fields, J.S.; Fonteno, W.C.; Jackson, B.E.; Heitman, J.L.; Owen, J.S. Hydrophysical properties, moisture retention, and drainage profiles of wood and traditional components for greenhouse substrates. *HortScience* **2014**, *49*, 827–832.
96. De Pascale, S.; Barbieri, G.; Rouphael, Y.; Gallardo, M.; Orsini, F.; Pardossi, A. Irrigation management: Challenges and opportunities. In *Good Agricultural Practices for Greenhouse VEGETABLE production in the South East European Countries for Greenhouse Vegetable*; Plant Production and Protection Paper 230; FAO: Rome, Italy, 2013; pp. 79–105.
97. Raviv, M.; Wallach, R.; Silber, A.; Medina, S.; Krasnovsky, A. The effect of hydraulic characteristics of volcanic materials on yield of roses grown in soilless culture. *J. Am. Soc. Hortic. Sci.* **1999**, *124*, 205–209.
98. Hosseini, S.M.M.M.; Ganjian, N.; Pisheh, Y.P. Estimation of the water retention curve for unsaturated clay. *Can. J. Soil Sci.* **2011**, *91*, 543–549. [CrossRef]
99. Bilderback, T.E.; Warren, S.L.; Owen, J.S.; Albano, J.P. Healthy Substrates Need Physicals Too! *Horttechnology* **2005**, *15*, 747–751. [CrossRef]
100. Nowak, J.S. Changes of Physical Properties in Rockwool and Glasswool Slabs During Hydroponic Cultivation of Roses. *J. Fruit Ornam. Plant Res.* **2010**, *18*, 349–360.
101. Jones, H.G.; Tardieu, F. Modelling water relations of horticultural crops: A review. *Sci. Hortic.* **1998**, *74*, 21–46. [CrossRef]
102. De Jong van Lier, Q. Field capacity, a valid upper limit of crop available water? *Agric. Water Manag.* **2017**, *193*, 214–220. [CrossRef]
103. Snyder, R.L. *Irrigation Scheduling: Water Balance Method*; University of California, Department of Land, Air and Water Resources Atmospheric Science Davis: Berkeley, CA, USA, 2014; pp. 1–39.
104. Greenwood, D.J.; Zhang, K.; Hilton, H.W.; Thompson, A.J. Opportunities for improving irrigation efficiency with quantitative models, soil water sensors and wireless technology. *J. Agric. Sci.* **2010**, *148*, 1–16. [CrossRef]
105. Carmassi, G.; Bacci, L.; Bronzini, M.; Incrocci, L.; Maggini, R.; Bellocchi, G.; Massa, D.; Pardossi, A. Modelling transpiration of greenhouse gerbera (*Gerbera jamesonii* H. Bolus) grown in substrate with saline water in a Mediterranean climate. *Sci. Hortic.* **2013**, *156*, 9–18. [CrossRef]
106. Medrano, E.; Lorenzo, P.; Sánchez-Guerrero, M.C.; Montero, J.I. Evaluation and modelling of greenhouse cucumber-crop transpiration under high and low radiation conditions. *Sci. Hortic.* **2005**, *105*, 163–175. [CrossRef]
107. Andrew, L.; Enthoven, N.; Kaarsemaker, R. *Best Practice Guidelines for Greenhouse Water Management*; GRODAN & Priva: Roermond, The Netherlands, 2016; pp. 1–38.
108. Kittas, C. Solar radiation of a greenhouse as a tool to its irrigation control. *Int. J. Energy Res.* **1990**, *14*, 881–892. [CrossRef]
109. Zhang, Z.K.; Liu, S.Q.; Liu, S.H.; Huang, Z.J. Estimation of Cucumber Evapotranspiration in Solar Greenhouse in Northeast China. *Agric. Sci. China* **2010**, *9*, 512–518. [CrossRef]
110. Shin, J.H.; Park, J.S.; Son, J.E. Estimating the actual transpiration rate with compensated levels of accumulated radiation for the efficient irrigation of soilless cultures of paprika plants. *Agric. Water Manag.* **2014**, *135*, 9–18. [CrossRef]
111. Lee, A. Reducing or eliminating fruit physiological disorders with correct root zone management. *Practical Hydroponics & Greenhouses*, May/June 2010; 53–59.
112. Katsoulas, N.; Kittas, C. Greenhouse Crop Transpiration Modelling. In *Evapotranspiration—From Measurements to Agricultural and Environmental Applications*; Gerosa, G., Ed.; IntechOpen Limited: London, UK, 2011; pp. 311–328. ISBN 978-953-307-512-9.
113. Lovelli, S.; Perniola, M.; Arcieri, M.; Rivelli, R.; Di Tommaso, T. Water use assessment in muskmelon by the Penman-Monteith 'one-step' approach. *Agric. Water Manag.* **2008**, *95*, 1153–1160. [CrossRef]
114. Morille, B.; Migeon, C.; Bournet, P.E. Is the Penman-Monteith model adapted to predict crop transpiration under greenhouse conditions? Application to a New Guinea Impatiens crop. *Sci. Hortic.* **2013**, *152*, 80–91. [CrossRef]
115. Fazlil Ilahi, W.F. Evapotranspiration Models in Greenhouse. Master's Thesis, Irrigation and Water Engineering Group, Wageningen University, Wageningen, The Netherlands, 2009; pp. 1–52.
116. Simba, F.M. A Flexible Plant Based Irrigation Control for Greenhouse Crops. A Thesis submitted in partial fulfillment for the requirements of the Master of Science, University of Zimbabwe, Harare, Zimbabwe, 2010.

117. Allen, R.; Pereira, L.; Raes, D.; Smith, M. *Crop Evapotranspiration Guidelines for Computing Crop Water Requirements*; FAO Irrigation and Drainage Paper, 56; FAO: Rome, Italy, 1998; pp. 1–289.
118. Abdel-Razzak, H.; Wahb-Allah, M.; Ibrahim, A.; Alenazi, M.; Alsadon, A. Response of cherry tomato to irrigation levels and fruit pruning under greenhouse conditions. *J. Agric. Sci. Technol.* **2016**, *18*, 1091–1103.
119. Fernández, M.D.; Bonachela, S.; Orgaz, F.; Thompson, R.; López, J.C.; Granados, M.R.; Gallardo, M.; Fereres, E. Measurement and estimation of plastic greenhouse reference evapotranspiration in a Mediterranean climate. *Irrig. Sci.* **2010**, *28*, 497–509. [CrossRef]
120. Markou, M.; Papadavid, G. Norm input -output data for the main crop and livestock enterprises of Cyprus. *Agric. Econ.* **2007**, *46*, 0379–0827.
121. Christou, A.; Dalias, P.; Neocleous, D. Spatial and temporal variations in evapotranspiration and net water requirements of typical Mediterranean crops on the island of Cyprus. *J. Agric. Sci.* **2017**, *1*, 1188–1197. [CrossRef]
122. Liu, H.J.; Cohen, S.; Tanny, J.; Lemcoff, J.H.; Huang, G. Estimation of banana (*Musa* sp.) plant transpiration using a standard 20 cm pan in a greenhouse. *Irrig. Drain. Syst.* **2008**, *22*, 311–323. [CrossRef]
123. Blanco, F.F.; Folegatti, M.V. Evaluation of evaporation measuring-equipments for estimating evapotranspiration within a greenhouse evapotranspiranspiration greenhouse. *Revista Brasileira de Engenharia Agrícola e Ambiental.* **2004**, *8*, 184–188. [CrossRef]
124. Sabeh, N.C. Evaluating and minimizing water use by greenhouse evaporative cooling systems in a semi-arid climate. In *Partial Fulfillment of the Requirements for the Degree of Doctor of Philosophy*; University of Arizona: Tucson, AZ, USA, 2007.
125. Seginer, I.; Kantz, D.; Levav, N.; Peiper, U.M. Night-time transpiration in greenhouses. *Sci. Hortic.* **1990**, *41*, 265–276. [CrossRef]
126. Mao, X.; Liu, M.; Wang, X.; Liu, C.; Hou, Z.; Shi, J. Effects of deficit irrigation on yield and water use of greenhouse grown cucumber in the North China Plain. *Agric. Water Manag.* **2003**, *61*, 219–228. [CrossRef]
127. Thompson, R.B.; Gallardo, M.; Valdez, L.C.; Fernández, M.D. Using plant water status to define threshold values for irrigation management of vegetable crops using soil moisture sensors. *Agric. Water Manag.* **2007**, *88*, 147–158. [CrossRef]
128. Buttaroa, D.; Santamariab, P.; Signoreb, A.; Cantorea, V.; Boari, F.; Montesano, F.F.; Parente, A. Irrigation Management of Greenhouse Tomato and Cucumber Using Tensiometer: Effects on Yield, Quality and Water Use. *Agric. Agric. Sci. Procedia* **2015**, *4*, 440–444. [CrossRef]
129. Contreras, J.I.; Alonso, F.; Cánovas, G.; Baeza, R. Irrigation management of greenhouse zucchini with different soil matric potential level. Agronomic and environmental effects. *Agric. Water Manag.* **2017**, *183*, 26–34. [CrossRef]
130. Depardieu, C.; Prémont, V.; Boily, C.; Caron, J. Sawdust and bark-based substrates for soilless strawberry production: Irrigation and electrical conductivity management. *PLoS ONE.* **2016**, *11*, e0154104. [CrossRef] [PubMed]
131. Gurovich, L.A.; Ton, Y.; Vergara, L.M. Irrigation scheduling of avocado using phytomonitoring techniques. *Cienc. E Investig. Agrar.* **2006**, *33*, 117–124.
132. Seelig, H.-D.; Stoner, R.J.; Linden, J.C. Irrigation control of cowpea plants using the measurement of leaf thickness under greenhouse conditions. *Irrig. Sci.* **2012**, *30*, 247–257. [CrossRef]
133. Sarlikioti, V.; Meinen, E.; Marcelis, L.F.M. Crop Reflectance as a tool for the online monitoring of LAI and PAR interception in two different greenhouse Crops. *Biosyst. Eng.* **2011**, *108*, 114–120. [CrossRef]
134. Ozcep, F.; Yıldırım, E.; Tezel, O.; Asci, M.; Karabulut, S. Correlation between electrical resistivity and soil-water content based artificial intelligent techniques. *Int. J. Phys. Sci.* **2010**, *5*, 47–56.
135. Morimoto, T.; Hashimoto, Y. An Intelligent Control Technique Based on Fuzzy Controls, Neural Networks and Genetic Algorithms for Greenhouse Automation. *IFAC Artif. Intell. Agnculture* **1998**, *31*, 61–66. [CrossRef]
136. Mohapatra, A.G.; Lenka, S.K. Neural Network Pattern Classification and Weather Dependent Fuzzy Logic Model for Irrigation Control in WSN Based Precision Agriculture. *Phys. Procedia* **2016**, *78*, 499–506. [CrossRef]
137. Saylan, L.; Kimura, R.; Caldag, B.; Akalas, N. Modeling of Soil Water Content for Vegetated Surface by Artificial Neural Network and Adaptive Neuro-Fuzzy Inference System. *Ital. J. Agrometeorol.* **2017**, *22*, 37–44.
138. Pérez-Castro, A.; Sánchez-Molina, J.A.; Castilla, M.; Sánchez-Moreno, J.; Moreno-Úbeda, J.C.; Magán, J.J. cFertigUAL: A fertigation management app for greenhouse vegetable crops. *Agric. Water Manag.* **2017**, *183*, 186–193. [CrossRef]

139. Sánchez-Molina, J.A.; Rodríguez, F.; Guzmán, J.L.; Ramírez-Arias, J.A. Water content virtual sensor for tomatoes in coconut coir substrate for irrigation control design. *Agric. Water Manag.* **2015**, *151*, 114–125. [CrossRef]
140. Ben Ali, R.; Bouadila, S.; Mami, A. Development of a Fuzzy Logic Controller applied to an agricultural greenhouse experimentally validated. *Appl. Therm. Eng.* **2018**, *141*, 798–810. [CrossRef]
141. Guirado-Clavijo, R.; Sanchez-Molina, J.A.; Wang, H.; Bienvenido, F. Conceptual Data Model for IoT in a Chain-Integrated Greenhouse Production: Case of the Tomato Production in Almeria (Spain). *IFAC-PapersOnLine* **2018**, *51*, 102–107. [CrossRef]
142. Rodríguez, F.; Castilla, M.; Sánchez, J.A.; Pawlowski, A. Semi-virtual Plant for the Modeling Control and Supervision of batch-processes. An example of a greenhouse irrigation system. *IFAC-PapersOnLine* **2015**, *48*, 123–128. [CrossRef]

6

Agronomic Management for Enhancing Plant Tolerance to Abiotic Stresses—Drought, Salinity, Hypoxia and Lodging

Luigi Mariani [1,2] and Antonio Ferrante [2,*]

1 Lombardy Museum of Agricultural History, via Celoria 2, 20133 Milan, Italy; luigi.mariani@unimi.it
2 Department of Agricultural and Environmental Sciences, Università degli Studi di Milano, via Celoria 2, 20133 Milan, Italy
* Correspondence: antonio.ferrante@unimi.it

Academic Editors: Alessandra Francini and Luca Sebastiani

Abstract: Abiotic stresses are currently responsible for significant losses in quantity and reduction in quality of global crop productions. In consequence, resilience against such stresses is one of the key aims of farmers and is attained by adopting both suitable genotypes and management practices. This latter aspect was reviewed from an agronomic point of view, taking into account stresses due to drought, water excess, salinity, and lodging. For example, drought tolerance may be enhanced by using lower plant density, anticipating the sowing or transplant as much as possible, using grafting with tolerant rootstocks, and optimizing the control of weeds. Water excess or hypoxic conditions during winter and spring can be treated with nitrate fertilizers, which increase survival rate. Salinity stress of sensitive crops may be alleviated by maintaining water content close to the field capacity by frequent and low-volume irrigation. Lodging can be prevented by installing shelterbelts against dominant winds, adopting equilibrated nitrogen fertilization, choosing a suitable plant density, and optimizing the management of pests and biotic diseases harmful to the stability and mechanic resistance of stems and roots.

Keywords: drought; lodging; hypoxia; salinity

---

1. Introduction

Crop yield and quality are the result of the interaction between a genotype's potential expression and the environment, which is modified by agronomic management in order to meet the objectives of the farmer. Different genotypes have varying yield capabilities depending on their adaptation abilities [1]. Agricultural systems are continuously evolving due to innovation in agronomic tools and the identification of high-performance cultivars coming from traditional or biotechnological genetic improvements [2]. Abiotic stresses such as low water availability, high salinity, high or low temperatures, hypoxia/anoxia, and nutrient deficiency are among the major causes of crop failure. Plants are able to perceive environmental stimuli and adapt to different environments; however, the degree of tolerance and adaptability to abiotic stresses varies among species and varieties. Crops exposed to abiotic stresses respond by activating defense mechanisms. Therefore, crops in an early stage of stress do not show visible symptoms but their physiology can undergo significant changes [3]. The energy used to counteract or cope with abiotic stresses is called "fitness cost" and does not contribute to crop production. Crops have to balance the resource allocation between productivity and defense actions [4].

In this review, agronomic strategies aimed at optimizing the resilience of crops exposed to abiotic stresses are covered. The work has been divided into two parts and the second part will be related to stress due to nutrients, high and low temperatures, and light excess or deficiency. Every agronomic strategy presented and discussed hereafter is sustainable not only socially and environmentally but also economically, because agriculture is an economic activity that cannot be done without adequate remuneration.

2. Role and Impact of Abiotic Limitations to Crop Yield

The general scheme proposed in Figure 1 describes the plant response to an abiotic stress with given features (duration, severity, etc.). The stress acts on a crop with given phenotypic characters to result in the final effects on growth, development, and mortality rate [5]. The basic scheme of Figure 1 is the result of a complex causal chain, which involves plant hormones and acts at the molecular and physiological level. The knowledge of this causal chain is substantially increasing thanks to transcriptomics, metabolomics, proteomics, and other integrated research approaches [3]. The conceptual scheme illustrated in Figure 2, widely adopted in crop yield simulation [6,7], shows crop production as the final result of a dry matter cascade triggered by solar radiation intercepted by plant canopies, which provides energy for the photosynthetic process.

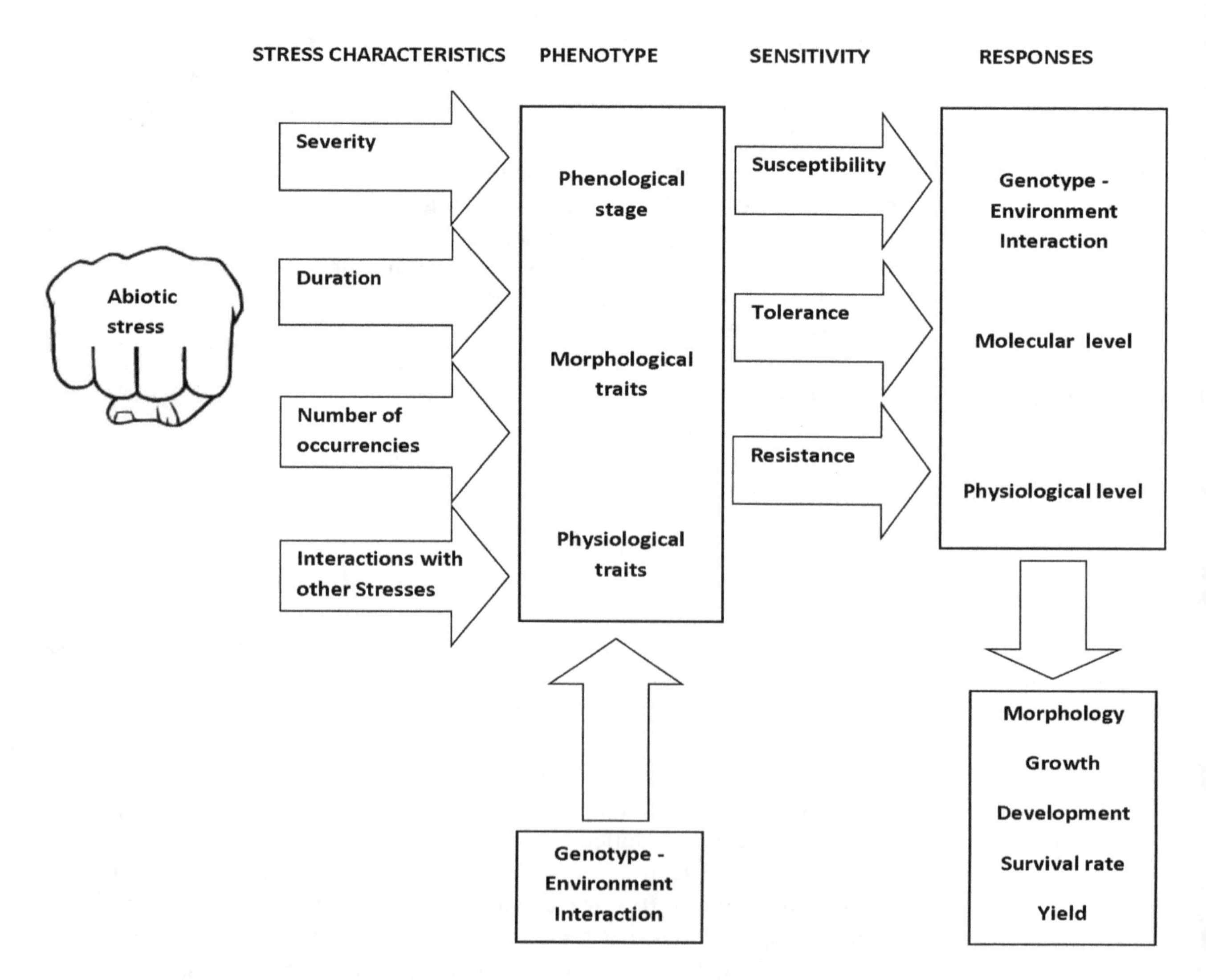

Figure 1. Scheme of factors that determine crop response to abiotic stresses.

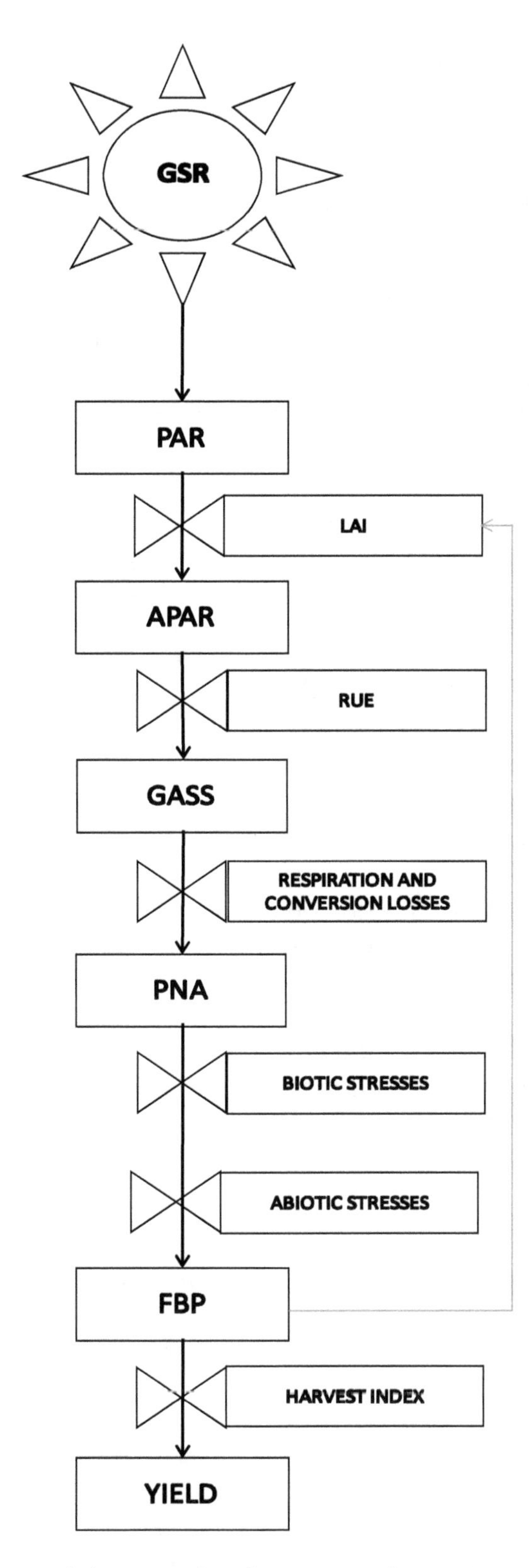

Figure 2. Conceptual scheme of the cascade of energy and organic matter that links global solar radiation and final yield. The role of abiotic stress as a rate variable that rules the conversion from potential net assimilation (PNA) to the final biomass production (FBP) is highlighted, with global solar radiation (GSR), photosynthetically active radiation (PAR), absorbed PAR (APAR), and gross photosynthetic assimilation (GASS) each contributing to the outcome. Rectangular boxes are state variables, and valves are rate variables.

Photosynthesis gives rise to a potential production of dry matter, gradually curtailed by different losses up to final production. Important losses occur by conversion from global solar radiation (GSR) to photosynthetically active radiation (PAR), efficiency of photosynthesis, translocation from photosynthetic to storage organs, maintenance and production respiration, and limitations by biotic (pests, fungi, bacteria, weeds, and so on) and abiotic stresses (temperature, soil water excess or shortage, nutrients, wind, etc.).

An example of a grain maize (class 700 FAO) field crop with Radiation Use Efficiency (RUE) of 4 g of glucose per MJ of Absorbed Photosynthetically Active Rradiation (APAR) and a harvest index (HI) of 0.6 cultivated on a flat plain at 45° North is useful to give an idea of the strength of the effect of different limitations on crop production. This field received 4337.0 $MJ \cdot m^{-2}$ of GSR in the period 1 April–30 September (data for Piacenza San Damiano, Italy—mean of the period 1993–2013); 50% of the GSR was useful for photosynthesis (PAR = 2168.5 MJ) and 80% of the PAR was intercepted by the canopy (APAR = 1734.8 $MJ \cdot m^{-2}$). By consequence, the Gross Assimilation (GASS) was 1734.8 $MJ \cdot m^{-2} \cdot 4\ g \cdot MJ^{-1} \cdot 10{,}000\ m^2 \cdot 1/1{,}000{,}000\ t \cdot g^{-1} = 69.4\ t \cdot ha^{-1}$ of glucose.

If we consider a loss of 35% related to maintenance and production respiration and translocation of photosynthetic products from the green organs to the storage tissues [8], there was a total net production without limitations (Potential Net Assimilation, PNA) of $69.4 \times (1 - 0.35) = 45.1\ t \cdot ha^{-1}$, which, multiplied by HI, gives a potential net grain production of 27.1 $t \cdot ha^{-1}$. This latter value was 48% higher than ordinary production and 33% higher than the maximum production attainable in ideal field conditions (Table 1). These gaps give an idea of the weight of biotic and abiotic stress factors in ordinary field conditions on the Po plain (Italy). In our experience, the weight of biotic factors (mainly effects of the European corn borer *Ostrinia nubilaris* and some fungal diseases) is about 15%, as can be inferred by comparing Italian production trends with those of the USA, where GMO BT maize largely eliminates the effects of biotic stresses from European corn borer and related fungal diseases. As a consequence, the effect of abiotic stresses on maize yield is about 33% (15–48%) in ordinary field conditions and drops to 18% (15–33%) in ideal field conditions (Table 1). Global values of abiotic limitations were simulated by Mariani [8] with a physiological-process-based crop simulation model driven by 1961–1990 monthly climate data from a global FAO dataset and applied to four crops (Wheat, Maize, Rice, and Soybeans (WMRS)) that account for 64% of global caloric consumption by humans. The model simulated only temperature and water limitations.

Table 1. The gap between net potential assimilation and final yield for maize cultivated in the Po valley in Italy [8].

Maize (Class 700 FAO) Yield in Field Conditions on the Po Plain (Italy)	Total Biomass Production ($t \cdot ha^{-1}$)	Harvest Index (%)	Grain Yield ($t \cdot ha^{-1}$)
Potential net assimilation (PNA)	45.1	0.6	27.1
Ordinary farmer objective in field conditions	21.7	0.6	13
Highest yield reachable in field conditions	30	0.6	18

As stated by Mariani [8], (i) the simulation was carried out on a global map with a pixel of 0.50×0.50 degrees in geographic coordinates (about 60×60 km at the equator), (ii) thermal and water limitations at the different latitudes were estimated only for the cells where the selected crops were present, (iii) water limitation for rice was estimated for rainfed crops and water, (vi) thermal limitations were obtained with suitable response curve models, and (v) the final weight of limitations on crop production was obtained by adopting a multiplicative approach.

The results in Figure 3 show the strength of global abiotic limitation and substantially agree with the results given by Buchanan et al. [5]. Moreover, the latitudinal distribution of abiotic limitations is shown in Figure 4.

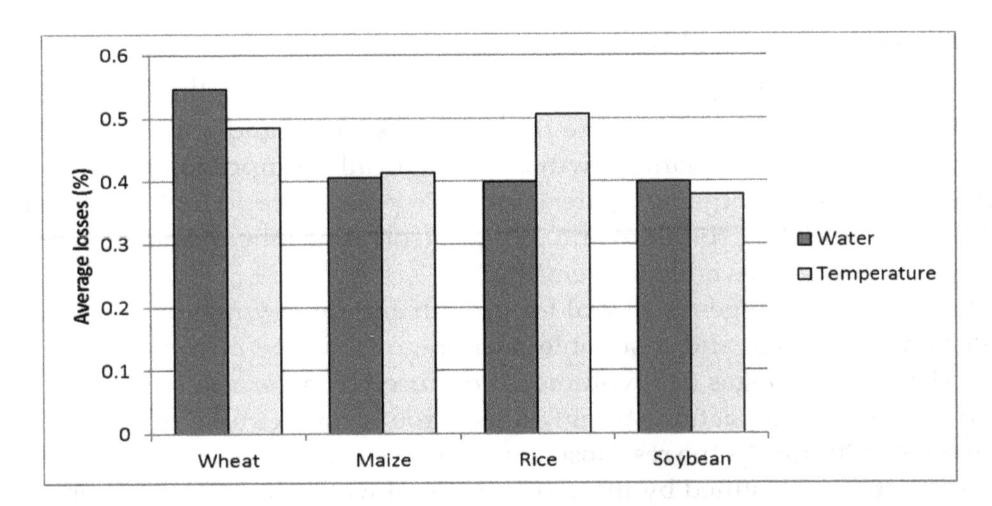

Figure 3. Mean global water and temperature stress losses for the four selected crops, wheat, maize, rice, and soybeans (WMRS) (% on potential net assimilation. PNA). Values above 1 of cumulative water and temperature stresses are the result of non-additive effects of multiple stresses [8].

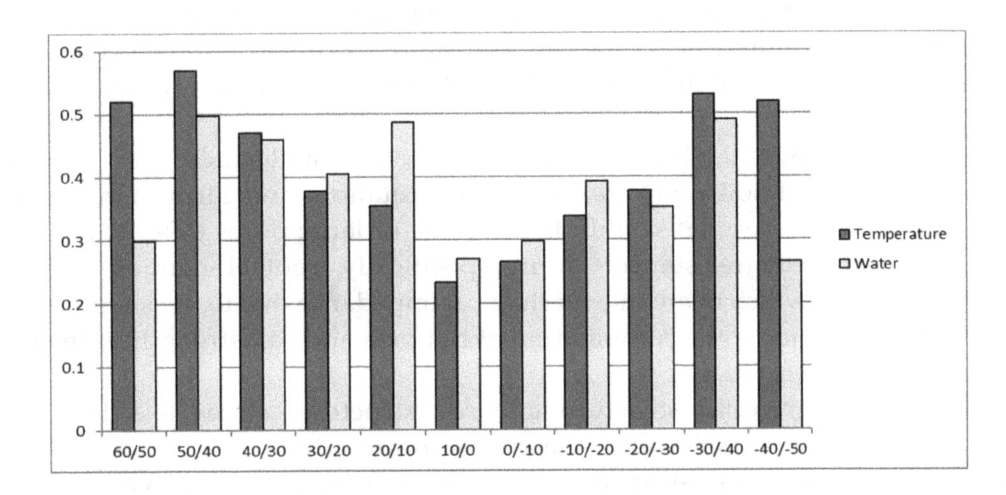

Figure 4. Mean latitudinal water and temperature stress for the four selected crops (WMRS). Values above 1 of cumulative water and temperature stresses are admissible due to the non-additive effects of multiple stresses [8].

Abiotic stresses interact not only among themselves but also with biotic stresses. For example, a crop that has undergone abiotic stress often shows greater susceptibility to attacks from insects, fungi, or mites, and a crop prone to these attacks shows greater sensitivity to water stress because stomata regulation is altered. Oerke [9] provided a global evaluation of the weight of biotic stresses due to weeds, pests, and pathogens for maize and wheat (two staple crops responsible for about 45% of global caloric intake by humans) and cotton (a commodity fundamental for the production of consumer goods). Data from three reference periods (1964–1965, 1988–1990, and 2001–2003) showed weights of 23.9%, 34.0%, and 28.2% for wheat; 34.8%, 38.3%, and 31.2% for maize; and 24.6%, 37.7%, and 28.8% for cotton, respectively.The global weight of biotic stresses on yield losses was estimated to be 70% by Boyer [10] and 13–94% by Farooq et al. [11]. Other data that illustrates both global yield losses and the weight of abiotic stresses have been reported by Cramer [4]. All the above mentioned estimates state the relevant gap between potential and real crop production induced by both biotic and abiotic stresses.

3. Soil Texture, Structure, and Field Hydraulic Arrangements

Plant resilience against abiotic stresses is at a maximum if soil conditions are suitable for plant growth and development. Soil is a disperse three-phase system, and a medium-textured soil at a condition considered optimal for plant growth is schematically composed (by volume) of 50% solid matter (textural particles and organic matter) and 50% pore space. The latter is equally divided between 25% liquid (circulating solution) and 25% gas (soil atmosphere) and an increase in liquid is associated with a decrease in gas and vice versa [12].

A suitable presence of oxygen is crucial for growth and deployment of roots that give an easy access to water and nutrients, and a suitable anchorage [13]. So, a relevant deviation from the abovementioned solid–liquid–gas (slg) volume ratio caused by an excess or shortage of water or by pore volume decrease by compaction strongly affects root growth and deployment, thus affecting crop production [14]. Obviously, the response to unsuitable slg volume ratios is strongly influenced by species and varieties, as testified by the case of ruderal weeds like *Plantago* spp. and *Polygonum aviculare* that are able to colonize compacted soils.

In medium- and fine-textured soils, a volume ratio suitable for most crops is generally attained if colloids (clay and organic matter) aggregate solid elements of texture in structural particles with a diameter of 0.3–3 mm, giving rise to a so-called granular structure, with pores characterized by a 50–50 ratio between macropores (large soil pores generally greater than 0.08 mm in diameter, which, after a saturating rainfall or irrigation, are rapidly drained and re-occupied by air) and micropores (small soil pores with diameters less than 0.08 mm that are mainly found within structural aggregates) [15].

As stated by Valentine et al. [16], who worked on 34 farms located in eastern Scotland that represented a wide range of soil types, textures, crop rotation, and farm management practices, root elongation was directly correlated with the volume of large pores (60–300 μm) and inversely correlated with penetrometer resistance. More specifically, root elongation was enhanced by low-resistance macropores, which overcome mechanical impedance due to the strength of the bulk soil and are limited by hypoxia (or some combination of hypoxia and soil strength) if the rate of oxygen diffusion to the root surface is too low.

Degradation of the granular structure towards structural states less favorable to crops (columnar, blocky, prismatic, massive, etc.) can result from various natural and man-made factors such as the effect of heavy rain on naked soil or the effect of preparatory or tillage carried out under unfavorable conditions of soil water content. Particularly damaging are field activities carried out with excessive soil water content. This explains, at least in part, the importance of hydraulic agricultural systems that aim to avoid excessive water due to rainy or saturated irrigation or high ground.

A fundamental presupposition to reach and maintain the granular structure is to avoid soil water excess; this can be attained by means of suitable soil field hydraulic arrangements. The basic rules for the field management of precipitation are that (1) soil water reservoir must be refilled until field capacity, (2) water excess must be quickly removed from field because it is harmful to most crops, and (3) the speed of this removal must be compatible with the need to avoid harmful erosion phenomena. Field hydraulic arrangements (ditches and drainage systems) should follow these general rules. According to Bonciarelli [17], primary field ditches should be sized with reference to the heavy and frequent rainfall that was quantified for Italy at 50/70 mm in 24 h, which means a reference volume for primary field ditch volumes of 250 to 350 $m^3 \cdot ha^{-1}$. This basic volume should be significantly reduced for very permeable soils. A quantitative analysis carried out by Mariani et al. [18] on rainfall data of 98 stations from throughout Italy, which mainly belongs to the Köppen-Geiger climate types Csa and Cfa, taking into account the 8th absolute 24 h rainfall maximum for the 1993–2012 period, highlighted that most of the territory needed a ditch volume of 250 to 350 $m^3 \cdot ha^{-1}$, as indicated by Bonciarelli [17], with some significant exceptions.

Table 2 lists factors that determine the conservation and improvement of granular structures. Minimum tillage and no tillage aim to promote the self-healing capacity of soils [19–21] and are particularly effective in soils with a sufficient content of good-quality clays.

A significant improvement of structure is also observed with soil organic or inorganic mulching [22] and with soil tillage carried out at a moisture content that maximizes the number of small aggregates [23]. Granular structure can be maintained or enhanced by amendments with organic matter like manure, slurry, compost, crop residues [24,25], or macromolecule polymers acting as soil conditioners [26,27]; or by mixing two or more soil layers in order to reach a more equilibrated texture [28]. Amendments improve soil physical properties, including increasing the content of water-stable aggregates, improving soil porosity and soil penetrability, improving water retention, decreasing soil bulk density and evaporation, and decreasing runoff amount and velocity.

Significant effects on soil structure are played by different tillage systems (moldboard ploughing, minimum tillage, or no tillage systems) with significant effects on macro- and microporosity. For example, Pagliai et al. [19] showed that minimum tillage significantly increased macroporosity, giving rise to a higher hydraulic conductivity and a less pronounced tendency to form a surface crust. Similar effects were highlighted for tropical soils under a no tillage system for 12 years [21].

Soil degradation factors are also listed in Table 2. Soil degradation due to rain [29] first involves the splash erosion that occurs when raindrops hit bare soil and is followed by runoff with sheet erosion (soil removal in thin layers by shallow surface flow), rill erosion (shallow drainage lines less than 30 cm deep), gully erosion (channels deeper than 30 cm that cannot be removed by normal cultivation), and tunnel erosion, which occurs when surface water moves into and through dispersive subsoils [30,31].

Heavy traffic of agricultural machinery is responsible for surface and subsurface soil compaction [28,32], which can be prevented by the adoption of machinery tracked or wheeled with low-pressure tires.

A relevant soil degradation factor is given by freezing–thawing cycles [33], such as the triggering factor for the peculiar type of gully erosion (named "calanchi" in Italian) that is typical of the Apennine clayey hills. The "calanchi" are more frequent in slopes exposed to the south, which are prone to freezing–thawing cycles.

Table 2. List of factors that determine conservation and improvement or degradation of granular soil structure.

Structure Conservation and Improvement Factors	Structure Degradation Factors
Self-healing capacity of soils [20] Organic or inorganic soil mulching [22] Minimum tillage and no tillage [19,21] Tillage at moisture content at which the largest number of small aggregates is produced [23] Amendments: - organic matter (e.g., manure, slurry, crop residues) [24,25] - soil conditioners (polymers) [26,27] - mixing of two or more soil layers in order to reach a more equilibrated texture [28]	Impact of rain and irrigation drops on bare soil surface [29] Runoff effect of rain and irrigation [29] Heavy traffic of agricultural machinery with high-volume pneumatics [28,32] Clods' exposure to freezing–thawing cycles [33]

4. The Impacts of Individual Stress Factors on Crops

4.1. Hypoxia/Anoxia Stress

Crops exposed to limited oxygen conditions must modify their physiology, biochemistry, and transcript profiles to adapt to stressful environments. Crop adaptability can allow survival if exposed to extreme environments or adverse seasons. Several physiological pathways are modified or activated, and many others are repressed [34], to allow plant survival. Specific stress activates a target gene or cluster of genes that may work as signals for cascade activation events and secondary responses [35]. Hypoxia or anoxia is quite a common event that can occur during plant life. During rainy seasons, plants can undergo long periods of flooding and suffer from waterlogging. However, waterlogging can also occur after excessive irrigation in soils with poor drainage. Crops under flooding conditions suffer from low oxygen availability at the root level, which causes a reduction of oxygen in tissues and leads to hypoxia/anoxia responses. In agriculture, oxygen limitation can cause a reduction in crop yield [36]. In fact, yields can dramatically decline after a long period of waterlogging: damages depend on adaptability to hypoxic conditions, soil properties, and drainage. In general, surviving oxygen deprivation depends on which plant tissue type is involved, the developmental stage, and the genotype, as well as the severity and duration of the stress light levels and temperature [37].

The agronomic strategies that can be applied to avoid flooding during the winter provide adequate drainage and soil arrangements, especially where there are structural problems [38]. Another important strategy is nitrate supply. Nitrate fertilization is quite uncommon in winter as plants have low metabolic activity and the frequent rain can increase the nitrate leaching. However, it has been proven that nitrate supply increased plant survival during the winter. Subsequently, several experiments demonstrated that increasing nutrient supply in waterlogged wheat increased plant growth and performance [39]. Among the different nutrients, the most important was nitrogen and, in particular, nitrate. In experiments on tomatoes, plants exposed to oxygen deprivation had delayed anoxia symptoms if nitrate was supplied [40]. The main role is not played by nitrate but the nitrate reductase enzyme. This enzyme in hypoxia conditions has been demonstrated to be involved in NO formation, which plays an important role in the hemoglobin oxygenation/reduction cycle [41].

4.2. *Drought Stress and Dry Farming*

Drought can drastically reduce crop productivity, especially if it occurs in the most critical stage of plant development. The crop tolerance to water stress depends on the ability of the plant to undergo physiological, biochemical, and morphological changes to enhance water use efficiency (WUE). Crop physiology is regulated by soil water availability and environmental conditions. With optimal soil water availability, plant transpiration is regulated by environmental variables around the leaves. The reduction of water in soil induces in the plant a regulation of the transpiration rate by reduction of stomatal conductance to equilibrate the amount of water uptake and maintain the crop water balance. It means that crops absorb and transpire the same amount of water; in this situation, soil water availability defines the crop yield. The prolonged decline of water availability induces the plant to produce compounds that enhance crop tolerance. Water movement occurs along gradients of water potentials so during drought conditions, crop plants accumulate osmolytes that are used for cell osmoregulation or osmotic adjustment that maintain water uptake [42]. Plants adapted to low water availability also show (i) morphological changes such as reduced leaf area and increased root biomass for exploring a wide volume of soil to find water, and (ii) phenological changes due to the need to reach maturity or ripening as the primary goal. The reduction of water availability induces several physiological and metabolic changes that lead plants to invest their energy to modify their morphology or produce osmolytes, but that reduce yield.

Only a quarter of land receives enough rain to meet crop water requirements. This explains the birth [43] of dry farming, an ensemble of cropping practices that can be adopted in areas without irrigation and where the annual average precipitation is between about 250 to 500 mm, or where rainy events are highly discontinuous and concentrated in limited periods of the year, as for example in Mediterranean environments (Csa type of Köppen-Geiger climate classification), where less than 30% of yearly precipitation falls during the summer. In the Mediterranean, water availability during the summer is often the main limiting factor for agriculture. The water shortage reduces yields and production can be only achieved by using efficient agronomic water management strategies, while climate variability and change have resulted in more sustainable management of water resources.

Dry farming aims in particular at the conservation of water resources by enhancing the storage of useful water in the soil reservoir and limiting water consumption. These objectives can be achieved by increasing the storage of water useful for crops in the soil reservoir, increasing the depth of the soil layer explored by roots, reducing water loss from the soil reservoir, and improving the crop Water Use Efficiency—WUE (water consumption per unit of dry matter produced).

Increased water storage for crops in the soil reservoir can be obtained in the following ways:

- adopting hydraulic arrangements that slow down the speed of surface water runoff and allow water penetration into soil, refilling the underground water reservoir. The infiltration of surface water can also be enhanced by appropriate tillage (e.g., by surface tillage that increases the roughness of the soil). Different tillage methods have a different effect on soil porosity and water infiltration. Ploughing to the depth of 20 cm is the tillage system that gives the highest porosity and water storage capacity, while the lowest was obtained in no-tillage systems [44];
- enhancing the drainage of infiltrated water in order to reach the entire profile explored by roots (e.g., by deep tillage carried out with rippers);
- favoring seasonal flooding of soils by water courses;
- catching runoff water and directing it towards compliant areas suitable for agriculture.

The depth of the layer explored by the roots can be increased in the following ways:

- breaking waterproof or compacted layers by means of ripping or ploughing;
- intervening with drainage or filled ditches to contain the winter rise of the water table, which imposes on crops a superficial root system, enhancing their sensitivity to summer drought.

The soil reservoir can be enhanced by:

- adopting management practices useful for promoting a granular structure with a good equilibrium between macro- and microporosity;
- increasing soil organic matter with organic fertilizers or green manure (the positive effect of organic matter on the soil reservoir is more significant in sandy soils or clay soils with low-quality clays—kaolinite). Organic matter increases the water storage ability of soils. Therefore, higher organic matter concentration means higher water availability for a crop [45];
- adjusting soil texture by mixing surface horizons with excess of sand with lower layers richer in clay (the presupposition for this activity is a suitable analysis of the soil profile and horizon distribution);
- adopting fallow techniques (field plowed and harrowed but left unseeded for one year) aimed at accumulating water in the soil during the "rest" period. For example, the biennial rotation fallow—wheat can be a solution for crop areas where yearly rainfall is insufficient for continuous cultivation. In this context, while the naked fallow should be avoided due to erosion problems and negative impacts on humus content, the most rational form of fallow is that of early autumn plowing (before the rainy season) and superficial work (harrowing) during the following spring and summer, whenever the soil appears encrusted or covered by weeds that waste a relevant quantity of water.

The loss of water from the soil reservoir can be limited by:

- minimizing tillage works (minimum tillage, sod seeding) in order to limit water evaporation from clods exposed to air. Superficial tillage usually limits water losses through evaporation since the capillarity is interrupted and water does not reach the soil surface. Water evaporation can be 70% higher in untilled soils compared with conventional ones [45].

- breaking the soil surface layer or soil crust by means of light soil work (harrows or weeders) in order to eliminate soil cracks that increase the exchange surface with the atmosphere and to interrupt the continuity between soil and atmosphere, reducing water flow towards this latter.
- implementing rational control of weeds, which are strong competitors for water. This is very important during the early stages of the crop cycle and the most critical phases in terms of the water deficit;
- reducing evaporation and transpiration loss through shading, windbreaks, mulching, and anti-hail nets. For example, on tendon-grapevines in Apulia (Italy), plastic films and/or anti-hail nets are used as cover in order to anticipate or delay the harvest and to reduce respiratory losses thanks to the shading effect and the limitation of the transpirational flow. Mulching with plastic or biodegradable films is commonly used in vegetable production to reduce water losses by evaporation and irrigation requirements [46,47]. A reduction of 45% of water need has been demonstrated with combined drip irrigation and mulching, in comparison with overhead sprinkler systems [48].
- using antitranspirants (mostly restricted to nurseries, to avoid excessive transpiration in newly transplanted crops). Antitranspirants are wax or plastic compounds that create a film on the leaves, covering the stomata. The effect is the reduction of water losses by transpiration and the reduction of photosynthesis, which means lower water use and improved tolerance of crops to drought stress [49]. Recent studies report that an antitranspirant sprayed on soybeans under a regular or low irrigation rate was able to improve WUE, acting on stomata and leaf gas exchange [50].
- adopting increased distances between rows and along the row in sowing and transplanting, reducing plant density and competition among plants;
- performing pruning or leaf removal in order to reduce the leaf area index;
- implementing a rational use of fertilizers. In this context, organic fertilization is generally useful for positive effects on soil water retention, and phosphate fertilization is often useful because it stimulates radicle growth, while nitrogen fertilization should be limited to avoid increasing concentrations of the circulating solution with greater difficulty in water supply for plants;
- delivering the water supply strictly needed to restore the useful water soil reservoir;
- choosing more efficient irrigation techniques (considering efficiency to be the ratio between water transpired by a crop to water distributed into the field, the mean efficiency is about 80–90% for drip, subsurface, center pivot or linear irrigation, 60–70% for sprinkler irrigation, and 30–40% for surface irrigation).
- selecting species and varieties so that the stage of maximum sensitivity to water stress does not coincide with the period of maximum dryness for the selected environment;
- choosing early sowing that enhances deep soil colonization by roots and in some species/varieties anticipates harvest. In this sense, autumn sowing is preferable for winter crops (wheat, barley, oat, canola, etc.) while early spring sowing is preferable for summer crops. This choice is obviously suitable only for zones that are not prone to frost risk or where early sowing is compatible with the harvest of previous crops;
- using biostimulants that can improve root development or enhance the biosynthesis of osmotic compounds [51]. These metabolites are able to improve crop tolerance and include plant hormones (abscisic acid) proline, sugars, amino acids, etc. The application of biostimulants can be carried out by soil drench or spray.

WUE can be enhanced by a suitable choice of crop species and varieties. An example is soft wheat (*Triticum aestivum* L.), in which the selection of varieties for the European environment led to varieties suitable for Mediterranean or transitional climates (Csa and Cfa of the Köppen-Geiger climate classification) with the length of the flowering–ripening period reduced with respect to that of varieties suitable for Oceanic environments (Cfb of Köppen-Geiger). Other examples are grapevine rootstocks from *Vitis rupestris* Scheele, which are more resistant than those of *Vitis Riparia* Michx., and sorghum (*Sorghum bicolor* L.), which in environments prone to drought is preferred to maize (*Zea mays* L.) because it is able to resume vegetation without excessive production damage after a drought event.

4.2.1. Precision Farming and Variable-Rate Irrigation

Soil water content is crucial for managing irrigation and can be measured by means of suitable sensors [52,53] or estimated by water balance models based on the continuity equation (conservation of water applied to the soil reservoir). These technologies allow for selecting irrigation time and volumes.

In recent years, these tools have been used to drive variable-rate irrigation (VRI) [54], which allows for the distribution of different amounts of water in different parts of the same field as a function of soil and crop heterogeneity. This approach can be particularly important for fields characterized by a high variability of soil characteristics (texture, structure, depth, and fertility). In these situations, the combination of precision farming and VRI can reduce water losses and improve the WUE of field crops.

4.2.2. Grafting as an Agronomic Tool to Improve Drought Tolerance

WUE can also be improved by means of suitable rootstocks The increase of WUE in plants by grafting can be a reliable agronomic strategy for enhancing crop adaptation and performance in dry environments and, in the past, grafting was widely used in vegetable crops to limit the effects of soil pathogens [55]; more recently, it has been used to induce tolerance against abiotic stresses, such as organic pollutants [56], boron and salinity [57,58], and thermal and water stress. About this latter, it has been shown that the grafting of scions susceptible to water stress onto tolerant rootstocks increased the resistance of grafted trees to this stress [59,60]. Sanders and Markhart [61] have shown that the osmotic potential of dehydrated scions of grafted beans (*Phaseolus vulgaris* L.) was determined by the rootstocks, while the osmotic potential of non-stressed scions was governed by the shoot [60]. Drought tolerance, provided by either the rootstock or the scion, resulted in enhanced nitrogen fixation in soybeans (*Glycine max* L.) [61]. Other grafting experiments on the effect of drought on fruit crops, such as kiwis and grapes, proved that drought-tolerant rootstocks are available and useable under commercial conditions [62,63]. In contrast, only a few studies exist with grafted fruit vegetables. Because eggplants are more effective at water uptake than tomato root systems, it could be useful to study the effects of their grafting on WUE under water-stress conditions. Grafting mini-watermelons onto a commercial rootstock (PS 1313: *Cucurbita maxima* Duchesne × *Cucurbita moschata* Duchesne) revealed a >60% higher marketable yield when grown under conditions of deficit irrigation compared to non-grafted melons [64].

In response to water stress, the hormone that serves as a link between the rootstock and scion seems to be abscisic acid (ABA), as observed in different species such as *Citrus* sp. [65] and *Vitis* sp. [66], but also in cucumber grafts on luffa (*Luffa* sp. Mill. 1754) [67]. Grafting experiments with ABA-deficient mutants of tomato showed that stomata can close independently of the leaf water status, suggesting that there is a chemical signal produced by the roots that controls stomata conductance [68]. Therefore, the selection of rootstocks with adequate biosynthesis and perception of the ABA could improve the efficient use of water and the drought tolerance of many horticultural varieties. Since Cucurbita is one of the genera most used as rootstocks of different species of Cucurbitaceae (watermelon, melon, cucumber), the identification of mutants with a greater ability to synthesize ABA, or hypersensitivity to ABA, could improve tolerance to plant water stress. This increase in ABA biosynthesis also occurred in ethylene-insensitive mutants [69]. On the other hand, it is known that ABA function in the closure of the stomata is mediated by Reactive Oxygen Species (ROS) production [70]. Therefore, used as rootstocks, ethylene-insensitive and ROS-tolerant mutants could also improve water stress tolerance of vegetables, including post-harvest dehydration of fruit.

4.3. Salinity Stress

Soil salinity is determined by the accumulation of soluble salts, which mainly include Cl^-, SO_4^{2-}, HCO_3^-, Na^+, Ca^{2+}, and Mg^{2+}. The accumulation of these ions derives from low-quality irrigation water and poor soil drainage [71]. Salinity reduces plant growth and yield when the concentration of salts reaches 4 dS/m. The reduction of growth and yield depends on crops' tolerance (Table 3).

Table 3. Tolerance thresholds expressed as electric conductivity (EC) and critical EC values for yield loss. Tolerance degree expressed as: S = sensitive; MS = moderately sensitive; MT = moderately tolerant; T = tolerant [72].

Vegetable Crops	Soil Salinity			Salinity in Irrigation Water			Tolerance
	Threshold (CEe) ($dS{\cdot}m^{-1}$)	Slope (%/$dS{\cdot}m^{-1}$)	Yield 0% ($dS{\cdot}m^{-1}$)	Threshold (CEi) ($dS{\cdot}m^{-1}$)	Slope (%/$dS{\cdot}m^{-1}$)	Yield 0% ($dS{\cdot}m^{-1}$)	
Artichoke (*Cynara scolymus* L.)	4.8	10.9	14	2.7	14.4	9.6	MT
Asparagus (*Asparagus officinalis* L.)	4.1	2	54.1	2.7	3.0	36	T
Bean (*Faseolus vulgaris* L.)	1	19	6.3	0.7	28.5	4.2	S
Broad bean (*Vicia faba* L.)	1.6	9.6	12	1.1	14.5	8	MS
Broccoli (*Brassica oleracea* var. *italica* Plenck)	2.8	9.2	13.7	1.9	13.8	9.2	MS
Brussels sprouts (*Brassica oleracea* var. *gemmifera* DC.)	-	-	-	-	-		MS
Cabbage hood (*Brassica oleracea* var. *capitata* L.)	1.8	9.7	12.1	1.2	14.6	8.1	MS
Carrot (*Daucus carota* L.)	1	14	8.1	0.7	21.0	5.5	S
Cauliflower (*Brassica oleracea* L. var. *botrytis* L.)	-	-	-	-	-	-	MS
Celery (*Apium graveolens* L. var. *dulce* [Mill.] Pers.)	1.8	6.2	17.9	1.2	9.3	12	MS
Cowpea (*Vigna unguiculata* [L.] Walpers subsp. *unguiculata*)	4.9	12	13.2	3.3	18.2	8.8	MT
Cucumber (*Cucumis sativus* L.)	2.5	13	10.2	1.7	19.5	6.8	MS
Eggplants (*Solanum melongena* L.)	1.1	6.9	15.6	0.7	10.3	10.4	MS
Funnel (*Foeniculum vulgare* Miller var. *azoricum* [Mill.] Thell.)	1.5	15	8.2	1.1	18.0	6.7	MS
Garlic (*Allium sativum* L.)	1.7	10	11.7	1.1	14.9	7.8	MS
Lettuce (*Lactuca sativa* L.)	1.3	13	9	0.9	19.5	6	MS
Melon (*Cucumis melo* L.)	1	8.4	12.9	0.7	12.7	8.6	MS
Onion (*Allium cepa* L.)	1.2	16	7.5	0.8	24.0	5	S
Pea (*Pisum sativum* L.)	-	-	-	-	-	-	S
Pepper (*Capsicum annuum* L.)	1.5	14	8.6	1.0	21.0	5.8	MS
Potato (*Solanum tuberosum* L.)	1.7	12	10	1.1	18.0	6.7	MS
Radish (*Raphanus sativus* L.)	1.2	13	8.9	0.8	19.5	5.9	MS
Rapa (*Brassica rapa* L. var. *rapa*)	0.9	9	12	0.7	13.5	8.1	MS
Spinach (*Spinacia oleracea* L.)	2	7.6	15.2	1.3	11.4	10.1	MS
Straberry (*Fragaria x ananassa* Duch.)	1	33	4.0	0.7	49.5	2.7	S
Swiss chard (Beta *vulgaris* L. var. *conditiva* Alef.)	4	9	15.1	2.7	13.5	10.1	MT
Tomato (*Lycopersicon esculentum* Mill.)	2.5	9.9	12.6	1.7	15.0	8.4	MS
Water melon (*Citrullus lanatus* [Thunberg] Matsumura et Nakai)	-	-	-	-	-	-	MS
Zucchini (*Cucurbita pepo* L.)	4.7	9.4	15.3	3.1	14.1	10.2	MT

High salinity is often a problem in areas located along the sea, especially in Mediterranean areas with intensive agriculture [73] and, in particular, for vegetable farms. Vegetable crops usually have short production cycles and require substantial amounts of water in short periods, increasing salinity problems. The majority of high salinity soil and irrigation water occurs in summer because the reduction of water availability increases seawater infiltration in the groundwater. In these conditions farms that use underground water pumped from soils for irrigation increase seawater infiltration. It has been estimated that half of irrigated agricultural lands are affected by salinity.

Agronomic strategies to reduce salinity stress during cultivation can act on soil or crops. At the soil level, the simplest strategy is to maintain high water availability, with frequent irrigation if possible. It is important to use irrigation systems with high efficiency, such as drip irrigation. The reduction of soil water content increases the salt concentration and crops suffer from osmotic stress. Summer is the most critical period for salinity stress; usually increased plant survival can be obtained by calcium nitrate or chloride [74]. The application of calcium has a beneficial effect on both the soil structure and plant tolerance. The calcium ions move the sodium ions from soil colloids and these can be leached by irrigation. The same effect can be reached by magnesium application. The mitigation effect is considered in the calculation of the sodium adsorption rate (SAR), which takes into consideration the concentration of sodium, calcium, and magnesium, in the following equation:

$$\mathrm{SAR} = \frac{Na^{+}}{\sqrt{\frac{Ca^{2+}+Mg^{2+}}{2}}}. \qquad (1)$$

Therefore, fertilizers containing calcium and magnesium in sodic soils improve the structure of soils and provide a better environment for roots and plant growth. At the crop physiology level, cytosolic calcium inhibits sodium channels in membranes, called salt overlay sensitivity (SOS), and reduces salt accumulation in the cells, alleviating salinity stress. The mutation of a gene encoding for a SOS plasma membrane Na^{+}–H^{+} antiporter increased salt sensitivity, while the overexpression of this gene increased salt tolerance [75]. Moreover, nitrates, if calcium nitrate is used, are in competition with sodium ions for accumulation in the vacuoles. Therefore, the supply of nitrates may reduce salt uptake. However, this particular aspect needs further investigation.

Positive effects have been reported for the application of plant growth-promoting bacteria (PGPB) in increasing crop tolerance to salinity [76]. The mechanism of action has not been elucidated yet but it seems that the bacteria help the roots avoid the excessive uptake of sodium. An analogous effect can be obtained using arbuscular-mycorrhizal fungi, which can improve the uptake of mineral nutrients and reduce salt stress, enhance osmotic adjustment, and have a direct effect on plant hormone biosynthesis. The application of *Glomus* species in lettuce stressed with sodium chloride improved photosynthetic activity, stomatal conductance, and WUE [77].

4.4. Lodging

The process by which shoots of winter or summer cereals are displaced from their vertical orientation is named lodging. In cereals such as wheat and barley, lodging is most likely to occur during the two or three months preceding harvest, usually after ear or panicle emergence, with the result that shoots permanently lean or lie horizontally on the ground. Lodging can be caused by the buckling of stems (stem lodging) or displacement of roots within the soil (root lodging) [78]. In stem lodging, roots are held firm in a strong soil where the wind force buckles one of the lower internodes of the shoot. Root lodging becomes more likely when the anchorage strength is reduced by weak soil or poorly developed anchorage roots. The effect is a reduction in crop yield by up to 80%, with further losses in grain quality, greater drying costs, and an increase in the time taken for harvesting [79].

The main factors that contribute to the lodging process are strong winds (e.g., foehn winds or downbursts associated with thunderstorms), heavy rain, crop pests (e.g., *Diabrotica vergifera* larvae bore deep into the roots, destroying them and giving rise to root lodging), diseases (e.g., fungal diseases that attack the basal part the stems of winter cereals give rise to stem lodging), and an excess of nitrogen fertilization (nitrogen enhances the vegetative growth of stems, with excessive elongation associated with a lower elasticity and increased weakness).

Lodging can be prevented by installing shelterbelts against dominant winds, adopting an equilibrated nitrogen fertilization, choosing a suitable plant density, and optimizing the management of pests and biotic diseases harmful to the stability and mechanic resistance of stems and roots. The lodging risk for crops can also be reduced by the introduction of semi-dwarf varieties (e.g., the wheat varieties produced by Nazareno Strampelli and Norman Borlaug in the 20th century) or by the adoption of plant growth regulators (PGRs).

Wind breaks are barriers given by plantations (trees and shrubs) or non-living material (walls, fences and so all), established in order to protect field crops from dominant winds. A crucial decision in order to optimize the effect of windbreaks is influenced by their porosity, because a very dense or low-density row of trees have low effectiveness, while the most effective are medium-density rows [80]. Numerous studies describe the effect of wind breaks on various atmospheric variables such as temperature (for example, the risk of frost can be increased) and evapotranspiration. Furthermore, shelter belt plantations show a more or less strong competition with crops for light, water, and nutrients [64]. All these effects, which are generally a function of the distance from the windbreak, should be considered in the design of these artifacts.

5. Conclusions

This review discussed agronomic strategies that can be adopted to cope with the effects of abiotic stress on crops, offering a series of ideas based on suitable cultivation techniques. Often defense against abiotic stress is only sought at the genetic level by the identification of tolerant genotypes. This is a correct approach, but agronomic tools can often offer an adequate and rapid solution for reducing crop yield losses. The interaction between genetics and management was a crucial factor of the 20th-century green revolution and is destined to receive increasing attention in the coming years due to the need to increase global agricultural production while respecting the quality requirements of the market. Agronomic management strategies have been considered static for a long time and have not been adequately reconsidered for controlling crop performance. Instead, agronomic management has to be continuously revised, considering innovations in crop tolerance and genetic improvements.

References

1. Zandalinas, S.I.; Mittler, R.; Balfagón, D.; Arbona, V.; Gómez-Cadenas, A. Plant adaptations to the combination of drought and high temperatures. *Physiol. Plant.* **2017**. [CrossRef] [PubMed]
2. Vinocur, B.; Altman, A. Recent advances in engineering plant tolerance to abiotic stress: Achievements and limitations. *Curr. Opin. Biotechnol.* **2005**, *16*, 123–132. [CrossRef] [PubMed]
3. Cramer, G.R.; Urano, K.; Delrot, S.; Pezzotti, M.; Shinozaki, K. Effects of abiotic stress on plants: A systems biology perspective. *BMC Plant Biol.* **2011**, *11*, 163. [CrossRef] [PubMed]
4. Atkinson, N.J.; Urwin, P.E. The interaction of plant biotic and abiotic stresses: From genes to the field. *J. Exp. Bot.* **2012**, *63*, 3523–3543. [CrossRef] [PubMed]
5. Buchanan, B.B.; Gruissem, W.; Russell, L.J. (Eds.) *Biochemistry and Molecular Biology of Plants*, 2nd ed.; Wiley: Hoboken, NJ, USA, 2015; 1280p, ISBN 978-0-470-71421-8.
6. Boogaard, H.; Wolf, J.; Supit, I.; Niemeyer, S.; van Ittersum, M. A regional implementation of WOFOST for calculating yield gaps of autumn-sown wheat across the European Union. *Field Crops Res.* **2013**, *143*, 130–142. [CrossRef]
7. Singh, R.; Van Dam, J.C.; Feddes, R.A. Water productivity analysis of irrigated crops in Sirsa district, India. *Agric. Water Manag.* **2006**, *82*, 253–278. [CrossRef]
8. Mariani, L. Carbon plants nutrition and global food security. *Eur. Phys. J. Plus* **2017**, *132*, 69. [CrossRef]
9. Oerke, E.C. Crop losses to pests. Centenary review. *J. Agric. Sci.* **2006**, *144*, 31–43. [CrossRef]

10. Boyer, J.S. Plant productivity and environment. *Science* **1982**, *218*, 443–448. [CrossRef] [PubMed]
11. Farooq, M.; Wahid, A.; Kobayashi, N.; Fujita, D.; Basra, S.M.A. Plant drought stress: Effects, mechanisms and management. *Agron. Sustain. Dev.* **2009**, *29*, 185–212. [CrossRef]
12. Hillel, D. *Introduction to Environmental Soil Physics*; Elsevier: Amsterdam, The Netherlands, 2013; 495p.
13. Rich, S.M.; Watt, M. Soil conditions and cereal root system architecture: Review and considerations for linking Darwin and Weaver (Darwin review). *J. Exp. Bot.* **2013**, *64*, 1193–1208. [CrossRef] [PubMed]
14. Passioura, J.B. Soil conditions and plant growth. *Plant Cell Environ.* **2002**, *25*, 311–318. [CrossRef] [PubMed]
15. Pearson, C.J.; Norman, D.W.; Dixon, J. *Sustainable Dryland Cropping in Relation to Soil Productivity—FAO Soils Bulletin 72*; Food & Agriculture Organization of the United Nations (FAO): Rome, Italy, 1995.
16. Valentine, T.A.; Hallett, P.D.; Binnie, K.; Young, M.W.; Squire, G.R.; Hawes, C.; Bengough, A.G. Soil strength and macropore volume limit root elongation rates in many UK agricultural soils. *Ann. Bot.* **2012**, *110*, 259–270. [CrossRef] [PubMed]
17. Bonciarelli, F. *Fondamenti di Agronomia Generale*; Edagricole: Bologna, Italy, 1989; 292p.
18. Mariani, L.; Cola, G.; Parisi, S. Dimensioning of field ditches in function of heavy and frequent precipitations. In Proceedings of the AIAM 2013 17th Annual Meeting of the Italian Agrometeorological Association Agrometeorology for Environmental and Food Security, Florence, Italy, 4–6 June 2013. (supplement to the *Italian Journal of Agrometeorology* 2013, 101–102).
19. Pagliai, M.; Vignozzi, N.; Pellegrini, S. Soil structure and the effect of management practices. *Soil Tillage Res.* **2004**, *79*, 131–143. [CrossRef]
20. Kibblewhite, M.G.; Ritz, K.; Swift, M.J. Soil health in agricultural systems. *Philos. Trans. R. Soc. Lond. B Biol. Sci.* **2008**, *363*, 685–701. [CrossRef] [PubMed]
21. Carmeis Filho, A.C.A.; Crusciol, C.A.C.; Guimarães, T.M.; Calonego, J.C.; Mooney, S.J. Correction: Impact of amendments on the physical properties of soil under tropical long-term no till conditions. *PLoS ONE* **2017**, *12*. [CrossRef] [PubMed]
22. Mulumba, N.L.; Lal, R. Mulching effects on selected soil physical properties. *Soil Tillage Res.* **2008**, *98*, 106–111. [CrossRef]
23. Wagner, L.E.; Ambe, N.M.; Barnes, P. Tillage-induced soil aggregate status as influenced by water content. *Trans. ASAE* **1992**, *35*, 499–504. [CrossRef]
24. Liu, Z.; Rong, Q.; Zhou, W.; Liang, G. Effects of inorganic and organic amendment on soil chemical properties, enzyme activities, microbial community and soil quality in yellow clayey soil. *PLoS ONE* **2017**, *12*, e0172767. [CrossRef] [PubMed]
25. Tejada, M.; Gonzalez, J.L. Influence of organic amendments on soil structure and soil loss under simulated rain. *Soil Tillage Res.* **2017**, *93*, 197–205. [CrossRef]
26. Wu, S.F.; Wu, P.T.; Feng, H.; Bu, C.F. Influence of amendments on soil structure and soil loss under simulated rainfall China's loess plateau. *Afr. J. Biotechnol.* **2010**, *9*, 6116–6121.
27. Su, L.; Wang, Q.; Wang, C.; Shan, Y. Simulation Models of Leaf Area Index and Yield for Cotton Grown with Different Soil Conditioners. *PLoS ONE* **2015**. [CrossRef] [PubMed]
28. Pearson, C.J.; Norman, D.W.; Dixon, J. Physical aspects of crop productivity. In *Sustainable Dryland Cropping in Relation to Soil Productivity, FAO Soils Bulletin 72*; Food & Agriculture Organization of the United Nations (FAO): Rome, Italy, 1995; Chapter 2; Available online: http://www.fao.org/docrep/V9926E/v9926e04.htm (accessed on 22 September 2017).
29. Mohamadi, M.A.; Kavian, A. Effects of rainfall patterns on runoff and soil erosion in field plots. *Int. Soil Water Conserv. Res.* **2015**, *3*, 273–281. [CrossRef]
30. NSW DPI Soils Advisory Office. Soil Erosion Solutions, Fact Sheet 1: Types of Erosion. 2017. Available online: https://www.dpi.nsw.gov.au/__data/assets/pdf_file/0003/255153/fact-sheet-1-types-of-erosion.pd (accessed on 3 September 2017).
31. Taguas, E.V.; Guzman, E.; Guzman, G.; Vanwalleghem, T.; Gomez, J.A. Characteristics and Importance of Rill and Gully Erosion. *Cuadenos de Investigacion Geografica* **2015**, *41*, 107–126. [CrossRef]
32. Soracco, C.G.; Lozano, L.A.; Villarreal, R.; Palancar, T.C.; Collazo, D.J.; Sarli, G.O.; Filgueira, R.R. Effects of compaction due to machinery traffic on soil pore configuration. *Rev. Bras. Ciênc. Solo* **2015**, *39*. [CrossRef]
33. Dagesse, D.F. Freezing cycle effects on water stability of soil aggregates. *Can. J. Soil Sci.* **2013**, *93*, 473–483. [CrossRef]

34. Tennis, E.S.; Dolferus, R.; Ellis, M.; Rahman, M.; Wu, Y.; Hoeren, F.U.; Grover, A.; Ismond, K.P.; Good, A.G.; Peacock, W.J. Molecular strategies for improving waterlogging tolerance in plants. *J. Exp. Bot.* **2000**, *51*, 89–97.
35. Drew, M.C. Oxygen deficiency and root metabolism: Injury and acclimation under hypoxia and anoxia. *Ann. Rev. Plant Physiol. Plant Mol. Biol.* **1997**, *48*, 223–250. [CrossRef] [PubMed]
36. Linkemer, G.; Board, J.E.; Musgrave, M.E. Waterlogging effects on growth and yield components in late-planted soybean. *Crop Sci.* **1998**, *38*, 1576–1584. [CrossRef] [PubMed]
37. Fukao, T.; Bailey-Serres, J. Plant responses to hypoxia—Is survival a balancing act? *Trends Plant Sci.* **2004**, *9*, 449–456. [CrossRef] [PubMed]
38. MacEwan, R.J.; Gardner, W.K.; Ellington, A.; Hopkins, D.G.; Bakker, A.C. Tile and mole drainage for control of waterlogging in duplex soils of south-eastern Australia. *Aust. J. Exp. Agric.* **1992**, *32*, 865–878. [CrossRef]
39. Huang, B.; Johnson, J.W.; Nesmith, S.; Bridges, D.C. Growth physiological and anatomical responses of two wheat genotypes to waterlogging and nutrient supply. *J. Exp. Bot.* **1994**, *45*, 193–202. [CrossRef]
40. Allègre, A.; Silvestre, J.; Morard, P.; Kallerhoff, J.; Pinelli, E. Nitrate reductase regulation in tomato roots by exogenous nitrate: A possible role in tolerance to long-term root anoxia. *J. Exp. Bot.* **2004**, *55*, 2625–2634. [CrossRef] [PubMed]
41. Igamberdiev, A.U.; Hill, R.D. Nitrate, NO and haemoglobin in plant adaptation to hypoxia: An alternative to the classic fermentation pathways. *J. Exp. Bot.* **2004**, *55*, 2473–2482. [CrossRef] [PubMed]
42. Serraj, R.; Sinclair, T.R. Osmolyte accumulation: Can it really help increase crop yield under drought conditions? *Plant Cell Environ.* **2002**, *25*, 333–341. [CrossRef] [PubMed]
43. Widtsoe, J.A. *Dry Farming, a System of Agriculture for Countries under a Low Rainfall*; The Mcmillan Company: New York, NY, USA, 1920; p. 501. Available online: http://archive.org/details/dryfarmingasyst01widtgoog (accessed on 15 September 2017).
44. Lipiec, J.; Kuś, J.; Słowińska-Jurkiewicz, A.; Nosalewicz, A. Soil porosity and water infiltration as influenced by tillage methods. *Soil Tillage Res.* **2006**, *89*, 210–220. [CrossRef]
45. Bond, J.J.; Willis, W.O. Soil water evaporation: Surface residue rate and placement effects. *Soil Sci. Soc. Am. J.* **1969**, *33*, 445–448. [CrossRef]
46. Gupta, S.; Larson, W.E. Estimating soil water retention characteristics from particle size distribution, organic matter percent, and bulk density. *Water Resour. Res.* **1979**, *15*, 1633–1635. [CrossRef]
47. Lament, W.J. Plastic mulches for the production of vegetable crops. *HortTechnology* **1993**, *3*, 35–39.
48. Kasirajan, S.; Ngouajio, M. Polyethylene and biodegradable mulches for agricultural applications: A review. *Agron. Sustain. Dev.* **2012**, *32*, 501–529. [CrossRef]
49. Clough, G.H.; Locascio, S.J.; Olson, S.M. Continuous use of polyethylene mulched beds with overhead or drip irrigation for successive vegetable production. In Proceedings of the 20th National Agriculture Plastics Congress, Portland, OR, USA, 25–27 August 1987; pp. 57–61.
50. Davenport, D.; Hagan, R.; Martin, P. Antitranspirants uses and effects on plant life. *Calif. Agric.* **1969**, *23*, 14–16.
51. Shasha, J.I.; Ling, T.O.N.G.; Fusheng, L.I.; Hongna, L.U.; Sien, L.I.; Taisheng, D.U.; Youjie, W.U. Effect of a new antitranspirant on the physiology and water use efficiency of soybean under different irrigation rates in an arid region. *Front. Agric. Sci. Eng.* **2017**, *4*, 155–164.
52. Bulgari, R.; Cocetta, G.; Trivellini, A.; Vernieri, P.; Ferrante, A. Application of biostimulants for improving yield and quality of vegetables and floricultural crops. *Biol. Agric. Hortic.* **2015**, *31*, 1–17. [CrossRef]
53. Alvino, A.; Marino, S. Remote sensing for irrigation of horticultural crops. *Horticulturae* **2017**, *3*, 40. [CrossRef]
54. Dukes, M.D.; Perry, C. Uniformity testing of variable-rate center pivot irrigation control systems. *Precis. Agric.* **2006**, *7*, 205. [CrossRef]
55. Schwarz, D.; Rouphael, Y.; Colla, G.; Venema, J.H. Grafting as a tool to improve tolerance of vegetables to abiotic stresses: Thermal stress, water stress and organic pollutants. *Sci. Hortic.* **2010**, *127*, 162–171. [CrossRef]
56. Edelstein, M.; Ben-Hur, M.; Plaut, Z. Grafted melons irrigated with fresh or effluent water tolerate excess boron. *J. Am. Soc. Hortic. Sci.* **2007**, *132*, 484–491.
57. Edelstein, M.; Plaut, Z.; Ben-Hur, M. Sodium and chloride exclusion and retention by non-grafted and grafted melon and Cucurbita plants. *J. Exp. Bot.* **2011**, *62*, 177–184. [CrossRef] [PubMed]

58. Garcia-Sanchez, F.; Syvertsen, J.P.; Gimeno, V.; Botia, P.; Perez-Perez, J.G. Responses to flooding and drought stress by two citrus rootstock seedlings with different water-use efficiency. *Biol. Plant* **2007**, *130*, 532–542. [CrossRef]
59. Satisha, J.; Prakash, G.S.; Bhatt, R.M.; Sampath Kumar, P. Physiological mechanisms of water use efficiency in grape rootstocks under drought conditions. *Int. J. Agric. Res.* **2007**, *2*, 159–164.
60. Sanders, P.L.; Markhart, A.H., III. Interspecific grafts demonstrate root system control of leaf water status in water stressed Phaseolus. *J. Exp. Bot.* **1992**, *43*, 1563–1567. [CrossRef]
61. Serraj, R.; Sinclair, T.R. Processes contributing to N_2-fixation insensitivity to drought in the soybean cultivar Jackson. *Crop Sci.* **1996**, *36*, 961–968. [CrossRef]
62. Clearwater, M.J.; Lowe, R.G.; Hofstee, B.J.; Barclay, C.; Mandemaker, A.J.; Blattmann, P. Hydraulic conductance and rootstock effects in grafted vines of kiwifruit. *J. Exp. Bot.* **2004**, *55*, 1371–1381. [CrossRef] [PubMed]
63. Rouphael, Y.; Cardarelli, M.; Colla, G.; Rea, E. Yield, mineral composition, water relations, and water use efficiency of grafted mini-watermelon plants under deficit irrigation. *HortScience* **2008**, *43*, 730–736.
64. Allario, T.; Brumos, J.; Colmenero-Flores, J.M.; Iglesias, D.J.; Pina, J.A.; Navarro, L.; Talon, M.; Ollitrault, P.; Morillon, R.; Morillon, R. Tetraploid Rangpur lime rootstock increases drought tolerance via enhanced constitutive root abscisic acid production. *Plant Cell Environ.* **2013**, *36*, 856–868. [CrossRef] [PubMed]
65. Serra, I.; Strever, A.; Myburgh, P.A.; Deloire, A. The interaction between rootstocks and cultivars (*Vitis vinifera* L.) to enhance drought tolerance in grapevine. *Aust. J. Grape Wine Res.* **2014**, *20*, 1–14. [CrossRef]
66. Liu, S.; Li, H.; Lv, X.; Ahammed, G.J.; Xia, X.; Zhou, J.; Zhou, Y. Grafting cucumber onto luffa improves drought tolerance by increasing ABA biosynthesis and sensitivity. *Sci. Rep.* **2016**, *6*, 20212. [CrossRef] [PubMed]
67. Holbrook, N.M.; Shashidhar, V.R.; James, R.A.; Munns, R. Stomatal control in tomato with ABA-deficient roots: Response of grafted plants to soil drying. *J. Exp. Bot.* **2002**, *53*, 1503–1514. [PubMed]
68. Corbineau, F.; Xia, Q.; Bailly, C.; El-Maarouf-Bouteau, H. Ethylene, a key factor in the regulation of seed dormancy. *Front. Plant Sci.* **2014**, *5*, 539. [CrossRef] [PubMed]
69. Mustilli, A.C.; Merlot, S.; Vavasseur, A.; Fenzi, F.; Giraudat, J. Arabidopsis OST1 protein kinase mediates the regulation of stomatal aperture by abscisic acid and acts upstream of reactive oxygen species production. *Plant Cell* **2002**, *14*, 3089–3099. [CrossRef] [PubMed]
70. Bernstein, L. Effects of salinity and sodicity on plant growth. *Ann. Rev. Phytopathol.* **1975**, *13*, 295–312. [CrossRef]
71. Acosta-Motos, J.R.; Ortuño, M.F.; Bernal-Vicente, A.; Diaz-Vivancos, P.; Sanchez-Blanco, M.J.; Hernandez, J.A. Plant Responses to Salt Stress: Adaptive Mechanisms. *Agronomy* **2017**, *7*, 18. [CrossRef]
72. FAO. Annex 1. Crop Salt Tolerance Data. Available online: http://www.fao.org/docrep/005/y4263e/y4263e0e.htm (accessed on 12 September 2017).
73. Jaleel, C.A.; Manivannan, P.; Sankar, B.; Kishorekumar, A.; Gopi, R.; Somasundaram, R.; Panneerselvam, R. Water deficit stress mitigation by calcium chloride in *Catharanthus roseus*: Effects on oxidative stress. proline metabolism and indole alkaloid accumulation. *Colloids Surf. B Biointerfaces* **2007**, *60*, 110–116. [CrossRef] [PubMed]
74. Yang, Q.; Chen, Z.Z.; Zhou, X.F.; Yin, H.B.; Li, X.; Xin, X.F.; Gong, Z. Overexpression of SOS (Salt Overly Sensitive) genes increases salt tolerance in transgenic Arabidopsis. *Mol. Plant* **2009**, *2*, 22–31. [CrossRef] [PubMed]
75. Mayak, S.; Tirosh, T.; Glick, B.R. Plant growth-promoting bacteria confer resistance in tomato plants to salt stress. *Plant Physiol. Biochem.* **2004**, *42*, 565–572. [CrossRef] [PubMed]
76. Ruiz-Lozano, J.M.; Azcon, R.; Gomez, M. Alleviation of salt stress by arbuscular-mycorrhizal Glomus species in *Lactuca sativa* plants. *Physiol. Plant.* **1996**, *98*, 767–772. [CrossRef]
77. Berry, P.M. Understanding and reducing lodging in cereals. *Adv. Agron.* **2004**, *84*, 217–271.
78. Tams, A.R.; Mooney, S.J.; Berry, P.M. The Effect of Lodging in Cereals on Morphological Properties of the Root-Soil Complex. In Proceedings of the SuperSoil 2004: 3rd Australian New Zealand Soils Conference, Sydney, Australia, 5–9 December 2004. Available online: http://www.regional.org.au/au/asssi/supersoil2004/s9/oral/1998_tamsa.htm (accessed on 20 September 2017).

79. Bean, A.; Alperi, R.W.; Federer, C.A. A method for categorizing shelterbelts porosity. *Agric. Meteorol.* **1975**, *14*, 417–429. [CrossRef]
80. Campi, P.; Palumbo, A.D.; Mastrorilli, M. Effect of tree windbreaks on microcliamte and wheat productivity in a Mediterrranean environment. *Eur. J. Agron.* **2009**, *30*, 220–227. [CrossRef]

7

Yield, Quality, Antioxidants and Elemental Composition of New Leek Cultivars under Organic or Conventional Systems in a Greenhouse

Nadezhda A. Golubkina [1,*], Timofey M. Seredin [1], Marina S. Antoshkina [1], Olga V. Kosheleva [2], Gabriel C. Teliban [3] and Gianluca Caruso [4]

1 Agrochemical Research Center, Federal Scientific Center of Vegetable Production, Odintsovo District, Vniissok, Selectsionnaya 14, Moscow 143072, Russia; timofey-seredin@rambler.ru (T.M.S.); limont_m@mail.ru (M.S.A.)
2 Federal Research Centre of Nutrition, Biotechnology and Food Safety, Ustinsky pr., 2/14, Moscow 109240, Russia; ok-ion-vit@yandex.ru
3 Department of Horticulture Technology, University of Agriculture Sciences and Veterinary Medicine, 3 M. Sadoveanu, 700490 Iasi, Romania; gabrielteliban@uaiasi.ro
4 Department of Agricultural Sciences, University of Naples Federico II, 80055 Portici, Naples, Italy; gcaruso@unina.it
* Correspondence: segolubkina45@gmail.com.

Abstract: Leek (*Allium porrum*) is known for its high antioxidant activity and the ability to accumulate significant amounts of potassium and iron. We assessed yield, quality indicators, antioxidants and elemental composition of nine leek cultivars grown in greenhouses under organic or conventional systems in the Moscow region. The management system did not affect yield, which attained the highest value with the cultivar Giraffe and the lowest with Premier and Cazimir. Pseudo-stem dry matter and sugars were higher with organic management, whereas nitrate concentration was higher with conventional management. The cultivars Vesta and Summer Breeze showed the highest dry matter and total sugar content, whereas Goliath had the highest antioxidant, selenium and potassium concentrations. Among the antioxidants, ascorbic acid attained higher values with organic management. The antioxidant system of leek was characterized by highly significant positive correlations between: Se and polyphenols, Se and ascorbic acid, Se and K, ascorbic acid and polyphenols, ascorbic acid and K, polyphenols and K (r = 0.94, 0.94, 0.95, 0.94, 0.95, 0.96, respectively, at $P \leq 0.001$). Negative correlations were recorded between leaf and pseudo-stem Se and between leaf and pseudo-stem polyphenols (r = −0.922 and −0.976, respectively, at $P \leq 0.001$). Among the mineral elements, only K was significantly affected by the management system, showing a higher content in organically grown pseudo-stems. Varietal differences in pseudo-stem element composition showed strong positive correlations of: Al with As, Co, Li, Pb and V; Cr with I, Mg, Si, Ca; V with As, Co and Fe; negative correlations of Se with Cr and I. Compared to related species such as garlic (A. sativum), leek accumulated levels equal to garlic of K, Mg, P, Cd, Cu, Mn, Se, Zn, lower levels of Si and significantly higher amounts of Ca, Na, Al, As, Cr, Ni, Pb, Sr, V, Sn, B, Co, Fe, I, Li. The strong relationships between quality, antioxidant and mineral components in leek plants may give wide possibilities in breeding programs for both conventional and organic management systems in greenhouses.

Keywords: *Allium porrum*; organic management; production; sugars; selenium; antioxidants; minerals

---

1. Introduction

Among *Allium* species, leek (*A. porrum*), garlic (*A. sativum*) and onion (*A. cepa*) are the most widely used for human consumption. The greatest production of leek is in Indonesia and Turkey, with France and Belgium being major producers in Europe.

Popularity of leek is connected not only with its high nutritional value but also with its wide spectrum of biological activities, primarily due to a high antioxidants content [1–5]. Leek shows antimicrobial, cardio-protective, hypocholesteremic, hypoglycemic and anticancer activities [6,7]. Leek consumption is known to improve liver and gastro-intestinal tract functioning, quicken metabolic processes, to be useful in rheumatism treatment, to decrease blood pressure, to protect against anemia, to enhance brain activity, to inhibit platelet aggregation and to prevent neural tube defects [3]. Moreover, consumption of *A. porrum* edible parts reportedly decreases the risk of prostate, colon, stomach and breast cancers [8,9]. Antimicrobial effects of leek have been recorded both against gram-positive (*Bacillus subtilis, Streptococcus pneumonia, Staphylococcus aureus*) and gram-negative bacteria (*Escherichia coli, Proteus vulgaris, Pseudomonasaeruginosa*) [10]; and, antifungal activity has also been reported [11,12].

The most important biologically active compounds contained in leek are polyphenols [5,13], glucosinolates, S-alkenyl-L-cysteine sulfoxides, and pectic polysaccharides, each showing immune stimulating activity [1,14,15]. Polyphenol content in *A. porrum* is comparable with the *A. ascalonicum* (shallot) and significantly exceeds that of *A. sativum, A. cepa* [16], *A. schoenoprasum* (chive) and *A. tuberosum* (garlic chive) [13]. The nutritional value of leek is also correlated with the high content of potassium and iron [17].

Protected cultivation is reportedly preferred for organic horticulture [18], as the latter is more susceptible to the environmental unbalances caused by less intensive management and, in addition, organic vegetables have higher market prices than the conventional ones, which results in a higher income for farmers.

The evaluation of varietal differences in accumulation of biologically active compounds as well as macro- and micro-elements by leek plants is of interest, both for identifying the characteristic interrelations between the components and for carrying out breeding for high concentrations of biologically active compounds. Despite several studies devoted to assessing leek biochemical characteristics and their role in disease treatment, there is some variation among studies concerning varietal differences in biologically active compounds [19] and their elemental composition [17].

Due to research lacking on the aforementioned topics, we have carried out a study to assess the effect of both cultivar and conventional versus organic management on leek yield, quality, antioxidant content and elemental composition of leek grown in greenhouses.

2. Materials and Methods

2.1. Plant Material and Growth Conditions

Leek was grown in a greenhouse at the experimental fields of the Federal Scientific Center of Vegetable Production, in Odintsovo (Moscow, Russia, 55°39.51′ N, 37°12.23′ E) in 2015 and 2016 on a clay-loam soil, with a pH 6.8, 2.1% organic matter, 108 mg kg^{-1} N, 450 mg kg^{-1} P_2O_5, 357 mg kg^{-1} K_2O, and exchangeable bases sum ≤95.2%. Monthly mean temperature and relative humidity values from May to October were 13.0, 16.1, 19.8, 18.6, 12.3, 6.4 °C and 59.1, 63.8, 69.7, 72.4, 79.1, 81.0% respectively. The experimental protocol was based on a factorial combination of two management systems (organic, conventional) and nine cultivars (Goliath, Summer Breeze, Premier, Casimir, Kalambus, Camus, Vesta, Giraffe, Bandit), using a split-plot design with three replicates of each cultivar by management system.

Seed was sown on 5 December in 8 × 8 cm trays and the plantlets were transplanted to the field on 14 May, spaced 15 cm within rows, with rows 40 cm apart. Leek crops were preceded by organically-grown vegetables in the previous four years, including carrot, bean, rape and pea. Prior to planting, the soil was ploughed to a 30 cm depth, hoed to 15 cm, and fertilized with 180 kg ha^{-1} N,

80 kg ha^{-1} P_2O_5 and 120 kg ha^{-1} K_2O. During plant production, 40 kg ha^{-1} N were supplied three times at two-week intervals and, in the last N application, 7 kg ha^{-1} of P_2O_5 and of K_2O were also provided. Half of each fertilizer dose was applied just before transplanting and the remaining 50% by sidedressing at two week intervals. Drip irrigation was started at 80% soil available water. The organic management practices complied with EC Regulations 834/2007 and 889/2008. Plant protection was achieved by applying copper oxychloride against rust, and deltamethrin or azadirachtin in the conventional or organic systems, respectively, against aphids.

Harvests of mature plants were performed from 5 to 10 October, when the pseudo-stems had reached their maximum growth, and the leaf blades were trimmed to a 15 cm length for obtaining a marketable product. In each plot, determinations were made of the weight of the marketable product (pseudo-stems with 15 cm long leaf blades) and the mean pseudo-stem (with 15 cm long leaf blades) weight from twenty-plant samples. Further plant samples were collected, gently washed with water to remove surface contaminants and dried with filter paper. Pseudo-stems and leaves were separated, cut with a plastic knife, dried to a constant weight and homogenized. The resulting powders were subjected to laboratory analyses.

2.2. *Dry Matter*

The dry matter content in leaves and pseudo-stems of *A. porrum* was assessed after drying the fresh samples in an oven at 70 °C, until they reached constant weight.

2.3. *Total Soluble Solids (TSS) and Sugars*

Determination of total soluble solids was performed in filtered water extracts of leek leaves and pseudo-stems (1 g of dry sample per 100 mL) using a TDS-3 conductometer (HM Digital, Inc., Seoul, Korea).

Monosaccharides were determined using the ferricyanide colorimetric method, based on the reaction of monosaccharides with potassium ferrycianide [20]. Total sugars were determined after acidic hydrolysis of 50 mL of filtered water extracts with 5 mL of 20% hydrochloric acid. Fructose was used as an external standard.

2.4. *Polyphenols*

The concentrations of total phenolics in each sample of leaves and pseudo-stems were determined in filtered 70% ethanol extract (0.5 g of dry sample in 25 mL; 1 h at 80 °C) using the Folin-Ciocalteu colorimetric method, according to Golubkina et al. [21] using a Unico 2804 UV (Unico Inc., Wixom, MI, USA) spectrophotometer. The phenolic contents were calculated by using a calibration curve of gallic acid constructed with five concentrations of this compound (0–90 μg/mL). Phenolic contents were expressed as milligrams of gallic acid equivalents per 100 g of dry weight (mg GAE/100 g d.w.).

2.5. *Ascorbic Acid*

Ascorbic acid content of leek leaves and pseudo-stems was assessed by visual titration of fresh plant extracts in 6% trichloracetic acid with Tillmans reagent [22]. Five grams of fresh leek leaves were homogenized in porcelain mortar with 5 mL of 6% trichloracetic acid and quantitatively transferred to measuring cylinder. The volume was brought to 80 mL using trichloracetic acid, and the mixture was filtered through filter paper 15 min later. The ascorbic acid concentration was determined from the volume of Tillmans reagent which went into titration of the sample.

2.6. *Antioxidant Activity*

The antioxidant activity of leek leaves and pseudo-stems was assessed using a redox titration method [23], via titration of 0.01 N $KMnO_4$ solution with ethanolic extracts of leaves and pseudo-stems used for polyphenol determination (see Section 2.4). Reduction of $KMnO_4$ to colorless Mn^{+2} in this process reflects the concentration of antioxidants dissolvable in 70% ethanol. The values were expressed in mg GAE/100 g d.w. The use of $KMnO_4$ acidic solution is known to be successfully used for the determination of *Ocimum basilicum* antioxidant potential [24] and antioxidant capacity of serum [25].

2.7. Nitrates

Nitrate was assessed in fresh pseudo-stems using an ion selective electrode using a ionomer Expert-001 (Econix, Moscow, Russia). Five grams of fresh homogenized leek pseudo-stems were mixed with 50 mL of distilled water. Forty-five mL of filtered extract were mixed with 5 mL of 0.5 M potassium sulfate background solution (necessary for ionic strength regulation) and analyzed through the ionomer for nitrate determination.

2.8. Elemental Composition

Al, As, B, Ca, Cd, Co, Cr, Cu, Fe, Hg, I, K, Li, Mg, Mn, Na, Ni, P, Pb, Si, Sn, Sr, V and Zn contents of leek pseudo-stems were assessed using an ICP-MS on quadruple mass-spectrometer Nexion 300D (Perkin Elmer Inc., Shelton, CT 06484, USA) equipped with the 7-port FAST valve and ESI SC DX4 autosampler (Elemental Scientific Inc., Omaha, NE 68122, USA) in the Biotic Medicine Center (Moscow, Russia). Rhodium ^{103}Rh was used as an internal standard to eliminate instability during measurements. Quantitation was performed using external standard (Merck IV, multi-element standard solution), potassium iodide for iodine calibration and Perkin-Elmer standard solutions for P, Si and V. All the standard curves were obtained at 5 different concentrations.

For quality control purposes, internal controls and reference materials were tested together with the samples on a daily basis. Microwave digestion of samples was achieved according to the standard method [26] with sub-boiled HNO_3 diluted 1:150 with distilled deionized water (Fluka No. 02650 Sigma-Aldrich, Co., Saint Louis, MO, USA) in the Berghof SW-4 DAP-40 microwave system (Berghof Products + Instruments GmbH, 72800 Eningen, Germany). Trace levels of Hg in samples were not taken into account and, accordingly, they were excluded from the Tables below.

The instrument conditions and acquisition parameters were: plasma power and argon flow, 1500 and 18 L min^{-1} respectively; aux argon flow, 1.6 L min^{-1}; nebulizer argon flow, 0.98 L min^{-1}; sample introduction system, ESI ST PFA concentric nebulizer and ESI PFA cyclonic spray chamber (Elemental Scientific Inc., Omaha, NE 68122, USA); sampler and slimmer cone material, platinum; injector, ESI Quartz 2.0 mm I.D/; sample flow, 637 μL/min; internal standard flow, 84 μL/min; dwell time and acquisition mode, 10–100 ms and peak hopping for all analytes; sweeps per reading, 1; reading per replicate, 10; replicate number, 3; DRC mode, 0.55 L min^{-1} ammonia (294993-Aldrich Sigma-Aldrich, Co., St. Louis, MO 63103, USA) for Ca, K, Na, Fe, Cr, V, optimized individually for RPa and RPq; STD mode, for the rest of analytes at RPa = 0 and RPq = 0.25.

Se content of leek leaves and pseudo-stems was analyzed using the fluorimetric method previously described for tissues and biological fluids [27]. The method includes digestion of dried homogenized samples via heating with a mixture of nitric-chloral acids, subsequent reduction of Se^{+6} to Se^{+4} with a solution of 6 N HCl, and formation of a complex between Se^{+4} and 2,3-diaminonaphtalene. The Se concentration was assessed in triplicate by recording piazoselenol fluorescence values in hexane at 519 nm λ emission and 376 nm λ excitation. The results precision was checked using a reference standard-lyophilized cabbage at each determination with 150 μg/Kg Se concentration (Institute of Nutrition, Moscow, Russia).

2.9. Statistical Analysis

Data were processed by analysis of variance and mean separations were performed using the Duncan multiple range test, $\alpha = 0.05$, using SPSS software version 21. The data expressed as a percentage were subjected to angular transformation before processing.

3. Results and Discussion

As the year of research had no significant effect on yield, quality and antioxidant variables examined, both as main factor or in interaction with the experimental factors management system or cultivar, the results are reported as average values of the two years of investigation.

3.1. *Yield, Dry Matter, Sugars and Nitrates of Pseudo-Stems*

The management system showed no significant effects on leek pseudo-stem yield and mean weight (Table 1). The experimental factor cultivar significantly affected the mean pseudo-stem weight and, accordingly, yield which ranged from 23.8×10^3 to 40.2×10^3 kg ha^{-1} , as an average management across systems; the variety Giraffe showed the highest yield, Premier and Kalambus the lowest. The coefficient of variation was rather low and reached 18.3%, which suggests a low genetic effect on this variable (Table 1).

Table 1. Yield, mean pseudo-stem weight, and content of dry matter, sugars and nitrates in leek.

Treatment	Yield 10^3 kg ha^{-1}	Mean Pseudo-Stem Weight g	Dry Matter %	Sugars g/100 g d.w.		Nitrates mg/kg f.w.
				Mono-	Total	
Crop management						
Organic	30.9	185.5	19.7	3.6	11.7	76
Conventional	31.2	187.0	17.8	3.2	10.4	102
	n.s.^{z}	n.s.	*	*	*	*
Cultivar						
Goliath	31.1 ^{d,e,y}	188.0 ^{d,e}	12.4 ± 0.4 ^{e}	4.8 ± 0.3 ^{a}	7.3 ± 0.5 ^{e}	77 ± 3 ^{d}
Premier	23.8 ^{g}	142.4 ^{g}	15.2 ± 0.5 ^{d}	4.4 ± 0.3 ^{a,b}	10.5 ± 0.7 ^{c}	74 ± 3 ^{d}
Bandit	35.0 ^{bc}	209.8 ^{b,c}	15.3 ± 0.6 ^{d}	3.5 ± 0.2 ^{c}	8.6 ± 0.5 ^{d}	103 ± 4 ^{a}
Kalambus	24.0 ^{g}	143.6 ^{g}	17.6 ± 0.6 ^{c}	3.9 ± 0.2 ^{b,c}	10.3 ± 0.7 ^{c}	70 ± 3 ^{d}
Cazimir	26.9 ^{f}	160.9 ^{f}	18.9 ± 0.6 ^{c}	2.8 ± 0.2 ^{d}	10.7 ± 0.7 ^{b,c}	90 ± 4 ^{b,c}
Giraffe	40.2 ^{a}	242.5 ^{a}	20.5 ± 0.8 ^{b}	3.4 ± 0.2 ^{c}	11.0 ± 0.7 ^{bc}	86 ± 3 ^{c}
Camus	33.3 ^{c,d}	201.7 ^{c,d}	21.4 ± 0.7 ^{b}	2.5 ± 0.2 ^{d}	12.2 ± 0.8 ^{b}	105 ± 5 ^{a}
Vesta	29.0 ^{e,f}	175.0 ^{e,f}	23.4 ± 0.8 ^{a}	2.6 ± 0.2 ^{d}	14.3 ± 0.8 ^{a}	98 ± 4 ^{ab}
Summer Breeze	36.7 ^{b}	221.3 ^{b}	24.3 ± 0.9 ^{a}	2.8 ± 0.2 ^{d}	15.1 ± 0.9 ^{a}	98 ± 4 ^{ab}
M	31.1	187.2	18.8	3.86	11.11	89
SD	5.7	34.8	3.2	1.23	1.84	11
CV (%)	18.3	18.6	17.3	31.9	16.8	12.4
Concentration range	23.8–40.2	142.4–242.5	12.4–24.3	2.5–8.4	7.3–15.1	70–105

^{z} n.s. not significant; * significant at $P \leq 0.05$. ^{y} Within each column, means followed by different letters are significantly different according to Duncan's Multiple Range Test at $P \leq 0.05$.

Organic management resulted in a higher dry matter and carbohydrate content in pseudo-stems, compared to the conventional managementsystem (Table 1). These results may be due to the enhancement of microbial biomass and activity leading to organic compound synthesis and, in this respect, similar trends were recorded in previous research carried out in greenhouses [28]. Moreover, dry matter content of the nine leek cultivars ranged from 12 to 24% (Table 1) which was much wider than leek grown in the Czech Republic (9–11%) [14]. These results identified varieties with high dry matter content (Summer Breeze and Vesta), which show a long shelf-life and areuseful for dry spice production, and genotypes with low dry matter content (Goliath, Premier and Bandit) more suitable for salad production, though dry matter is also dependent on water regime [29].

Cultivar differences in total sugar content were 2-fold lower than the coefficient of variation for monosaccharide content (Table 1). Therefore, the existence of a multidirectional nature of mono- and disaccharide accumulation in leek pseudo-stems may exist.

A significant positive correlation was observed between dry matter and disaccharides ($r = 0.98$ at $P < 0.01$), but there was a negative correlation ($r = -0.86$ at $P < 0.05$) between dry matter and monosaccharides (Figure 1). Indeed, the negative correlation between monosaccharides and dry matter content shown in Figure 1 is consistent with that previously found in onion [30]. Interestingly, total sugar content in onion did not differ among cultivars with low versus high dry matter content [31], in contrast to leek where the maximum value was twice the minimum. As observed in Figure 1, the highest concentrations of mono- and total saccharides in leek were associated with cultivars with high dry matter content. However, contrasting carbohydrate trends were found with the two varieties: Goliath had low dry matter content and a significantly higher concentration of monosaccharides compared to disaccharides; in Summer Breeze, the disaccharide content exceeded the monosaccharide content by 4.4-fold , with the ratio between the related total sugar contents at 2.1 (Figure 1).

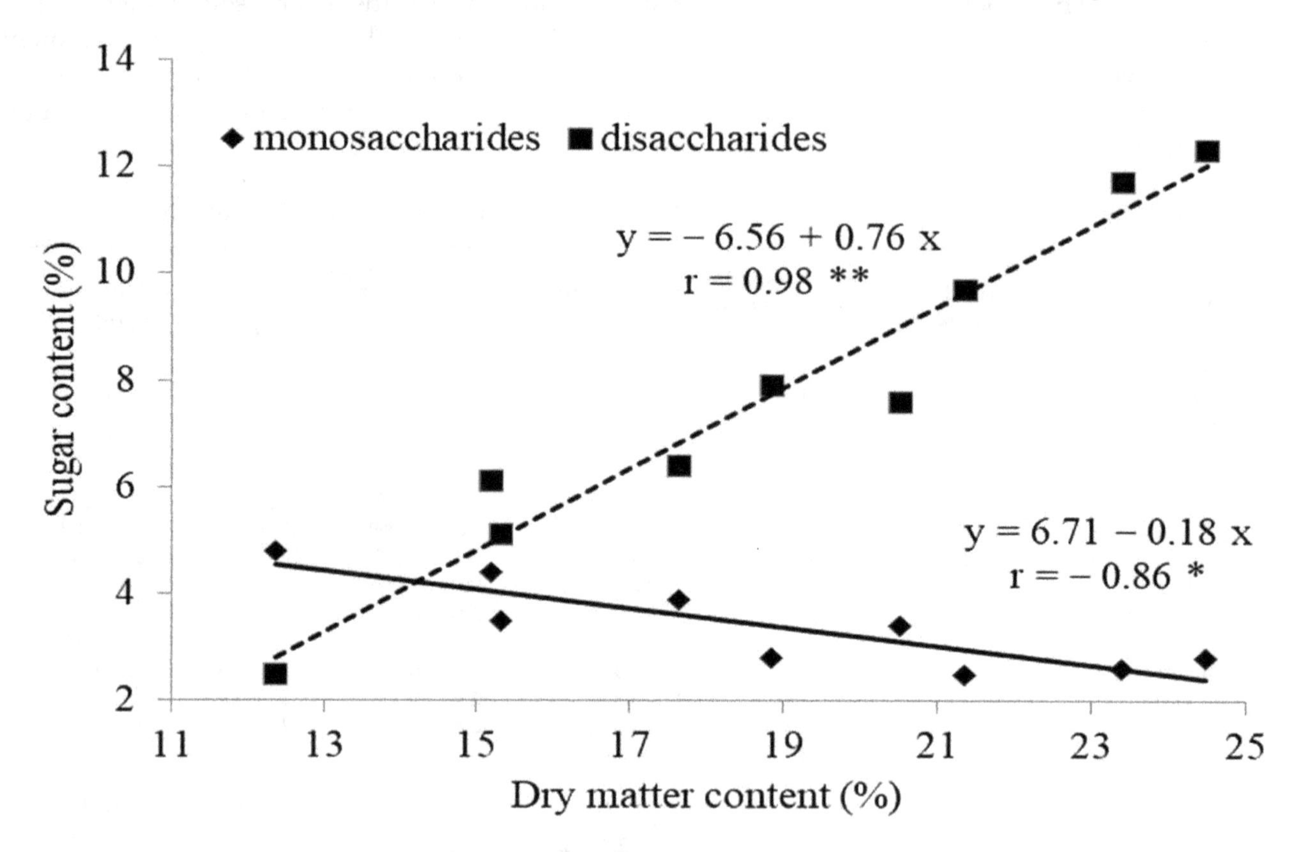

Figure 1. Correlations between dry matter and sugar content in leek pseudo-stems.

With regard to nitrate concentration in pseudo-stems (Table 1), lower values were recorded the organic management system compared to conventional management, consistent with previous reports [28]. Indeed, a slow nitrate release from organic fertilizers does not elicit abundant uptake of this anion in a short time period, which prevents its accumulation in plant parts [32].

Leek is characterized by a relatively high plant nitrate content [33], though it is much lower than in the top-accumulating species [32]. However, in our research the cultivars investigated had less than 105 mg nitrate per kg fresh weight (f.w.) and, moreover, the ascorbic acid content was sufficient to make the product safe and healthy, as ascorbic acid participates in producing essential nitrogen oxide for humans thus preventing nitrosamine formation [33]. Notably, low varietal differences in nitrate accumulation make this quality parameter the most stable among the cultivars grown in similar conditions.

3.2. *Antioxidants*

Polyphenols and vitamin C significantly affect plant antioxidant activity [34]. In our research, organic management resulted in higher vitamin C in leek pseudo-stems compared to conventional management, consistent with some previous investigations [35], but differently from others where no significant differences were observed between the two management systems [28]. The management system did not affect either polyphenol or selenium concentration in leek pseudo-stems and leaves.

There were significant varietal differences in ascorbic acid accumulation, in contrast to polyphenol content which was characterized by higher stability in both stems and especially in leaves (Table 2). Among the cultivars, Goliath showed the highest ascorbic acid content; notably, all of the cultivars from domestic selection had lower levels of polyphenols compared to prior reports, which may be related to different crop cycles and harvest times [36]. Similar findings have been reported from an investigation in Belgium on thirty leek cultivars [19], where the ascorbic acid concentration range was 90 to 350 mg/100 g d.w. and the polyphenol ranged from 7.3 to 11.3 mg GA/g d.w., whereas the domestic cultivars showed lower polyphenol values, i.e., 3.0 to 5.5 mg GA/g d.w. (Table 2).

Table 2. Concentrations of ascorbic acid, polyphenols and selenium in leek.

Treatment	Ascorbic Acid in Pseudo-Stems mg/100 g f.w.	Polyphenols mg GAE/100 g d.w.		Selenium µg/kg d.w.	
		Pseudo-Stems	Leaves	Pseudo-Stems	Leaves
Crop management					
Organic	47.2	375.4	696.7	76.3	64.4
Conventional	37.0	356.2	674.9	73.1	60.8
	* ^{z}	n.s.	n.s.	n.s.	n.s.
Cultivar					
Goliath	136.8 ± 8.1 ^{a,y}	555 ± 52 ^{a}	711 ± 19 ^{a}	107 ± 7 ^{a}	14 ± 1 ^{e}
Premier	58.6 ± 3.3 ^{b}	432 ± 34 ^{b}	650 ± 21 ^{b}	80 ± 5 ^{b}	65 ± 3 ^{c}
Bandit	40.5 ± 2.6 ^{c}	394 ± 26 ^{b,c}	647 ± 26 ^{b}	75 ± 5 ^{b,c}	48 ± 2 ^{d}
Kalambus	25.5 ± 2.3 ^{d,e}	319 ± 19 ^{d,e}	731 ± 46 ^{a}	72 ± 4 ^{b,c}	74 ± 3 ^{b}
Cazimir	30.2 ± 2.6 ^{d}	284 ± 20 ^{e}	728 ± 53 ^{a}	60 ± 3 ^{e}	76 ± 4 ^{a,b}
Giraffe	24.9 ± 2.0 ^{e,f}	331 ± 19 ^{d}	665 ± 41 ^{a,b}	73 ± 4 ^{b,c}	49 ± 2 ^{d}
Camus	21.1 ± 1.4 ^{f,g}	347 ± 21 ^{c,d}	684 ± 43 ^{a,b}	64 ± 3 ^{d,e}	81 ± 5 ^{a,b}
Vesta	19.2 ± 1.3 ^{g}	301 ± 19 ^{d,e}	740 ± 55 ^{a}	69 ± 3 ^{c,d}	84 ± 5 ^{a}
Summer Breeze	22.2 ± 1.9 ^{f,g}	329 ± 20 ^{d,e}	616 ± 42 ^{b}	72 ± 3 ^{b,c}	72 ± 4 ^{b,c}
M	42.1	317	686	74.7	62.5
SD	24.7	71	37	8.4	17
CV, %	58.7	22.4	5.4	11.2	27.2
Concentration range	19.2–136.8	284–555	616–740	60–107	14–84

^{z} n.s. not significant; * significant at $P \leq 0.05$. ^{y} Within each column, means followed by different letters are significantly different according to Duncan's Multiple Range Test at $P \leq 0.05$.

A significant positive correlation between ascorbic acid and polyphenol concentration in leek stems is of particular interest (r = 0.94 at $P < 0.01$). The lack of a similar correlation in a study using thirty leek cultivars conducted in Belgium [19] was presumably related to the heterogeneity of the cultivars, which were selected on the basis of morphological types (light-green summer type, dark-green winter type and intermediate autumn type). Indeed, the autumn-grown Belgium varieties resulted in a similar correlation to ours between ascorbic acid and polyphenol content (r = 0.71 at $P < 0.05$).

The ratio between leaf and stem polyphenol concentration in leek plants was of interest. In the nine cultivars in our research polyphenol content was always higher in leaves compared to stems, with a negative correlation between the tissues (Figure 2). Otherwise, as found by Ben Arfa et al. [5], polyphenol levels in *A. porrum* leaves may be both higher or lower than those in stems. In this respect, the total polyphenol content per plant should be more stable that polyphenol concentrations recorded in stems or leaves.

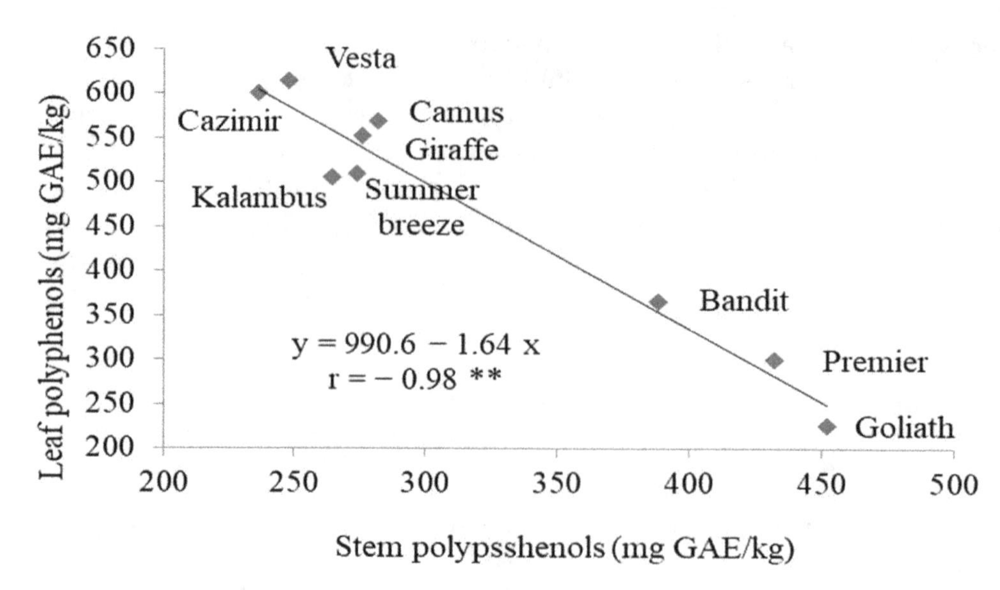

Figure 2. Correlation between leaf and stem polyphenol content.

Among the components of plant antioxidant systems, selenium plays a significant role. Indeed, though it is not an essential element for plants, selenium is able to provide a powerful antioxidant defense to plants against drought, salinity, frost, flooding, UV light and herbivores [37]. Notably, *Allium* species belong to the secondary selenium accumulators, which show a remarkable tolerance to high concentrations and consequent accumulation of this element due to Se ability to substitute for sulfur in natural compounds, as also reported by Turkish scientists in leek [17].

In our research, *A. porrum* grown in the Moscow region showed a Se accumulation range from 60 to 107 μg/kg d.w., which is much lower than the values recorded in Turkey [17]. This suggests the significant effect of selenium status in the environment on plant ability to concentrate the trace element. The negative correlation between selenium content in leaves and stems ($r = -0.95$ at $P < 0.01$; Figure 3), similar to that recorded for polyphenols, entails a rather stable level of selenium accumulation in the plant.

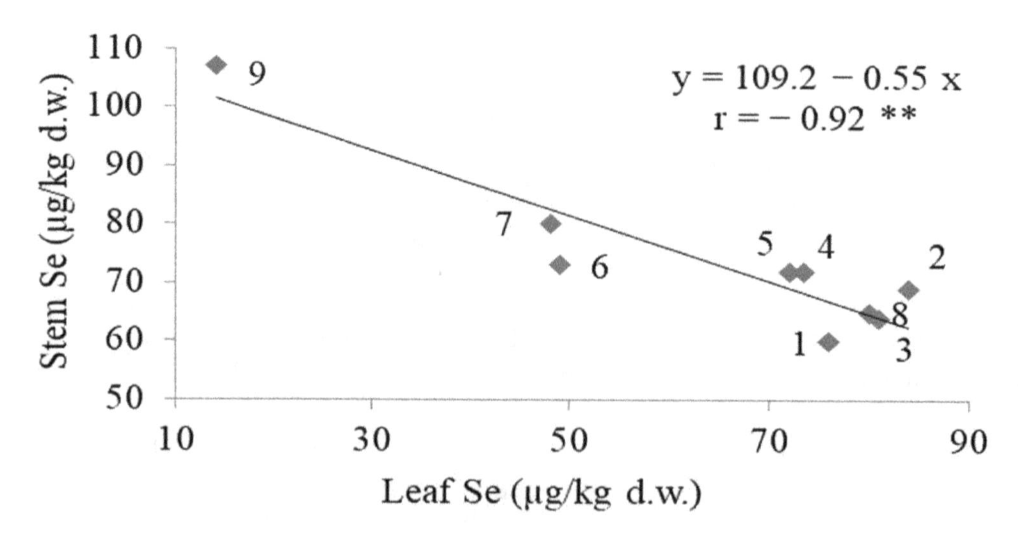

Figure 3. Correlation between leaf and stem selenium content in leek cultivars: (1) Cazimir, (2) Vesta, (3) Camus, (4) Kalambus, (5) Summer Breeze, (6) Giraffe, (7) Bandit, (8) Premier, (9) Goliath.

Reports relevant to selenium as plant secondary metabolites, as well as Se to polyphenols particularly in the absence of selenium loading are rather scarce and often controversial. In this respect, a positive correlation between selenium and polyphenol content was found in wheat [38] and an adverse correlation between quercetin and selenium was recorded in onion [30]. However, moderate doses of selenium are deemed to enhance the content of antioxidants such as polyphenols, flavonoids and carotenoids [39,40].

In our research, the leek genotypes investigated showed significant correlations between the components of the antioxidant system, i.e., selenium, ascorbic acid and polyphenols: Se and ascorbic acid (r = 0.93 at $P < 0.01$); Se and polyphenols (r = 0.92 at $P < 0.01$); ascorbic acid and polyphenols (r = 0.94 at $P < 0.01$). The latter correlations relevant to leek stems may be very useful for leek selection based on high antioxidant content.

3.3. Elemental Composition

The beneficial effect of many elements to human health has created unflagging interest in mineral composition of vegetable crops and in particular of leek [17,41]. Investigations of element content in leek have revealed that the plant is able to accumulate high concentrations of K and Fe. However, all investigations to date have restricted macro- and microelements (K, Ca, P, Na, Mg, Fe, Zn, Cu, Se) availability, giving no opportunity to evaluate either the leek total mineral profile or the varietal peculiarities of element accumulation.

Our research with nine leek cultivars has shown the existence of significant varietal differences in stem and leaf ash content (Figure 4). The ratio between leaf and stem ash content decreased as follows: Summer Breeze > Cazimir > Vesta > Giraffe > Bandit > Kalambus > Camus > Premier > Goliath.

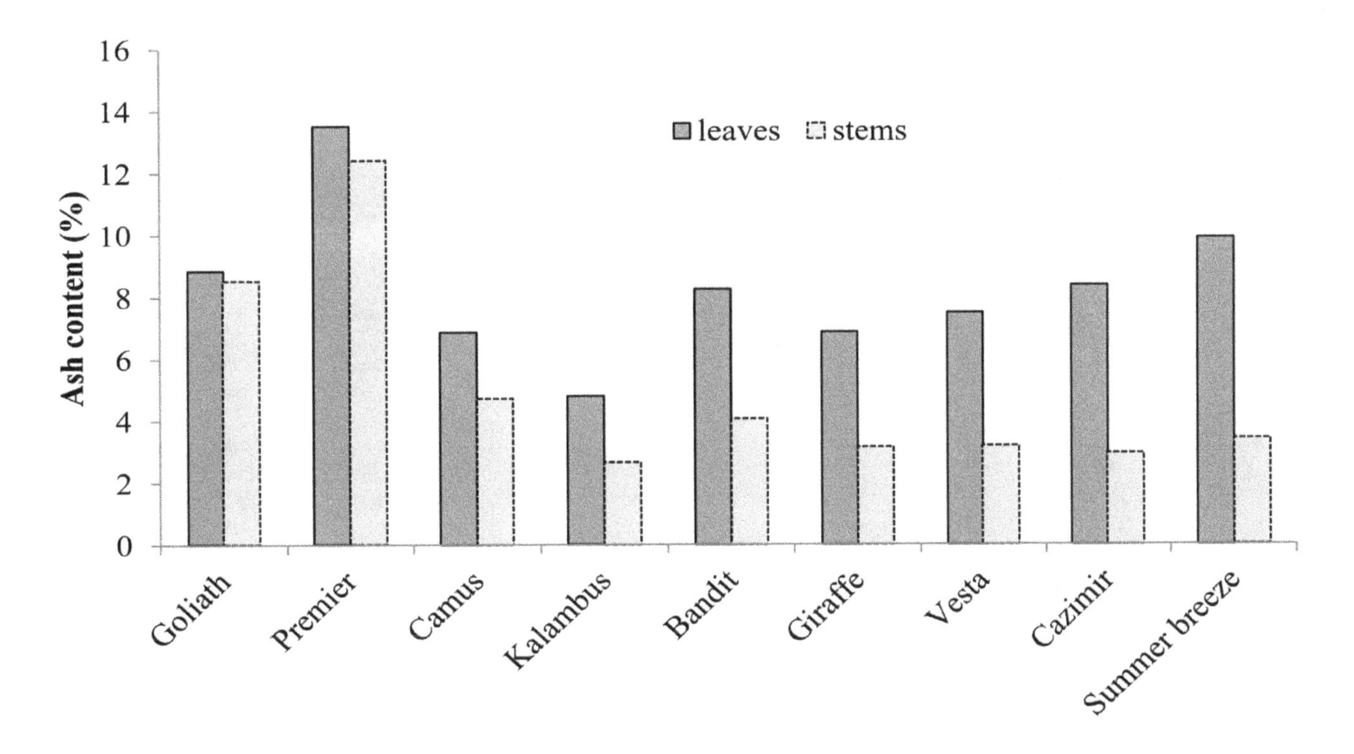

Figure 4. Varietal differences in leek leaf and stem ash content.

The ratio between leaf and stem element content was negatively correlated with stem polyphenol concentration (Figure 5), the latter being therefore related to both content and distribution of minerals in leek plants. As shown in Figure 5, the higher value of angular coefficient was associated with the water extract method, which provides higher mineral concentrations but lower polyphenol levels compared to an ethanol extract.

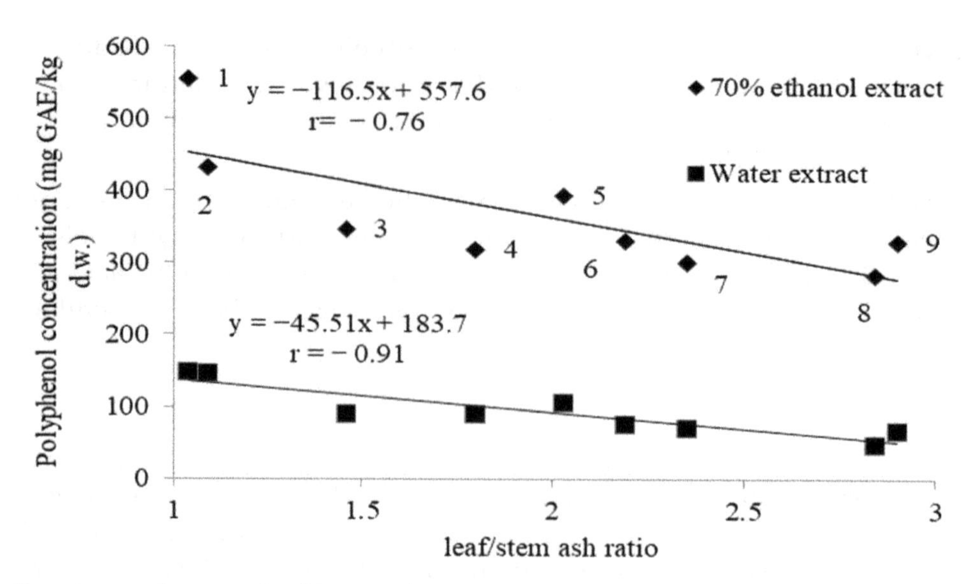

Figure 5. Correlations between leaf/stem ash content and stem polyphenol concentration: (1) Goliath; (2) Premier; (3) Camus; (4) Kalambus; (5) Bandit; (6) Giraffe; (7) Vesta; (8) Cazimir; 9) Summer Breeze.

The analysis of the content of twenty-five mineral elements in leek pseudo-stems has provided the opportunity to assess the varietal differences in elemental profile and has demonstrated that the ash content is directly connected with the concentration of K (Tables 3–5). The latter mineral showed a higher accumulation in pseudo-stems grown with organic management compared to those grown with conventional management (19.46 vs. 17.23 g/kg d.w.), whereas no significant differences were evident between the two management systems with regard to all the other elements analyzed.

Table 3. Macroelement concentration in *A. porrum* pseudo-stems (g/kg d.w.).

Element	Goliath	Cazimir	Premier	Vesta	Kalambus	Summer Breeze	Bandit	Giraffe	Camus
					Macro elements				
Ca	3.98 ^{b,c,z}	3.40 ^{c,d}	11.32 ^{a}	4.35 ^{b}	4.68 ^{b}	2.81 ^{d}	4.82 ^{b}	4.24 ^{b,c}	4.73 ^{b}
K	51.76 ^{a}	4.71 ^{e}	23.39 ^{b}	13.50 ^{c}	8.00 ^{d}	15.94 ^{c}	16.95 ^{c}	15.05 ^{c}	15.82 ^{c}
Mg	0.78 ^{c}	0.76 ^{c}	2.02 ^{a}	0.56 ^{d}	0.53 ^{d}	0.65 ^{c,d}	1.00 ^{b}	0.70 ^{c,d}	0.80 ^{c}
Na	0.34 ^{b,c}	0.39 ^{b}	0.81 ^{a}	0.16 ^{e}	0.16 ^{e}	0.12 ^{e}	0.17 ^{e}	0.19 ^{d,e}	0.28 ^{c,d}
P	2.95 ^{b}	2.61 ^{bc}	2.41 ^{c,d}	2.30 ^{c,d}	1.97 ^{d}	2.37 ^{cd}	3.85 ^{a}	2.64 ^{bc}	2.65 ^{b,c}

^{z} Within each row, means followed by different letters are significantly different according to Duncan's Multiple Range Test at $P \leq 0.05$.

Table 4. Microelements concentration in *A. porrum* pseudo-stems (mg/kg d.w.).

Element	Goliath	Cazimir	Premier	Vesta	Kalambus	Summer Breeze	Bandit	Giraffe	Camus
B	21.25 ^{a,z}	15.21 ^{b,c}	16.75 ^{b}	9.68 ^{d,e}	8.61 ^{e}	9.55 ^{d,e}	9.73 ^{d,e}	12.79 ^{c,d}	11.14 ^{d}
Co	0.070 ^{c}	0.050 ^{d}	0.091 ^{b}	0.034 ^{d}	0.290 ^{a}	0.035 ^{d}	0.035 ^{d}	0.035 ^{d}	0.099 ^{b}
Cu	4.81 ^{d,f}	4.52 ^{e,f}	3.46 ^{g}	5.90 ^{b,c}	6.66 ^{a,b}	5.15 ^{c,e}	7.18 ^{a}	4.08 ^{f,g}	5.53 ^{c,d}
Fe	221 ^{a}	116 ^{c}	178 ^{b}	101 ^{b,d}	77 ^{e}	84 ^{d,e}	98 ^{c,e}	104 ^{c,d}	235 ^{a}
I	0.060 ^{a}	0.040 ^{d}	0.353 ^{a}	0.042 ^{c,d}	0.055 ^{b,d}	0.057 ^{b,d}	0.038 ^{d}	0.071 ^{b,c}	0.073 ^{b}
Li	0.110 ^{b}	0.040 ^{c}	0.160 ^{a}	0.025 ^{c}	0.014 ^{c}	0.032 ^{c}	0.023 ^{c}	0.029 ^{c}	0.109 ^{b}
Mn	12.57 ^{c}	12.18 ^{c}	23.15 ^{a}	9.87 ^{c}	6.39 ^{d}	9.69 ^{c}	10.93 ^{c}	19.97 ^{b}	22.51 ^{a,b}
Si	14.62 ^{c}	10.74 ^{b}	28.78 ^{a}	9.50 ^{e}	13.43 ^{c,d}	13.86 ^{c,d}	11.41 ^{d,e}	20.00 ^{b}	16.17 ^{c}
Sn	0.160 ^{e}	0.240 ^{c,d}	0.023 ^{f}	0.171 ^{e}	0.519 ^{b}	0.193 ^{d,e}	0.574 ^{a}	0.245 ^{c}	0.247 ^{c}
Zn	23.97 ^{a,b}	27.27 ^{a}	11.96 ^{f}	18.45 ^{de}	16.26 ^{e}	19.58 ^{ce}	21.96 ^{b,c}	22.83 ^{b,c}	21.72 ^{b,d}

^{z} Within each row, means followed by different letters are significantly different according to Duncan's Multiple Range Test at $P \leq 0.05$.

Table 5. Heavy metal concentration in *A. porrum* pseudo-stems (mg/kg d.w.).

Element	Goliath	Cazimir	Premier	Vesta	Kalambus	Summer Breeze	Bandit	Giraffe	Camus
					Heavy metals				
Al	84.0 ^{c,z}	31.3 ^{d}	137.0 ^{a}	21.9 ^{d,f}	8.0 ^{g}	25.3 ^{d,e}	12.6 ^{f,g}	20.2 ^{e,f}	96.2 ^{b}
As	0.030 ^{b}	0.020 ^{c,d}	0.066 ^{a}	0.015 ^{c,e}	0.013 ^{d,e}	0.017 ^{c,e}	0.009 ^{e}	0.023 ^{b,c}	0.066 ^{a}
Cd	0.090 ^{c,d}	0.110 ^{b,c}	0.196 ^{a}	0.110 ^{b}	0.085 ^{d}	0.073 ^{d}	0.113 ^{b}	0.182 ^{a}	0.122 ^{b}
Cr	0.130 ^{c}	0.080 ^{g}	0.524 ^{a}	0.104 ^{d,f}	0.095 ^{e,g}	0.122 ^{c,d}	0.085 ^{f,g}	0.161 ^{b}	0.111 ^{c,e}
Ni	1.100 ^{a}	0.480 ^{c}	1.000 ^{a,b}	1.010 ^{a,b}	0.578 ^{c}	0.588 ^{c}	0.875 ^{b}	0.621 ^{c}	1.140 ^{a}
Pb	0.360 ^{b}	0.290 ^{b,c}	0.894 ^{a}	0.108 ^{e}	0.096 ^{e}	0.143 ^{d,e}	0.220 ^{c,d}	0.117 ^{e}	0.878 ^{a}
Sr	29.0 ^{a,b}	25.7 ^{c}	31.3 ^{a}	25.0 ^{c}	28.7 ^{a,b}	17.8 ^{d}	29.3 ^{a,b}	26.8 ^{b,c}	29.1 ^{a,b}
V	0.230 ^{b}	0.090 ^{c,d}	0.311 ^{a}	0.079 ^{c,d}	0.043 ^{e}	0.097 ^{c}	0.066 ^{d,e}	0.073 ^{c,d}	0.299 ^{a}

^{z} Within each row, means followed by different letters are significantly different according to Duncan's Multiple Range Test at $P \leq 0.05$.

The comparison between the nine leek cultivars, in terms of elemental composition, has indicated three cultivars with contrasting features: Premier, Goliath and Cazimir. Indeed, Premier preferably accumulated Co, I, Al, As, Cd, Ni, Pb and Sr, but low levels of Cu and Zn. Goliath was characterized by the highest content of Fe, B, Zn, Se and K, and the lowest of Cd. Cazimir showed the highest concentration of Zn and Na, but the lowest of I, Se, K, Cr and Ni.

In our research, the nine leek cultivars have shown higher concentrations of most elements, compared to twenty varieties belonging to related species such as garlic grown in the same geochemical conditions [42]: Ca, Na, As, Cr, Ni, Co, Fe, I, Li, Pg, Sr, V, B, Co, Fe, I, Li, Sn; equal amounts of K, Mg, P, Cd, Cu, Mn, Se and Zn; and lower concentrations of Si. It is worth noting that, in conditions of marginal selenium deficiency in the Moscow region, the average levels of selenium accumulation by leek and garlic did not differ from each other (Figure 6).

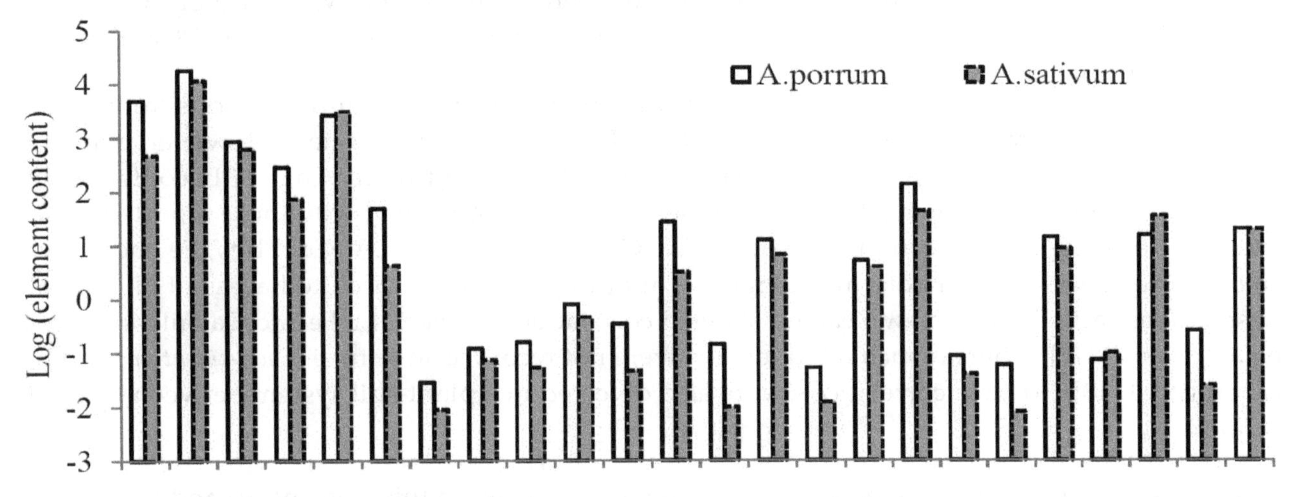

Figure 6. Comparative elemental profile of leek and winter garlic [42] grown in the same geochemical conditions of the Moscow region.

Interestingly, significant varietal differences in the content of most elements were evident (Figure 7). Though the values of the coefficient of variation (CV) relevant to macroelements were considerably lower than those of heavy metals and microelements, even among the former the CV reached 30 to 50%, with the exception of P having lower values. Among heavy metals, Al, Pb, V, As and Cr attained the highest coefficients of variation (40 to 80%), whereas among microelements Li, I, Sn, Co, Fe and Mn showed a CV > 40%.

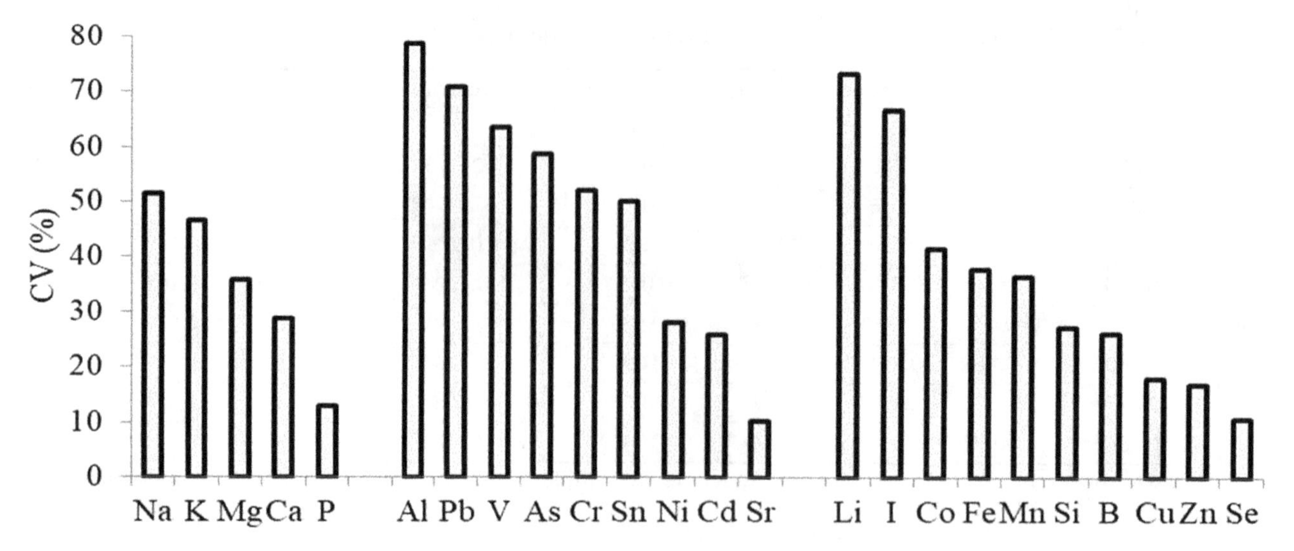

Figure 7. Coefficients of variation relevant to macro- and microelement content in *A.porrum*.

The several significant correlations recorded between the different elements revealed a higher complexity concerning the mineral dynamics in leek plants (Table 6), compared to related species such as garlic [42].

The highest number of significant correlations was recorded for Al, whose physiological role in plants has not been clearly determined so far, though this element is supposed to both activate some enzymes and control membrane permeability at low doses [43]. The significant intra- and interspecies variability in Al accumulation depends on plant tolerance thresholds to this element. The multiplicity of relationships between Al and other elements (Ca, Mg, Na, Co, Li, Fe, I, Cr, Mn, Cr, As, Pb, V) undoubtedly reflects the complex physiological functions of Al in leek. Among the minerals analyzed, the highest correlation coefficients were recorded between Al and As, Pb, V, Co and Li. As for heavy metals and As, highly significant correlations were found between V and Al, As, Co, Pb and Fe.

Following Al, Li showed wide varietal variation, consistent with previous reports [44]. According to the literature, leek is one of the least Li accumulating species. However, compared to Yalamanchali's [45] findings, our results suggest a wider concentration range of Li content in leek plants. The relationships between Li and other elements are in agreement with those reported, and, in particular, correlations between Li and Al, As, Ca, Co, Cr, Fe, I, Pb, Na, V, Co, Pb, V, Fe and Cr were found in five species grown both in ecological unpolluted and in oil polluted areas of Nigeria [46]. Investigations carried out in New Zealand showed correlations between Li, Fe and Ca only in ryegrass (*Lolium perenne*) [47], whereas such relationships were not recorded in lettuce (*Lactuca sativa*) and beet (*Beta vulgaris*) [48]. Positive correlations were also detected in a plant-soil system between Li and Fe, Al and Na.

Among the relationships involving Li, the interactions between this element, Al and Fe should be considered important, as the two latter minerals show similar atomic radius to lithium.

Although varietal differences in Se content in leek were rather low compared to other elements, the significant correlation between Se and K is a remarkable characteristic of this *Allium* species and it has been scarcely investigated so far.

Table 6. Correlation coefficients between mineral elements in leek.

	Al	As	B	Ca	Cd	Co	Cr	Cu	Fe	I	K	Li	Mg	Mn	Pb
As	0.93 *	1	0.38												
Ca	0.71 *	0.66	0.29	1											
Cd	0.47	0.54	0.26	0.72 *	1										
Co	0.95 ***	0.95 ***	0.53	0.55	0.37	1									
Cr	0.74 *	0.66	0.40	0.94 ***	0.73	0.53	1								
Fe	0.85 **	0.80 **	0.64	0.31	0.20	0.92 ***	0.30	−0.36	1						
I	0.77 **	0.70 *	0.37	0.96 ***	0.70 *	0.58	0.99 ***	−0.60	0.33	1					
K	0.50	0.23	0.75 e	0.13	−0.03	0.39	0.21	−0.23	0.63	0.17	1				
Li	0.99 ***	0.90 ***	0.68	0.70 *	0.46	0.93 ***	0.73 *	−0.59	0.86 **	0.76 *	0.56	1			
Mg	0.75 *	0.65	0.41	0.94 ***	0.69	0.59	0.93 ***	−0.50	0.36	0.94 ***	0.21	0.75 *	1		
Mn	0.74 *	0.86 **	0.36	0.58	0.81 **	0.77 *	0.62	−0.65	0.64	0.62	0.14	0.72 *	0.62	1	
Na	0.83 **	0.71 *	0.64	0.85 **	0.63	0.70	0.88 **	−0.59	0.50	0.90 ***	0.25	0.83 **	0.90 ***	0.61	
Ni	0.65	0.58	0.32	0.38	0.16	0.66	0.27	0.01	0.76 *	0.29	0.60	0.65	0.34	0.42	
Pb	0.92 ***	0.97 ***	0.38	0.67	0.47	0.97 ***	0.63	−0.42	0.80 **	0.68	0.22	0.90 ***	0.71 *	0.79 **	1
Se	0.35	0.05	0.69	0.14	−0.06	0.19	−0.72 *	−0.16	0.42	−0.71 *	0.95 ***	0.42	0.18	−0.03	0.04
Si	0.72 *	0.72 *	0.39	0.83 **	0.80 **	0.57	0.91 ***	−0.70	0.37	0.90 ***	0.23	0.72 *	0.81 **	0.77 *	0.64
Sn	−0.66	−0.57	−0.57	−0.37	−0.39	−0.53	−0.58	0.844 c	−0.47	−0.55	0.26	−0.64	−0.42	−0.52	−0.45
V	0.98 ***	0.94 ***	0.56	0.60	0.38	0.98 ***	0.61	−0.50	0.91 ***	0.65	0.51	0.97 ***	0.64	0.75 *	0.94 ***
Zn	−0.34	−0.33	0.21	−0.74 *	−0.31	−0.15	−0.69	0.05	0.07	−0.71 *	0.05	−0.31	−0.54	−0.14	−0.29

*** $P < 0.001$; ** $P < 0.01$; * $P < 0.05$.

In spinach (*Spinacia oleracea*) plants fertilization with sodium selenate increased K content in female but not male plants [21], whereas in other research [49] garlic biofortification led to selenium antagonistic activity towards K. Taking into account that K participates in plant protection against all forms of biotic and abiotic stresses along with Se and other components of antioxidant defense systems [50], the close relationship between the two minerals in leek suggests intensive interactions of all components of the defense system. The predominance of K in leek elemental composition and the significant correlation between polyphenol concentration and ash content (Figure 5) is in good agreement with the mentioned results. The correlation coefficient relevant to ash to K in leek was 0.78 at $P < 0.01$, whereas that related to K and polyphenols reached 0.96 at $P < 0.01$. The known ability of K to decrease the activity of polyphenol oxidase in plants and enhance polyphenol accumulation [51] may be a good explanation of the positive correlation between polyphenols and K in leek plants. The active participation of K in the antioxidant defense system of this *Allium* species was also characterized by a positive correlation of the element with the ascorbic acid content ($r = 0.95$ at $P < 0.01$). In this respect, the results of the present work revealed the close relationship between the main components of leek antioxidant systems including polyphenols, ascorbic acid as well as the macro- and trace elements Se and K.

In our research, the lowest negative correlation coefficients were Se with Cr and I. Se is known as an antagonist of Cr and its protective role towards Cr has been previously reported [24,52,53]. The interaction between Se and I is more complex; neither element is essential for plants, but at low concentrations, they may improve plant growth, development and protection from biotic and abiotic stresses [54]. However, the rather scarce and contradictory data about Se and I interactions in plants do not allow clear conclusions. Separate plant fortification with Se and I showed the possibility of mutual stimulation by the two elements in some but not all cases [40]. The selective accumulation of selenium in male spinach plants and of iodine in female spinach plants suggests the participation of phytohormones in the interactions between Se and I [21]. A negative correlation between Se and I in leek plants has not yet been reported.

4. Conclusions

From research carried out in the Moscow region with the aim of evaluating the performance of nine leek (*A. porrum*) cultivars grown in greenhouses under either organic or conventional management systems, interesting clues have been drawn. The varieties showed a uniform behavior under both management systems: no yield differences were recorded between organic and conventional systems. When cultivated with organic procedures, all cultivars attained higher dry matter, sugar, ascorbic acid and potassium content but lower nitrates in the pseudo-stems than for conventional management, but with the same ranking as for conventional management. Moreover, in contrast to related species, highly significant correlations between the antioxidants and mineral elements in leek plants provide opportunities for obtaining genotypes with improved quality features.

Author Contributions: N.A.G. and G.C. designed the experimental protocol; T.M.S. and G.C.T. were concerned with the crop organic management; N.A.G., M.S.A. and O.V.K. performed analytical measurements; G.C. and N.A.G. equally performed the data statistical processing and manuscript writing.

References

1. Ozgur, M.; Akpinar-Bayaziy, A.; Ozcan, T.; Afolayan, A.J. Effect of dehydration on several physic-chemical properties and the antioxidant activity of leeks (*Allium porrum* L.). *Not. Bot. Hort. Agrobot. Cluj-Napoca* **2011**, *39*, 144–151. [CrossRef]
2. Steinmetz, A.K.; Potter, D.J. Vegetables, fruit, and cancer prevention: A review. *J. Am. Diet. Assoc.* **1996**, *96*, 1027–1039. [CrossRef]
3. Fattorusso, E.; Lanzotti, V.; Taglialatela-Scafati, O.; Cicala, C. The flavonoids of leek, *Allium porrum*. *Phytochemistry* **2001**, *57*, 565–569. [CrossRef]

4. Galeone, C.; Pelucchi, C.; Levi, F.; Negri, E.; Fraceschi, S.; Talamini, R.; Giacosa, A.; La Vecchia, C. Onion and garlic use and human cancer. *Am. J. Clin. Nutr.* **2006**, *84*, 1027–1032. [CrossRef] [PubMed]
5. Ben Arfa, A.; Najjaa, H.; Yahia, B.; Tlig, A.; Neffati, M. Antioxidant capacity and phenolic composition as a function of genetic diversity of wild Tunisian leek (*Allium ampeloprasum* L.). *Acad. J. Biotechnol.* **2015**, *3*, 15–26. [CrossRef]
6. Griffiths, G.; Trueman, L.; Crowther, T.; Thomas, B.; Smith, B. Onions as global benefit to health. *Phytother. Res.* **2002**, *16*, 603–615. [CrossRef] [PubMed]
7. Radovanović, B.; Mladenović, J.; Radovanović, A.; Pavlović, R.; Nikolić, V. Phenolic composition, antioxidant, antimicrobial and cytotoxic activites of *Allium porrum* L. (Serbia) extracts. *J. Food Nutr. Res.* **2015**, *3*, 564–569. [CrossRef]
8. Bianchini, F.; Vainio, H. *Allium* vegetables and organosulfur compounds: Do they help prevent cancer? *Environ. Health Perspect.* **2001**, *109*, 893–902. [CrossRef] [PubMed]
9. Hsing, A.W.; Chokkalingam, A.P.; Gao, Y.T.; Madigan, M.P.; Deng, J.; Gridley, G.; Fraumeni, J.F. *Allium* vegetables and risk of prostate cancer: A population based study. *J. Natl. Cancer Inst.* **2002**, *94*, 1648–1651. [CrossRef] [PubMed]
10. Kyoung-Hee, K.; Hye-Joung, K.; Myung-Woo, B.; Hong-Sun, Y. Antioxidant and antimicrobial activities of ethanol extract from 6 vegetables containing different sulfur compounds. *J. Korean Soc. Food Sci. Nutr.* **2012**, *41*, 577–583.
11. Vergawen, R.; Van Leuven, F.; Van Laere, A. Purification and characterization of strongly chitin-binding chitinases from salicylic acid-treated leek (Allium porrum). *Physiol. Plant.* **1998**, *104*, 175–182. [CrossRef]
12. Yin, M.C.; Tsao, S.M. Inhibitory effect of seven Allium plants upon three Aspergillus species. *Int. J. Food Microbiol.* **1999**, *49*, 49–56. [CrossRef]
13. Mnayer, D.; Fabiano-Tixier, A.-S.; Petitcolas, E.; Hamieh, T.; Nehme, N.; Ferrant, C.; Fernandez, X.; Chemat, F. Chemical composition, antibacterial and antioxidant activities of six essential oils from the *Alliaceae* family. *Molecules* **2014**, *19*, 20034–20053. [CrossRef] [PubMed]
14. Lundegardh, B.; Botek, P.; Schulzo, V.; Hajšlo, V.J.; Stromberg, V.A.; Andersson, H.C. Impact of different green manures on the content of *S*-Alk(en)yl-L-cysteine sulfoxides and L-Ascorbic acid in leek (*Allium porrum*). *J. Agric. Food Chem.* **2008**, *56*, 2102–2111. [CrossRef] [PubMed]
15. Naem-Rana, K.; Had-Noora, A. The antimicrobial activity of *Allium porrum* water extract against pathogenic bacteria. *J. Kerbala Univ.* **2012**, *10*, 45–49.
16. Sekara, A.; Pokluda, R.; Del Vacchio, L.; Somma, S.; Caruso, G. Interactions among genotype, environment and agronomic practices on production and quality of storage onion (*Allium cepa* L.). *Rev. Hortic. Sci.* **2017**, *44*, 21–42. [CrossRef]
17. Koca, I.; Tasci, B. Mineral composition of leek. *Acta Hortic.* **2016**, *1143*, 147–151. [CrossRef]
18. Conti, S.; Villari, G.M.; Amico, E.; Caruso, G. Effects of production system and transplanting time on yield, quality and antioxidant content of organic winter squash (*Cucurbita moschata* Duch.). *Sci. Hortic.* **2015**, *183*, 136–143. [CrossRef]
19. Bernaert, N.; De Paepe, D.; Bouten, C.; De Clercq, H.; Stewart, D.; Van Bockstaele, E.; De Loose, M.; Van Droogenbroeck, B. Antioxidant capacity, total phenolic and ascorbate content as a function of the genetic diversity of leek (*Allium ampeloprasum* var. porrum). *Food Chem.* **2012**, *134*, 669–677. [CrossRef] [PubMed]
20. Swamy, P.M. *Laboratory Manual on Biotechnology*; Rastogi Publications: Meerut, India, 2008; 617p.
21. Golubkina, N.A.; Kosheleva, O.V.; Krivenkov, L.V.; Nadezhkin, S.M.; Dobrutskaya, H.G.; Caruso, G. Intersexual differences in plant growth, yield, mineral composition and antioxidants of spinach (*Spinacia oleracea* L.) as affected by selenium form. *Sci. Hortic.* **2017**, *225*, 350–358. [CrossRef]
22. AOAC. *The Official Methods of Analysis of AOAC (Association Official Analytical Chemists) International*; AOAC: Arlington, VA, USA, 2012; Volume 22.
23. Maximova, T.V.; Nikulina, I.N.; Pakhomov, V.P.; Shkarina, H.I.; Chumakova, Z.V.; Arzamastsev, A.P. Method of Antioxidant Activity Determination. RF Patent No. 2.170,930, 20 July 2001.
24. Srivastava, S.; Adholeya, A.; Conlan, X.A.; Cahill, D.M. Acidic potassium permanganate chemiluminescence for the determination of antioxidant potential in three cultivars of *Ocimum basilicum*. *Plant Foods Hum. Nutr.* **2015**, 70. [CrossRef] [PubMed]
25. Zhan, M.G.; Liu, N.; Liu, H. Determination of the total mass of antioxidant substances and antioxidant capacity per unit mass in serum using redox titration. *Bioinorg. Chem. Appl.* **2014**, 928595. [CrossRef]

26. Skalny, A.V.; Lakarova, H.V.; Kuznetsov, V.V.; Skalnaya, M.G. *Analytical Methods in Bioelementology*; Saint Petersburg-Science: St. Petersburg, Russia, 2009.
27. Alfthan, G.V. A micromethod for the determination of selenium in tissues and biological fluids by single-test-tube fluorimetry. *Anal. Chim. Acta* **1984**, *65*, 187–194. [CrossRef]
28. Caruso, G.; Villari, G.; Borrelli, C.; Russo, G. Effects of crop method and harvest seasons on yield and quality of green asparagus under tunnel in southern Italy. *Adv. Hort. Sci.* **2012**, *26*, 51–58.
29. Biswas, S.K.; Khair, A.; Sarker, P.K.; Alom, M.S. Yield and storability of onion (*Allium cepa* L.) as affected by various levels of irrigation. *Bangladesh J. Agric. Res.* **2010**, *35*, 247–255. [CrossRef]
30. Golubkina, N.A.; Kekina, H.G.; Antoshkina, M.S.; Agafonov, A.F.; Nadezhkin, S.M. Inter varietal differences in accumulation of biologically active compounds by *Allium cepa* L. *Messenger Russ. Agric. Sci.* **2016**, *2*, 51–55.
31. Sinclair, P.J.; Blakeney, A.B.; Barlow, E.W. Relationships between bulb dry matter content, soluble solids concentration and non-structural carbohydrate composition in the onion (*Allium cepa*). *J. Sci. Food Agric.* **1995**, *69*, 203–209. [CrossRef]
32. Caruso, G.; Conti, S.; La Rocca, G. Influence of crop cycle and nitrogen fertilizer form on yield and nitrate content in different species of vegetables. *Adv. Hort. Sci.* **2011**, *25*, 81–89.
33. Santamaria, P. Nitrates in vegetables: Toxicity content, intake and EC regulation. *J. Food Agric.* **2006**, *86*, 10–17. [CrossRef]
34. Proteggente, A.R.; Pannala, A.S.; Pagana, G.; Van Buren, L.; Wagner, E.; Wiseman, S. The antioxidant activity of regularly consumed fruit and vegetables reflects their phenolic and vitamin C composition. *Free Radic. Res.* **2002**, *36*, 217–233. [CrossRef] [PubMed]
35. Conti, S.; Villari, G.; Faugno, S.; Melchionna, G.; Somma, S.; Caruso, G. Effects of organic vs. conventional management system on yield and quality of strawberry grown as an annual or biennial crop in southern Italy. *Sci. Hortic.* **2014**, *180*, 63–71. [CrossRef]
36. Biesiada, A.; Kolota, E.; Adamczewska-Sowinska, K. The effect of maturity stage on nutritional value of leek, zucchini and kohlrabi. *Res. Bull.* **2007**, *66*, 39–45. [CrossRef]
37. Malagoli, M.; Schiavon, M.; dall'Acqua, S.; Pilon-Smits, E.A.H. Effects of selenium biofortification on crop nutritional quality. *Front. Plant Sci.* **2015**, *6*, 280. [CrossRef] [PubMed]
38. Lachman, J.; Miholová, D.; Pivec, V.; Jírů, K.; Janovská, D. Content of phenolic antioxidants and selenium in grain of einkorn (*Triticum monococcum*), emmer (*Triticum dicoccum*) and spring wheat (*Triticum aestivum*) varieties. *Plant Soil Environ.* **2011**, *57*, 235–243. [CrossRef]
39. Kavalcová, P.; Bystrická, J.; Trebichalský, P.; Volnová, B.; Kopernická, M. The influence of selenium on content of total polyphenols and antioxidant activity of onion (*Allium cepa* L.). *J. Microbiol. Biotechnol. Food Sci.* **2014**, *3*, 238–240.
40. Golubkina, N.; Kekina, H.; Caruso, G. Foliar biofortification of Indian mustard (*Brassica juncea* L.) with selenium and iodine. *Plants* **2018**, *7*, 80. [CrossRef] [PubMed]
41. Sajid, M.; Butt, M.S.; Shehzad, A.; Tanwer, S. Chemical and mineral analysis of garlic, a golden herb. *Pak. J. Food Sci.* **2014**, *24*, 108–110.
42. Seredin, T.M.; Agafonov, A.F.; Gerasimova, L.I.; Krivenkov, L.V. Element composition of winter garlic. *Veg. Crop.* **2015**, *3–4*, 81–85.
43. Ahn, S.-J.; Matsumoto, H. The role of the plasma membrane in the response of plant roots to aluminum toxicity. *Plant Signal Behav.* **2006**, *1*, 37–45. [CrossRef] [PubMed]
44. Kabata-Pendias, A.; Pendias, H. *Trace Elements in Soils and Plants*; CRC Press: Boca Raton, FL, USA, 2010.
45. Yalamanchali, R.C. Lithium, an Emerging Environmental Contaminant, Is Mobile in the Soil-Plant System. Ph.D. Thesis, Lincoln University, Lincoln, UK, 2012.
46. Essiett, U.A.; Effiong, G.S.; Ogbemudia, F.O.; Bruno, E.J. Heavy metal concentrations in plants growing in crude oil contaminated soil in AkwaIbom State, South-Eastern Nigeria. *Afr. J. Pharm. Pharmacol.* **2010**, *4*, 465–470.
47. Crush, J.R.; Evans, J.P.M.; Cosgrove, G.P. Chemical composition of ryegrass (*Lolium perenne* L.) and prairie grass (*Bromus willdenowii* Kunth) pastures. *N. Z. J. Agric. Res* **1989**, *32*, 461–468. [CrossRef]
48. Bozokalfa, K.; Yağmur, B.; Aşçıoğul, T.; Eşiyok, D. Diversity in nutritional composition of Swiss chard (*Beta vulgaris* subsp. L. var. cicla) accessions revealed by multivariate analysis. *Plant Genet. Resour.* **2011**, *9*, 557–566. [CrossRef]
49. Põldma, P.; Tõnutare, T.; Viitak, A.; Luik, A.; Moor, U. Effect of Selenium treatment on mineral nutrition, bulb size, and antioxidant properties of garlic (*Allium sativum* L.). *J. Agric. Food Chem.* **2011**, *59*, 5498–5503.

50. Wang, M.; Zheng, Q.; Shen, Q.; Guo, S. The critical role of potassium in plant stress response. *Int. J. Mol. Sci.* **2013**, *14*, 7370–7390. [CrossRef] [PubMed]
51. Mudau, F.N.; Soundy, P.; du Toit, E.S. Effects of nitrogen, phosphorus, and potassium nutrition on total polyphenol content of bush tea (Athrixia phylicoides L.) leaves in shaded nursery environment. *Hort Sci.* **2007**, *42*, 334–338.
52. Qing, X.; Zhao, X.; Hu, C.; Wang, P.; Zhang, Y.; Zhang, X.; Wang, P.; Shi, H.; Jia, F.; Qu, C. Selenium alleviates chromium toxicity by preventing oxidative stress in cab-bage (*Brassica campestris* L. ssp. Pekinensis) leaves. *Cotoxicol. Environ. Saf.* **2015**, *114*, 179–189. [CrossRef] [PubMed]
53. Belokobylsky, A.I.; Ginturi, E.I.; Kuchava, N.E.; Kirkesali, E.I.; Mosulishvili, L.M.; Frontasyeva, M.V.; Pavlov, S.S.; Aksenova, N.G. Accumulation of selenium and chromium in the growth dynamics of *Spirulina platensis*. *J. Radioanal. Nucl. Chem.* **2004**, *259*, 65–68. [CrossRef]
54. Pilon-Smits, E. Selenium in plants. In *Progress in Botany*; Luttge, U., Beyschlag, W., Eds.; Springer International Publishing: Basel, Switzerland, 2015; pp. 93–107.

8

The Effect of Environment and Nutrients on Hydroponic Lettuce Yield, Quality and Phytonutrients

William L. Sublett [1], T. Casey Barickman [1,*] and Carl E. Sams [2]

[1] North Mississippi Research and Extension Center, Department of Plant and Soil Sciences, Mississippi State University, Verona, MS 38879, USA; willsublett@gmail.com
[2] Department of Plant Sciences, The University of Tennessee, Knoxville, TN 37996, USA; carlsams@utk.edu
* Correspondence: t.c.barickman@msstate.edu.

Abstract: A study was conducted with green and red-leaf lettuce cultivars grown in a deep-water culture production system. Plants were seeded in rockwool and germinated under greenhouse conditions at 25/20 °C (day/night) for 21 days before transplanting. The experimental design was a randomized complete block with a 2 × 3 factorial arrangement of cultivar and nutrient treatments that consisted of six replications. Treatments consisted of two lettuce genotypes, (1) green (Winter Density) and (2) red (Rhazes), and three nutrient treatments containing electroconductivity (EC) levels of (1) 1.0; (2) 2.0; and (3) 4.0 mS·cm^{-1}. After 50 days, plants were harvested, processed, and analyzed to determine marketable yield, biomass, plant height, stem diameter, phenolics, and elemental nutrient concentrations. An interaction between growing season and lettuce cultivar was the predominant factor influencing yield, biomass, and quality. Nutrient solution EC treatment significantly affected biomass and water content. EC treatments significantly impacted concentrations of 3-O-glucoside and uptake of phosphorous, potassium, iron, boron, zinc, and molybdenum. Effects of growing season and cultivar on leafy lettuce yield and quality were more pronounced than the effect of nutrient solution EC treatment. Thus, greenhouse production of green and red-leaf lettuce cultivars in the south-eastern United States should be conducted in the spring and fall growing seasons with elevated nutrient solution EC of ≈4.0 mS·cm^{-1} to maximize yield and quality.

Keywords: electro-conductivity; polyphenols; phenolics; flavonoids

---

1. Introduction

In the United States, lettuce is a valuable vegetable crop and a staple food in the diet. Lettuce contributes a notable amount of polyphenolic compounds, vitamins A, C, and E, calcium, and iron [1]. Due to its raw consumption in relatively large quantities, it provides an important source of dietary antioxidants and possesses high radical scavenging activity, which is often credited with aiding in the prevention of many chronic illnesses such as cancer and cardiovascular disease [2,3]. Lettuce is a cool-season vegetable, which thrives in temperatures ranging from 7 to 24 °C. In the southern United States, field production typically occurs in the fall and winter months, allowing growers to take advantage of shorter days and cooler temperatures. However, the increasing consumer demand for high quality, locally sourced produce and off-season availability has fueled the expansion of greenhouse production over the past decade [4]. Due to the increased ability to precisely control the greenhouse environment and maintain year-round production, lettuce yield and quality is greater, compared to open field production per unit of space [5]. The high cost of greenhouse production leaves little room for error and must be offset by high gross returns.

Southern United States greenhouse growers have production advantages during the cool seasons, such as milder temperature, greater light intensity, and reduced energy costs. Lettuce production during late spring and summer often negatively affects yield and quality and threatens economic returns [6]. In the south-east United States, adverse temperatures and long days largely limit warm season production of lettuce. Consistent exposure to these supra-optimal conditions decreases lettuce quality. For example, lettuce subjected to 13 h of daylight and temperatures above 24 °C resulted in premature inflorescence initiation, otherwise known as bolting [7]. Crisphead lettuce subjected to heat stress for a 3 or 5 day period, two weeks after heading resulted in 46% of mature lettuce heads with rib discoloration [8]. Additionally, genotype determines the susceptibility of lettuce to tipburn, but the incidence is heavily influenced by environment. An analysis of 125 harvests of butterhead lettuce over a 3-year period found that high light intensity, fresh head mass, and elevated temperature were the predominant variables positively correlated with tipburn incidence [9].

In closed greenhouse hydroponic cultivation systems, fertilizers are dissolved in water, and the total amount of solutes in the solution are referred to as the electrical conductivity (EC). Numerous studies have examined the effect of differing EC levels on lettuce production. Previous research has indicated that increasing EC levels resulted in a reduction of lettuce yield and leaf nitrate in a floating system but increased total phenolic compounds and antioxidant activity [10]. Additionally, Scuderi et al. [11] found that increasing solution EC decreased lettuce yield and resulted in reduced leaf nitrate content. Conversely, three lettuce varieties subjected to increasing EC treatments also resulted in reduced total yield but showed no significant effect on leaf nitrate content. Moreover, increasing EC levels resulted in notable increases in leaf phosphorous (P), zinc (Zn), manganese (Mn), and iron (Fe) concentrations in greenhouse lettuce [12]. While lettuce is considered mildly sensitive to high EC levels, research indicates that moderate EC is associated with the biosynthesis of secondary metabolites, such as phenolic compounds [13]. Furthermore, red-leafed lettuce varieties are characterized by higher phenolic content than green-leafed varieties. Kim et al. [14] reported that phenolic content and antioxidants increased in romaine lettuce produced with long-term irrigation and relatively low EC concentration. However, green and red-leafed baby lettuce grown with increasing EC levels contained greater amounts of flavonoids, phenolic acids, and carotenoids in both varieties [15].

Information is lacking and inconclusive regarding the effects of environmental stress on greenhouse lettuce by altering the EC of the plant nutrient solution. However, Fallovo et al. [16] investigated the effect of macro and micronutrient proportions on lettuce yield and quality of 'Green Salad Bowl' during spring and summer production seasons. The results indicated that marketable yield, leaf area index, and shoot biomass were unaffected by the nutrient solution, and growing season played the most determinant role in plant yield and quality. A high amount of calcium (Ca) did result in increased quality parameters, such as chlorophyll, glucose, fructose, and leaf Ca concentrations. Moreover, green oakleaf lettuce produced during winter and summer seasons and grown in increasing EC concentrations reached maturity more quickly during summer, and yield was unaffected regardless of nutrient solution concentration [16]. More information is needed to determine the relationship between nutrient solution EC concentrations and growing season on lettuce yield and nutritional quality. Therefore, the purpose of this study was to determine the effect of increased nutrient solution EC and growing season on lettuce plant height and stem diameter, biomass accumulation, mineral nutrient uptake, yield, and polyphenolic content of green and red-leafed lettuce cultivars.

2. Materials and Methods

2.1. Plant Culture and Harvest

Three separate studies were conducted in the spring, summer, and fall of 2016 and 2017 to examine the effects of season and nutrient solution concentrations on green and red leaf lettuce growth, minerals nutrients, and secondary metabolites. Seeds of green-leaf, 'Winter Density' lettuce, and red-leaf, 'Rhazes' lettuce, (Johnny's Selected Seed, Waterville, ME, USA) were sown into rockwool

(3.81 cm × 3.81 cm; Hummert Int., Earth City, MO, USA) and germinated in greenhouse conditions (Verona, MS, USA; 34° N, 89° W) at 25/20 °C (day/night). The natural photoperiod and light intensity were not enhanced with any supplemental lighting. Daily light intensity readings of photosynthetic active radiation (PAR) were taken using the WatchDog 1000 Series plant growth micro station (Spectrum Technologies, Aurora, IL, USA), while temperature and relative humidity were monitored with a WatchDog A-Series data logger (Spectrum Technologies, Aurora, IL, USA). After 21 days (third leaf stage), three plantlets from each cultivar were transferred into a closed hydroponic system composed of 36, 11-L Rubbermaid© Roughneck plastic storage containers (Rubbermaid, Atlanta, GA, USA). Each tub was filled with 10-L of nutrient solution using a modified Hoagland formulation [17]. Elemental concentrations of modified half-strength nutrient solution consisted of ($mg \cdot L^{-1}$): N (105), P (91.5), K (117.3), Ca (80.2), Mg (24.6), S (32.0), Fe (1.0), B (0.25), Mo (0.005), Cu (0.01), Mn (0.25), and Zn (0.025). The experimental design was a randomized complete block in a 2 × 3 factorial arrangement of cultivar and EC treatments that consisted of six replications, with individual tubs representing an experimental unit. Treatments consisted of two lettuce genotypes, (1) green (Winter Density) and (2) red (Rhazes), and three nutrient treatments containing EC levels of (1) 1.0 $mS \cdot cm^{-1}$; (2) 2.0 $mS \cdot cm^{-1}$; and (3) 4.0 $mS \cdot cm^{-1}$. Electroconductivity readings were measured weekly with a portable pH/Conductivity meter (Accumet© AP85; Fisher Scientific, Hampton, NH, USA), and growth solutions were changed every two weeks. Water was added to the containers to maintain a 10 L level of nutrient solution to keep up with the transpiration losses by the lettuce plants. After 50 days, lettuce plants were harvested by replication and treatment. Plants were separated into roots and shoots, and the fresh weights and stem diameter were recorded. A 20–30 g subsample of leaf tissue from three lettuce plants per treatment was retained to be freeze-dried (Labconco Corp., Kansas City, MO, USA). The subsamples were taken from the first fully expanded leaf of the lettuce plants. Freeze dried leaf tissue was then ground by mortar and pedestal, placed in an ultra-low freezer (−80 °C) until further analyzed for nutritional quality. The remaining plant material and roots were dried in a forced-air oven at 80 °C then weighed again to determine plant biomass production. All subsamples for chemical analysis were taken from each cultivar and treatment (n = 3) from each of the six replications.

2.2. Flavonoid Analysis

Flavonoid analysis was conducted according to Neugart et al. [18] and modified for the analysis of lettuce by Becker et al. [19]. Freeze-dried lettuce leaf samples were ground using a mortar and pestle for homogenous sub-samples. A 0.04 g sub-sample was extracted in a 2 mL microcentrifuge tube by adding 1.0 mL of extraction solvent (60:37:3) consisting of methanol, de-ionized water, and formic acid. The samples were then vortexed for 1 min and centrifuged at 12,000 rpm for 15 min. After centrifugation, the samples were filtered through a 0.45 μm polytetrafluoroethylene (PTFE) syringe filter and collected in a 2-mL high-performance liquid chromatography (HPLC) vial for analysis. Separation parameters and flavonoid quantification were carried out with authentic standards using an Agilent 1260 series HPLC with a multiple wavelength detector (Agilent Technologies, Willington, DE, USA). Chromatographic separations were achieved using a 150 × 4.6 mm i.d., 2.6 μm analytical scale Kinetex F5 reverse-phase column (Phenomenex, Torrance, CA, USA), which allows for effective separation of chemically similar flavonoid compounds. The column was equipped with a Kinetex F5 12.5 × 4.6 mm i.d. guard cartridge and holder (Phenomenex), and it was maintained at 30 °C using a thermostat column compartment. All separations were achieved using mobile gradient phase of reverse osmosis (RO) water adjusted to pH 2.5 with trifluoroacetic acid and acetonitrile. Anthocyanin analysis was similar to the flavonoid determination procedure with slight modifications. Briefly, 0.04 g of red lettuce sub-samples were extracted in a 2 mL microcentrifuge tube by adding 1.0 mL of extraction solvent (50:40:10) consisting of water, methanol, and acetic acid. The samples were then vortexed for 1 min and centrifuged at 12,000 rpm for 15 min. After centrifugation, the samples were filtered through a 0.45 μm PTFE syringe filter and collected in a 2 mL HPLC vial for analysis.

2.3. Mineral Composition

Nutrient analysis was conducted according to Barickman et al. [20] with slight modifications. Briefly, a 0.5 g subsample of dried leaf tissue was combined with 10-mL of 70% HNO_3, was digested in a microwave digestion unit (Model: Ethos, Milestone Inc., Shelton, CT, USA). Leaves were collected and dried for 48 h in a forced air oven (model large; Fisher Scientific, Atlanta, GA, USA) at 65 °C. Dried samples were ground to homogeneity using liquid nitrogen, and a 0.5 g sub-sample was weighed for analysis. Nutrient analysis was conducted using an inductively coupled plasma mass spectrometer (ICP-MS; Agilent Technologies, Inc., Wilmington, DE, USA). The ICP-MS system was equipped with an octopole collision/reaction cell, Agilent 7500 ICP-MS ChemStation software, a Micromist nebulizer, a water-cooled quartz spray chamber, and a CETAC (ASX-510, CETAC Inc., Omaha, NE, USA) auto-sampler. The instrument was optimized daily in terms of sensitivity (lithium: Li, yttrium: Y, thallium: Tl), level of oxide, and doubly charged ion using a tuning solution containing 10 $\mu g \cdot L^{-1}$ of Li, Y, Tl, cerium (Ce), and cobalt (Co) in a 2% HNO_3/0.5% HCl (v/v) matrix. Tissue nutrient concentrations are expressed on a dry weight (DW) basis.

2.4. Statistical Analysis

Data were subjected to the GLIMMIXED procedure and mean separation using Tukey's Honest Significant Difference test ($P \leq 0.05$) with SAS statistical software (Version 9.4; SAS Institute, Cary, NC, USA).

3. Results

3.1. Season, Cultivar, and Treatment Effects on Plant Growth and Biomass Production

Cumulative light energy levels (Figure 1A–F) registered the highest average levels in the spring and summer in both project years. Additionally, the summer growing season produced the greatest day and nighttime average temperatures in 2016 (Figure 1A–C) and 2017 (Figure 1D–F).

Statistical analysis of the results indicated that there were no effects of year (2016 and 2017). Thus, data from 2016 and 2017 were pooled and analyzed together for each lettuce plant parameter. The growing season produced a significant effect on stem diameter (Figure 2), and the lettuce cultivar impacted stem diameter (Figure 3).

The spring season produced plants with the greatest stem diameter and was statistically different than lettuce plants produced in the summer and fall season. The stem diameters of lettuce produced in the summer and fall were 32.9% and 21.3% smaller, respectively, when compared with lettuce plants produced in the spring season. Green-leaf 'Winter Density' produced plants that averaged 13.11 mm and averaged 28.5% larger stem diameter compared to red-leaf 'Rhazes' lettuce.

There were significant interactions between growing seasons and EC treatments for lettuce leaf fresh mass (FM; Figure 4). The spring season produced the greatest leaf fresh mass and was significantly more lettuce FM was produced with high and medium (4.0 and 2.0 $mS \cdot cm^{-1}$) EC treatments. There was a 17.7% increase in leaf FM when comparing the spring season, high and medium EC treatments. Conversely, there was a significant difference between spring high EC treatment leaf FM compared to the summer and fall high EC treatments. Additionally, the summer and fall high EC treatment lettuce leaf FM decreased 35.4% and 40.0%, respectively. Overall, there were significant decreases in lettuce leaf FM as the seasons progressed and EC treatments were reduced. Also, there was a significant difference between lettuce cultivars for leaf fresh mass. The green cultivar 'Winter Density' produced more fresh mass compared to the red cultivar 'Rhazes' (Figure 5). When comparing the two lettuce cultivars, there was a 42.6% decrease in lettuce fresh mass between 'Winter Density' and 'Rhazes'.

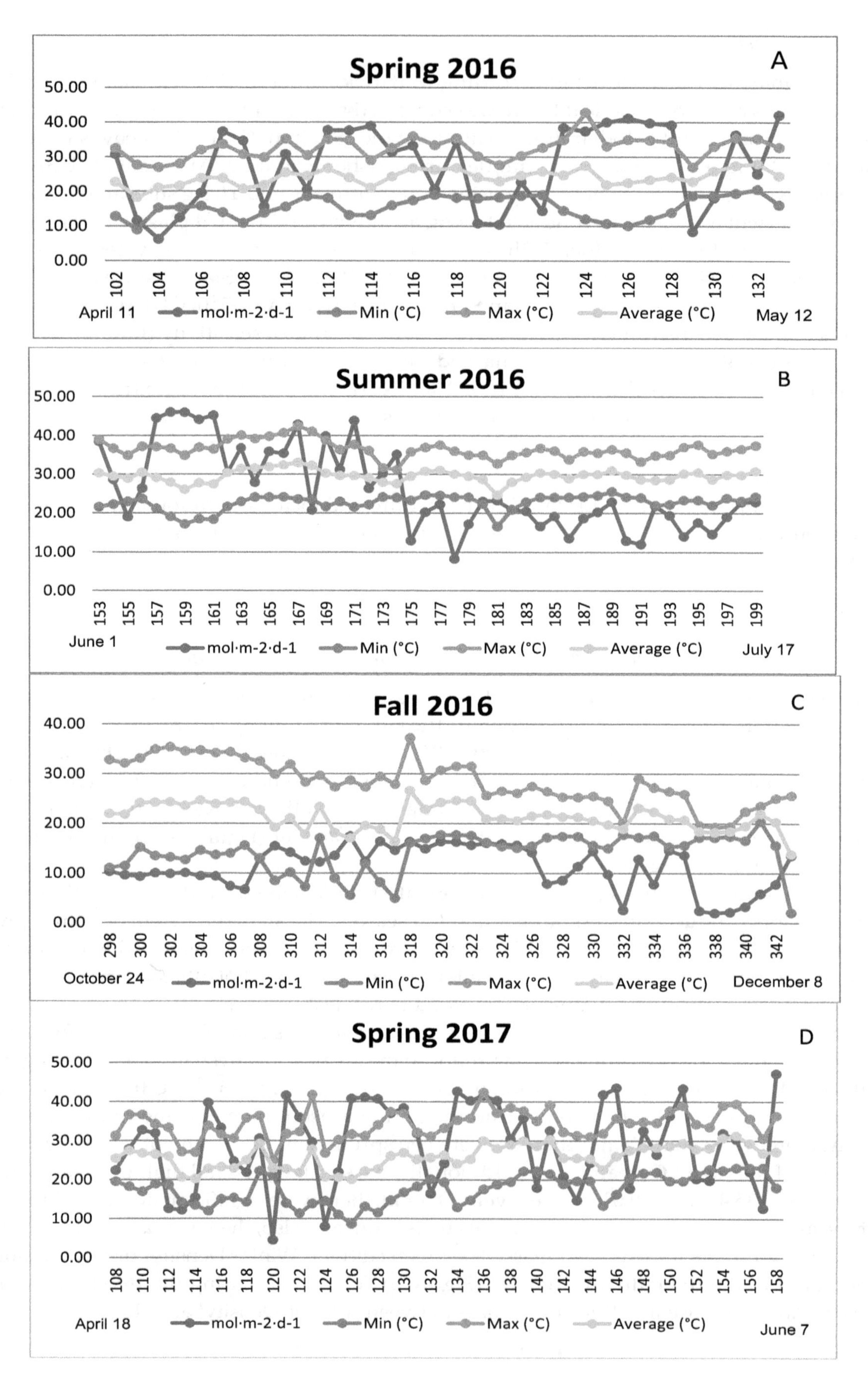

Figure 1. *Cont.*

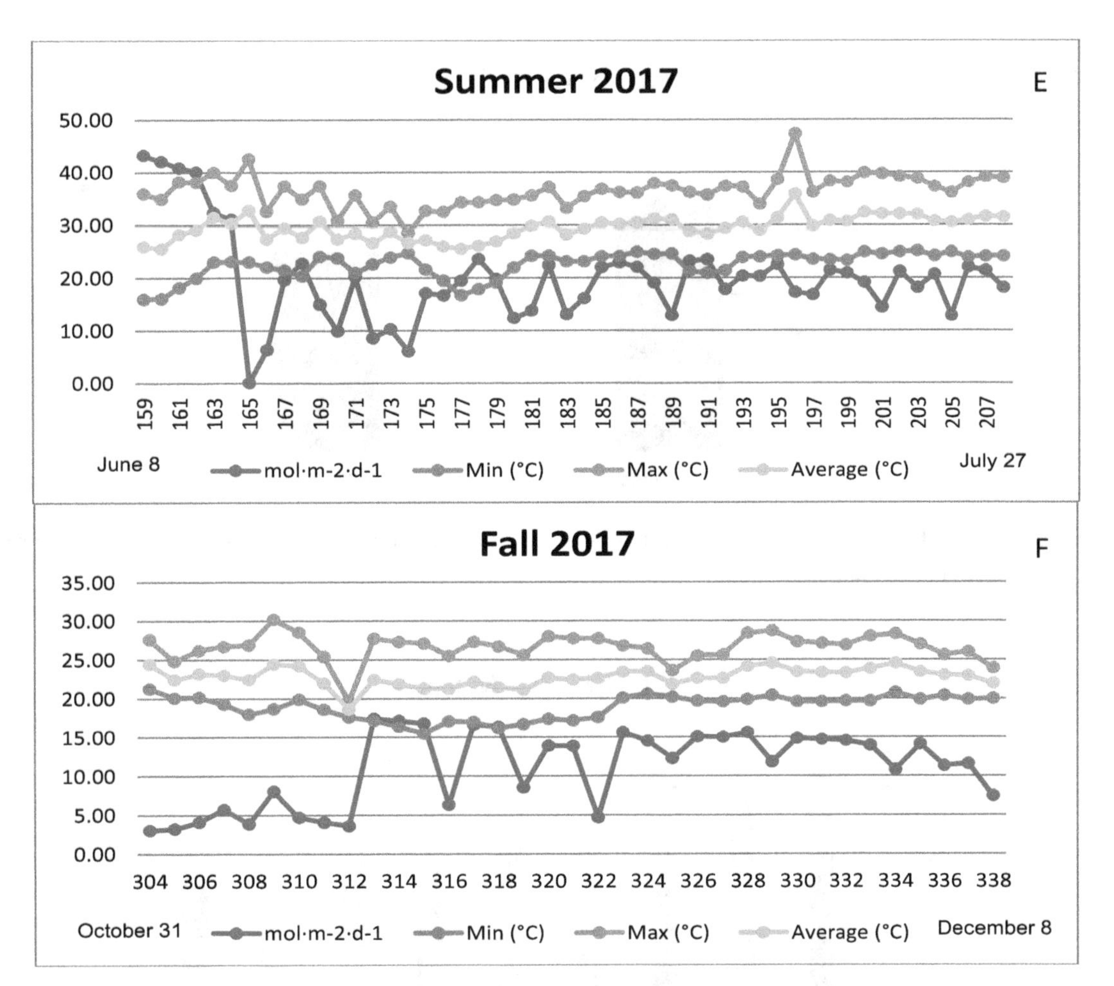

Figure 1. 2016 spring cumulative daily light energy (based on a 12 h d), and maximum, minimum, and average daily temperature (**A**); summer cumulative daily light energy, and maximum, minimum, and average daily temperature (**B**); fall cumulative daily light energy, and maximum, minimum, and average daily temperature (**C**); 2017 spring cumulative daily light energy, and maximum, minimum, and average daily temperature (**D**); summer cumulative daily light energy, and maximum, minimum, and average daily temperature (**E**); and fall cumulative daily light energy, and maximum, minimum, and average daily temperature (**F**).

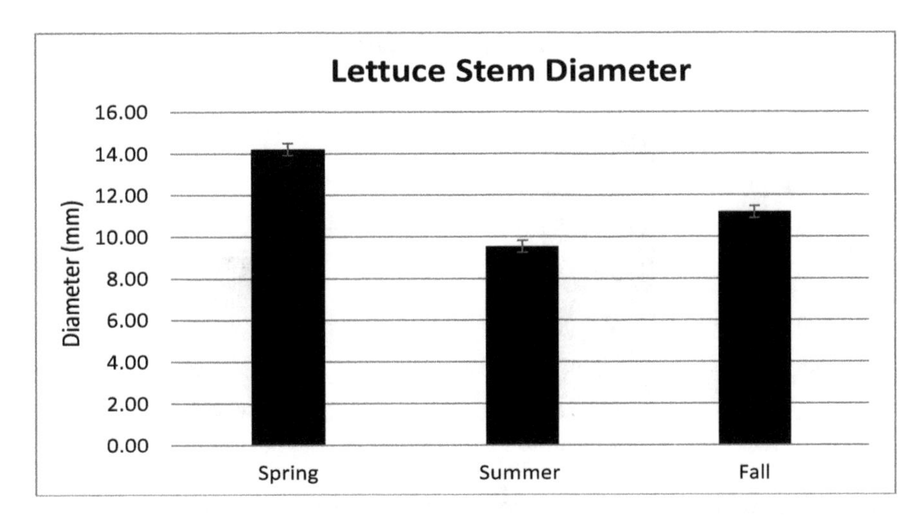

Figure 2. The effect of growing season on lettuce stem diameter. The standard error of the mean was: stem diameter $\pm$ 0.28.

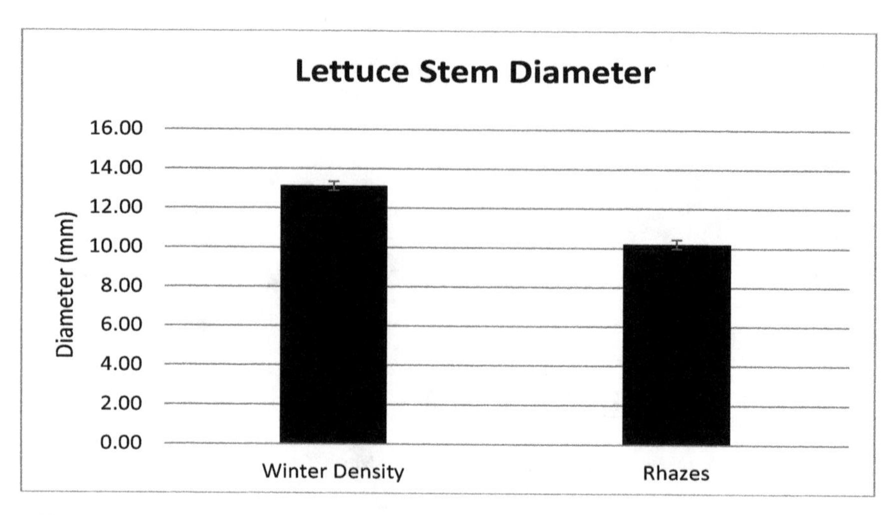

Figure 3. The effect of cultivar on greenhouse lettuce stem diameter. The standard error of the mean was: Stem diameter ± 0.23.

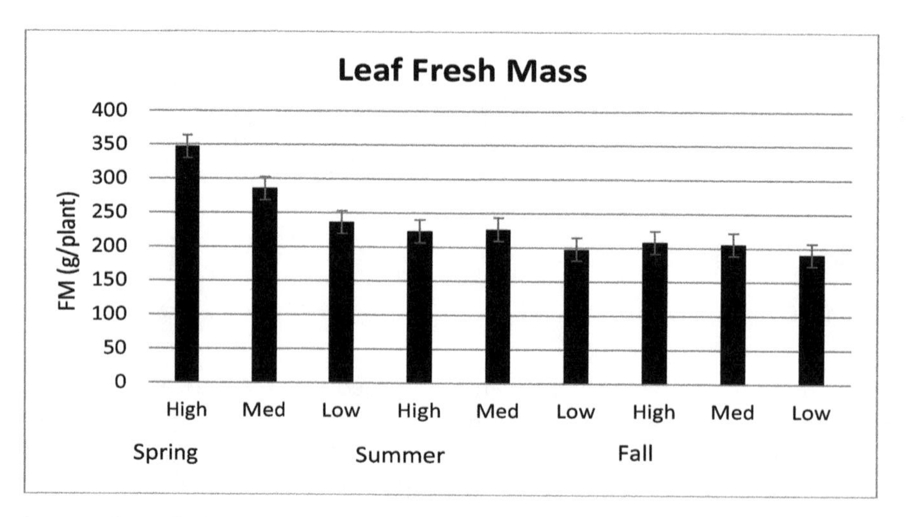

Figure 4. The interaction of growing season and electrical conductivity (EC) treatment on lettuce leaf fresh mass. The standard error of the mean: 12.65. The EC treatment: high = 4.0 mS·cm^{-1}, medium = 2.0 mS·cm^{-1}, and low = 1.0 mS·cm^{-1}.

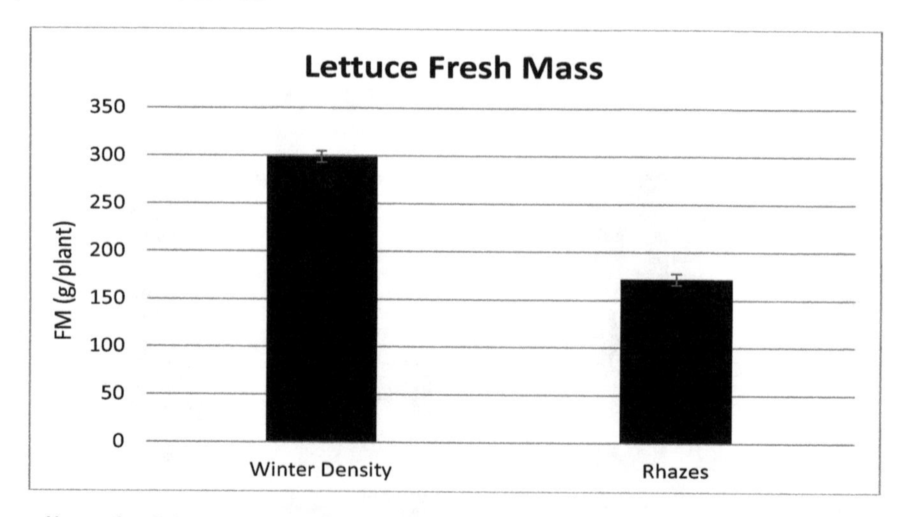

Figure 5. The effect of cultivar on greenhouse lettuce fresh mass. The standard error of the mean was: lettuce fresh mass ± 5.96.

There were no interactions between growing season, lettuce cultivar, and EC treatment for leaf dry mass (DM), DM:fresh mass (FM) ratio, and leaf water content (Table 1). Lettuce plants that were produced in the spring had significantly more leaf DM when compared to summer and fall lettuce plants. For example, there was a decrease in leaf DM by 19.4% and 33.2% when comparing the spring plants to summer and fall plants, respectively. The green-leafed lettuce cultivar 'Winter Density' produced 47.3% more leaf DM when compared to the red-leafed lettuce cultivar 'Rhazes'. Additionally, the high EC treatment produced the greatest leaf DM when compared to the medium and low EC treatments by 14.6% and 18.0%, respectively. The ratio of DM:FM was also significantly different for growing season, lettuce cultivar, and EC treatments. The summer growing season produced the greatest difference between DM:FM with a 10.2% and 10.8% increase compared to the spring and fall season, respectively. There were differences in cultivar and EC treatment DM:FM ratio. The leaf water content also saw similar trends as leaf DM in response to cultivar and EC treatment differences.

Table 1. The effect of growing season, lettuce cultivar, and EC treatment on leaf dry mass (DM) (g/plant), DM:fresh mass (FM) ratio, and lettuce leaf water content.

Treatments	Leaf DM (g)	DM:FM (g) [a]	Leaf Water %
Spring	12.37 a	0.0413 b	95.86 a
Summer	9.97 b	0.0460 a	95.41 b
Fall	8.26 c	0.0410 b	95.92 a
Winter Density	13.36 a	0.0444 a	95.56 b
Rhazes	7.04 b	0.0410 b	95.90 a
High [b]	11.44 a	0.0433 ab	95.61 b
Med	9.77 b	0.0408 b	95.93 a
Low	9.38 b	0.0440 a	95.64 b
***P*-Value** [c,d]			
Season	***	**	***
Cultivar	***	**	**
Electro-Conductivity	**	ns	*

[a] Lettuce DM:FM is reported in grams of dry mass to grams of fresh mass; [b] The EC treatment: high = 4.0 $mS \cdot cm^{-1}$, medium = 2.0 $mS \cdot cm^{-1}$, and low = 1.0 $mS \cdot cm^{-1}$. [c] The standard error of the mean was for growing season leaf DM ± 0.48; leaf DM:FM ± 0.0012; leaf water ± 0.13, cultivar standard error for leaf DM ± 0.40; leaf DM:FM ± 0.0011; leaf water ± 0.12, and EC treatment standard error for leaf DM ± 0.48; leaf DM:FM ± 0.0016; leaf water ± 0.13, [d] ns, *, **, *** indicate non-significant or significant at $P \leq 0.05, 0.01, 0.001$, respectively.

Lettuce root FM and DM peaked during spring production and was significantly reduced during the summer and fall concerning each cultivar (Table 2). Notably, spring green-leaf lettuce roots averaged 60.08 g FM, which was 93% greater than the root FM of red-leaf lettuce. Root biomass and water content were comparable between both cultivars produced in the fall season as well as between green-leaf lettuce grown in the summer and red-leaf lettuce grown in the spring (Table 2). Plant height (data not shown) and stem diameter were impacted and resulted in green and red-leaf summer lettuce achieving the greatest height, but smallest stem diameter, compared to their spring and summer counterparts. Rhazes lettuce growth in the fall was minimally impacted by season and cultivar and was 66% shorter compared to the Winter Density lettuce.

There were no interactions for EC treatments. Thus, EC treatments are presented as main effects. Low and high EC treatments resulted in comparable amounts of leaf DM. Conversely, lettuce leaf water content increased slightly by 0.7% when subjected to medium EC treatments. Additionally, season and treatment interactions significantly affected root biomass, water content, and stem diameter. Root biomass and water content had an inverse relationship when grown in different seasons and nutrient solution treatments. Root biomass in the spring and fall season increased by 25% and 20%, respectively, when the concentrations of the nutrient solution increased from low to high strength (data not shown). Conversely, root water content decreased 1% in the spring and fall and increased by 1% in summer with increasing nutrient solution strength. Lettuce stem diameter increased by 19% with respect to the spring season and increasing nutrient strength but decreased by 4% during the summer.

Fall production resulted in an increase of 1% from low to medium solution strength and decreased by 6% from medium to high strength.

Table 2. The effect of the interaction of season and greenhouse lettuce cultivars on root fresh mass, dry mass, dry mass to fresh mass ratio, and water content.

Season	Cultivar	Root FM (g)	Root DM (g)	Root DM:FM (g) [a]	Root Water %
Spring	Winter Density	60.08 a	2.42 a	0.04 c	0.96 a
Summer	Winter Density	30.00 b	1.47 b	0.05 b	0.95 b
Fall	Winter Density	23.73 c	1.22 c	0.05 b	0.95 b
Spring	Rhazes	31.11 b	1.32 bc	0.05 b	0.95 b
Summer	Rhazes	12.28 d	0.79 d	0.06 a	0.94 c
Fall	Rhazes	13.88 d	0.68 d	0.05 b	0.95 b
	***P*-Value [b,c]**	***	**	**	**

[a] Lettuce DM:FM is reported in grams of dry mass to grams of fresh mass. [b] The standard error of the mean was Root FM ± 2.14; Root DM ± 0.09; Root DM:FM ± 0.002; Root Water ± 0.002; [c] ns, **, *** indicate non-significant or significant at $P \leq 0.01$, 0.001, respectively.

3.2. Season, Cultivar, and Treatment Effect on Lettuce Quality

Growing season alone demonstrated a significant effect on chlorogenic acid content of greenhouse lettuce cultivars (Figure 6). Concentrations of chlorogenic acid were statistically comparable in the spring and summer seasons but significantly different from the fall. Chlorogenic acid levels were greatest in the spring, which was 73% higher compared to the fall.

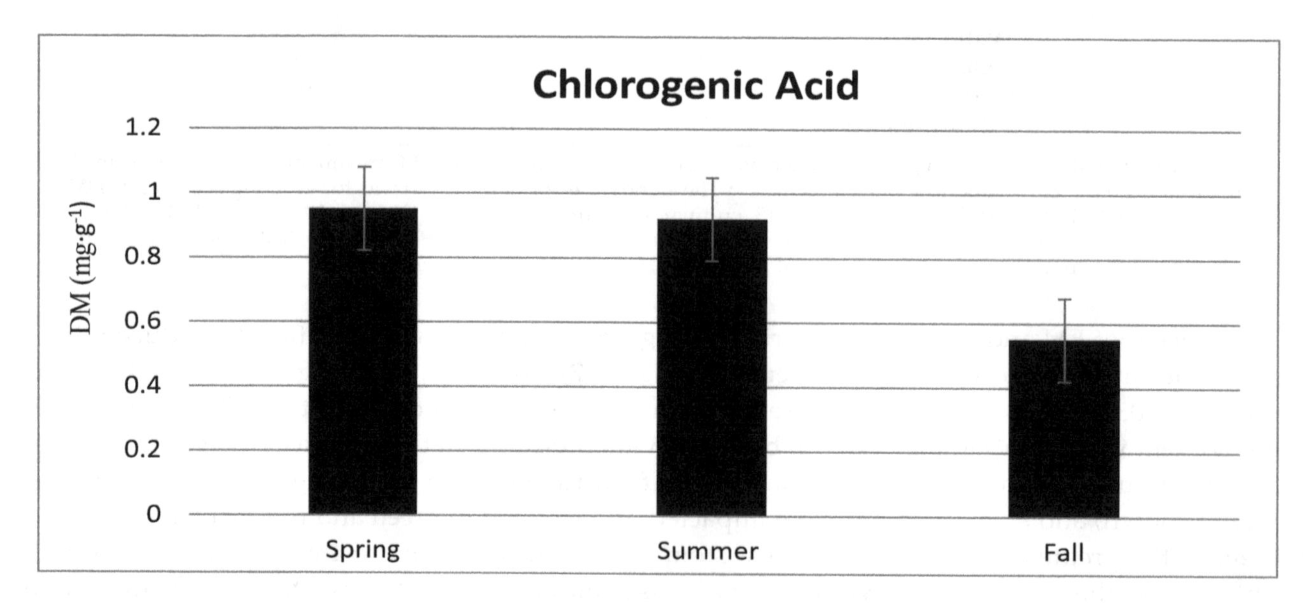

Figure 6. The effect of growing season on greenhouse lettuce chlorogenic acid content. The standard error of the mean was: Leaf DM ± 0.05. Different letters are significantly different at $P \leq 0.05$ according to Tukey's honest significant difference test.

Interactions between growing season and lettuce cultivars significantly affected chicoric acid and lettuce flavonoids (Table 3). Levels of chicoric acid increased from spring to summer to fall in both lettuce cultivars. The maximum concentration of chicoric acid, produced by red-leaf lettuce in the fall, was 131% greater compared to summer red-leaf lettuce and 175% greater than spring red-leaf lettuce.

Moreover, fall red-leaf lettuce contained 94% greater levels of chicoric acid compared to fall green-leaf lettuce. Concerning lettuce flavonoids, quercetin glucoside and quercetin glucuronide had an inverse relationship. Levels of quercetin glucoside increased from spring to summer but decreased from summer to fall in both cultivars. However, levels of quercetin glucuronide decreased from spring to summer before increasing in the fall. Spring red-leaf lettuce produced the highest concentration of luteolin (9.86 $mg \cdot g^{-1}$), although maximal concentrations in green-leaf lettuce (1.56 $mg \cdot g^{-1}$) were achieved in the fall. Interactions between season and cultivar resulted in increasing levels of quercetin malonyl from spring through the fall, and the greatest accumulation was present in substantially higher concentrations among red-leaf lettuce compared to green-leaf. The impact of nutrient solution treatment on lettuce phenolics was insignificant for all compounds except for quercetin glucoside, which at low-solution treatments were 69% greater than medium-solution treatments and 62% greater than high solution treatments (data not shown).

Table 3. The effect of seasons and cultivars on concentrations of greenhouse lettuce phenolics and flavonoids.

Season	Cultivar	Concentrations of Phenolics and Flavonoids ($mg \cdot g^{-1}$ DM) [a,b]					
		chlo	chic	qgluc	qglucor	luteolin	qmal
Spring	Winter Density	0.83 b	11.34 d	0.87 b	1.01 c	1.15 d	3.56 d
Summer	Winter Density	0.95 ab	15.26 d	1.51 b	0.55 c	0.91 d	4.63 d
Fall	Winter Density	0.52 c	33.85 b	1.05 b	1.20 c	1.56 d	5.93 cd
Spring	Rhazes	1.08 a	23.79 c	3.85 a	9.25 a	9.86 a	15.33 bc
Summer	Rhazes	0.90 ab	28.31 bc	3.98 a	5.19 b	5.40 c	18.93 b
Fall	Rhazes	0.58 c	65.52 a	1.34 b	6.50 b	7.73 b	46.90 a
	P-Value [c]	ns	**	*	*	*	***

[a] Abbreviations: chlo—chlorogenic acid; chic—chicoric acid; qgluc—quercetin glucoside; qglucor—quercetin glucuronide; qmal—quercetin malonyl; [b] The standard error of the mean was chlo $\pm$ 0.07; chic $\pm$ 3.06; qgluc $\pm$ 0.53; qglucor $\pm$ 0.64; lutein $\pm$ 0.67; qmal $\pm$ 3.57; [c] ns, *, **, *** indicate non-significant or significant at $P \leq 0.05$, 0.01, 0.001, respectively.

3.3. Season, Cultivar, and Treatment Effects on Leaf Mineral Content

Growing season exhibited an effect on leaf sulfur, copper, and zinc concentrations. While the largest amount of sulfur (S) was achieved in the summer, spring growing season resulted in comparable concentrations (data not shown). Spring and fall growing seasons resulted in similar concentrations of copper (Cu), which were, respectively, 68% and 37% larger compared to the summer. Fall production resulted in the highest concentrations of zinc (Zn), followed by summer, with the lowest concentrations in the spring. Zn levels in the fall growing season were 27% greater compared to the spring. Additionally, cultivar produced a significant effect on Cu and Zn micronutrients. Both nutrients were found in the highest concentrations in the red-leaf lettuce cultivar. Cu was 33% more concentrated in red-leaf lettuce, and Zn levels were 18% larger. The interaction between season and cultivar significantly impacted the macronutrients magnesium (Mg), phosphorous (P), potassium (K), and calcium (Ca) (Table 4). Concerning green-leaf lettuce, spring production resulted in the most accumulation of Mg and Ca, which declined by 16% and 17% in the summer and an additional 7% and 12% in the fall, respectively. P and K did not display any significant changes in relation to season. Concerning red-leaf lettuce, Mg and Ca concentrations increased from spring to summer by 12% and 2%, respectively, then declined in the fall by 3% for each cultivar. P and K had the lowest accumulation in the spring (5.66/48.13 $mg \cdot g^{-1}$) and steadily increased during the summer by 18% and 11% and fall season by 31% and 8%, respectively. The interaction between season and cultivar significantly impacted the micronutrients boron (B), manganese (Mn), and molybdenum (Mo) (Table 4).

Table 4. The effect of season and cultivar on the concentrations of elemental nutrients in freeze-dried greenhouse lettuce leaf tissue.

Season	Cultivar	Elemental Nutrient Concentrations [a] ($mg \cdot g^{-1}$) ($\mu g \cdot g^{-1}$)										
		Mg	P	S	K	Ca	Fe	B	Mn	Cu	Zn	Mo
Spring	Winter Density	5.30 a	6.22 bc	5.22 a	48.73 c	18.02 a	120.53 abc	37.16 a	58.31 ab	4.56 a	22.39 b	0.97 a
Summer	Winter Density	4.44 b	6.46 b	5.83 a	46.26 c	14.89 b	138.57 a	28.77 b	40.59 b	2.85 b	23.58 b	0.78 b
Fall	Winter Density	4.11 bc	6.49 b	1.36 b	48.87 c	13.08 c	108.08 bc	30.50 b	47.20 b	2.83 b	24.81 b	0.51 c
Spring	Rhazes	3.34 e	5.66 c	4.39 a	48.13 c	13.83 bc	98.68 c	34.43 a	40.55 b	5.37 a	22.90 b	0.64 bc
Summer	Rhazes	3.73 cd	6.68 b	5.86 a	53.56 b	14.12 bc	128.74 ab	30.02 b	55.39 ab	3.04 b	28.36 ab	0.54 c
Fall	Rhazes	3.62 de	8.73 a	1.32 b	57.60 a	13.73 bc	127.34 ab	37.22 a	74.98 a	5.25 a	32.53 a	0.51 c
	P-Value [b,c]	***	***	ns	**	***	ns	**	*	ns	ns	*

[a] Abbreviations: Mg—Magnesium; P—Phosphorous; S—Sulfur; K—Potassium; Ca—Calcium; Fe—Iron; B—Boron; Mn—Manganese; Cu—Copper; Zn—Zinc; Mo—Molybdenum; [b] The standard error of the mean was Mg $\pm$ 0.15; P $\pm$ 0.27; S $\pm$ 0.70; K $\pm$ 1.63; Ca $\pm$ 0.55; Fe $\pm$ 14.53; B $\pm$ 1.43; Mn $\pm$ 9.35; Cu $\pm$ 0.66; Zn $\pm$ 2.64; Mo $\pm$ 0.68; [c] ns,*, **, *** indicate non-significant or significant at $P < 0.05, 0.01, 0.001$, respectively.

Concerning green-leaf lettuce, B and Mn concentrations were greatest in the spring; whereas, summer and fall concentrations did not significantly differ. Mo concentrations were greatest in the spring (0.97 $\mu g \cdot g^{-1}$) and decreased during the summer by 20% and an additional 35% in the fall. Concerning red-leaf lettuce, B and Mn concentrations were greatest in the fall. However, B concentration decreased 13% from spring to summer, while Mn increased 37% from spring to summer. Molybdenum concentrations decreased 16% from spring to summer and an additional 6% from summer to fall. Increasing solution EC impacted leaf concentrations of P, K, Fe, B, Zn, and Mo. Each nutrient increased from treatment 1 to treatment three except for K, which reached a saturation point at treatment 2 and declined with the elevated EC of treatment 3. Additionally, this general trend was observed concerning the other mineral nutrients that were considered not statistically significant.

4. Discussion

4.1. Season, Cultivar, and Treatment Effect on Plant Growth and Biomass Production

The current study examines how the seasonal environment and increasing nutrient solution EC affect lettuce root and shoot mass, plant height and stem diameter, mineral nutrient content, and concentrations of selected phenolic compounds in green and red-leaf romaine cultivars. While season, cultivar, and EC treatments created significant differences in leaf fresh mass and stem diameter, it was the interaction between growing season and lettuce cultivar that demonstrated the most significant effect on root and shoot biomass. Spring growing season and highest EC treatment resulted in the greatest production of leaf and root FM in both cultivars. Greenhouse environmental data measured during 2016 and 2017 show that the spring growing seasons registered the highest levels of cumulative light energy. Light is known as a primary regulatory factor in plant growth and development, and previous research has indicated that daily light intensity significantly affects the production of shoot biomass. For example, Fu et al. [21] examined the effect of increasing light intensity (60, 140, and 220 $\mu mols \cdot m^{-2} \cdot s^{-1}$) and nitrogen concentrations (7, 15, and 23 $mmols \cdot L^{-1}$) on the growth and quality of hydroponic leaf lettuce. The results revealed that plants subjected to 220 $\mu mols \cdot m^{-2} \cdot s^{-1}$ light intensity and 7 $mmols \cdot L^{-1}$ of N produced the greatest amount of dry biomass. Similarly, lettuce plants grown during fall and spring seasons with 50 or 100 $\mu mols \cdot m^{-2} \cdot s^{-1}$ of supplemental white light produced more than 270% greater biomass production compared to control treatments [22].

The current study's results suggest that lettuce cultivar had the greatest influence on the production of leaf FM content in green and red-leafed cultivars. Lettuce leaf DM, DM:FM ratio, and leaf water content were influenced the most by growing season, lettuce cultivar, and EC treatments. There was an interaction between the growing season and lettuce cultivars that created the most consistent favorable conditions for the production of root biomass, root DM:FM ratio, and root water content. These results are mixed with other studies that demonstrate lettuce sensitivity to increasing EC concentrations [12]. In the spring, EC treatments were significantly greater compared to the summer and fall growing seasons. Consequently, the summer and fall growing season correspond to other studies. For example, Scuderi et al. [11] reported that increasing solution EC decreased yield and leaf nitrate content in lettuce planted at high densities in a deep-water culture production system. Furthermore, previous research demonstrated that increasing salinity treatments in three lettuce cultivars also resulted in reduced total yield [12]. Temperature is known to heavily influence the partitioning of photoassimilates in plants, and studies of lettuce [16,23], tomato [4], and zucchini [24] have indicated differences in plant biomass due to light and temperature interactions. Under suboptimal conditions, lettuce's resilience to common physiologically induced disorders such as tipburn [9,23,25], rib-discoloration [26], bolting [7], and the increase of bitterness compounds [27] is highly correlated to lettuce genotype.

4.2. Season, Cultivar, and Treatment Effect on Lettuce Quality

Previous research has demonstrated that despite the influence on lettuce yield, increasing EC levels caused greater production of flavonoid and phenolic compounds [14,15]. The results of the current study were inconsistent with these findings. Nutrient solution EC did not significantly affect flavonoid and phenolic concentration of any compounds except for quercetin glucoside, which was the highest flavonoid concentration in the leaf tissue and grown under the lowest EC treatment. However, season and the interaction between season and lettuce cultivar showed a significant impact on phenolic production. Chlorogenic acid is well studied in plants and acts as an antioxidant as well as protecting against ultra-violet radiation [28]. This corresponds with the results of the current study, indicating the greatest concentrations of chlorogenic acid in the spring and summer when greenhouse light intensity was at its peak. Furthermore, red-leaf lettuce cultivars contain higher concentrations of phenolic compounds than their green-leaf counterparts, and previous studies have shown great variability in the production of these compounds with respect to cultivar and growth environment. For example, Oh et al. [29] reported that exposing five-week-old lettuce plants to mild environmental stresses resulted in a two to three-fold increase in phenolic compounds in the leaf tissue. Specifically, the study found that decreasing temperature elevated concentrations of quercetin and luteolin glycosides. Moreover, increasing photosynthetic photon flux density (PPFD) from 43 to 410 $\mu mols \cdot m^{-2} \cdot s^{-1}$ also increased concentrations of quercetin, luteolin, and cyanidin glycosides [19], and increasing ultraviolet (UV) radiation in field grown lettuce resulted in a dose-dependent response of quercetin and luteolin glycosides and total phenolic acid concentrations [28]. These findings are consistent with the results of the current study, which demonstrated significant increases in flavonoids and phenolic content among red and green-leaf cultivars during spring and fall growing seasons where PPFD levels were higher and average daily temperatures were cooler, respectively, compared to summer.

4.3. Season, Cultivar, and Treatment Effects on Leaf Mineral Content

While climatic factors predominantly influenced the content of lettuce flavonoid and phenolic compounds, all production variables in the current study affected the uptake and concentration of leaf mineral nutrients. In field production, the uptake of mineral nutrients occurs when nutrients become available, which is dependent on soil pH, buffering capacity, and moisture [30]. It is generally accepted that increasing the nutrient supply when nutrients are already present in sufficient amounts will not improve plant growth, especially under extreme adverse environmental conditions [31]. However, in hydroponic production systems, plant roots are provided with a constant supply of purified water with a low buffering capacity. The pH of this water can be adjusted and held at the preferred range of 5.5 to 6.0, which allows maximum availability of nutrients to plant roots. Previous research indicated that even slight increases of pH to levels of 7.0 could significantly reduce lettuce FM and DM [32]. Several studies have examined the effect of increased nutrient solution EC on plant mineral nutrient content. Fallovo et al. [16] investigated the effect of growing season and increasing nutrient solution EC on yield and quality of hydroponic lettuce. The results of this study demonstrated that leaf mineral content of macroelements P, K, and Mg increased with increasing solution EC. Additionally, altering macro-anion and macro-cation nutrient solution proportions in spring and summer growing seasons significantly affected leaf concentrations of N, K, Mg, and Ca [16]. Furthermore, Barickman et al. [30] found that elevating K for greenhouse lettuce production resulted in higher concentrations of K in lettuce leaf tissue. However, a saturation point was reached before negative effects developed at higher levels of K fertilization. The results of these experiments are consistent with the findings of the current study where season, cultivar, and the interactions between the two demonstrated the most significant effect on leaf mineral nutrient content. Additionally, mineral nutrient concentrations increased with increasing solution EC except for K, which reached a saturation point and decreased in plants exposed to the highest solution concentration.

To develop a thorough understanding of the genotypical mechanisms and external contributing factors that produce variable results with respect to lettuce growth and development, secondary compound production, and sequestration of mineral nutrients, more information is required. While it is generally true that exposing lettuce to mild abiotic stresses, specifically elevated light irradiance and temperature, the effects of increasing growth solution EC are inconsistent concerning yield and quality. While the results of this study agree with previous work that suggested yield and quality are predominantly affected by growing season as opposed to increasing EC, all the tested leaf elemental nutrient concentrations increased as nutrient solution EC increased with statistical significance. Thus, the results of this study suggest that fall and spring production of greenhouse green and red-leaf cultivars with elevated EC solution of $\approx 4.0\ mS{\cdot}cm^{-1}$ should be used to maximize lettuce yield and nutritional quality.

Author Contributions: For this research article, T.C.B. conceived and designed the experiments; W.L.S. performed the experiments; T.C.B. and W.L.S. analyzed the data; T.C.B., W.L.S., and C.E.S. contributed sample analysis; W.L.S. wrote the manuscript; T.C.B. and C.E.S. edited the manuscript.

Acknowledgments: This publication is a contribution of the Mississippi Agriculture and Forestry Experiment Station and supported by the USDA NIFA Hatch S-294 Project MIS 146030. This research received no external funding.

References

1. Romani, A.; Pinelli, P.; Galardi, C.; Sani, G.; Cimato, A.; Heimler, D. Polyphenols in Greenhouse and Open-Air-Grown Lettuce. *Food Chem.* **2002**, *79*, 337–342. [CrossRef]
2. Husain, S.R.; Cilurd, J.; Cillard, P. Hydroxyl radical scavenging activity of Flavonoids. *Phytochemistry* **1987**, *26*, 2489–2491. [CrossRef]
3. Cartea, M.E.; Francisco, M.; Soengas, P.; Velasco, P. Phenolic Compounds in Brassica Vegetables. *Molecules* **2011**, *16*, 251–280. [CrossRef] [PubMed]
4. Gruda, N. Impact of environmental factors on product quality of greenhouse vegetables for fresh consumption. *Crit. Rev. Plant Sci.* **2005**, *24*, 227–247. [CrossRef]
5. Jensen, M. Hydroponics Worldwide. *Acta Hortic.* **1999**, *481*, 719–729. [CrossRef]
6. Zbeetnoff, C. *The North American Greenhouse Vegetable Industry. Farm Credit Canada and AgriSuccess*; Agro-Environmental Consulting: White Rock, BC, Canada, 2006.
7. Rappaport, L.; Wittwer, S.H. Night temperature and photoperiod effects on flowering of leaf lettuce. *Proc. Am. Soc. Hortic. Sci.* **1956**, *68*, 279–282.
8. Jenni, S. Rib Discoloration: A Physiological Disorder Induced by Heat Stress in Crisphead Lettuce. *HortScience* **2005**, *40*, 2031–2035.
9. Wissemeier, A.H.; Zühlke, G. Relation between climatic variables, growth and the incidence of tipburn in field-grown lettuce as evaluated by simple, partial and multiple regression analysis. *Sci. Hortic.* **2002**, *93*, 193–204. [CrossRef]
10. Zanin, G.; Ponchia, G.; Sambo, P. Yield and quality of vegetables grown in a floating system for readyto-eat produce. *Acta Hortic.* **2009**, *807*, 433–438. [CrossRef]
11. Scuderi, D.; Restuccia, C.; Chisari, M.; Barbagallo, R.N.; Caggia, C.; Giuffrida, F. Salinity of nutrient solution influences the shelf-life of fresh-cut lettuce grown in floating system. *Postharvest Biol. Technol.* **2011**, *59*, 132–137. [CrossRef]
12. Abou-Hadid, A.F.; Abd-Elmoniem, E.M.; El-Shinawy, M.Z.; Abou-Elsoud, M. Electrical conductivity effect on growth and mineral composition of lettuce plants in hydroponic system. *Acta Hortic.* **1996**, *434*, 59–66. [CrossRef]
13. Sgherri, C.; Perez-Lopez, U.; Micaelli, F.; Miranda-Apodaca, J.; Mena-Petite, A.; Munoz-Rueda, A.; Quartacci, M.F. Elevated CO_2 and salinity are responsible for phenolics-enrichment in two differently pigmented lettuces. *Plant Physiol. Biochem.* **2017**, *115*, 269–278. [CrossRef] [PubMed]

14. Kim, H.J.; Fonseca, J.M.; Choi, J.H.; Kubota, C.; Dae, Y.K. Salt in irrigation water affects the nutritional and visual properties of romaine lettuce (*Lactuca sativa* L.). *J. Agric. Food Chem.* **2008**, *56*, 3772–3776. [CrossRef] [PubMed]
15. Neocleous, D.; Koukounaras, A.; Siomos, A.S.; Vasilakakis, M. Assessing the salinity effects on mineral composition and nutritional quality of green and red 'baby' lettuce. *J. Food Qual.* **2014**, *37*, 1–8. [CrossRef]
16. Fallovo, C.; Rouphael, Y.; Cardarelli, M.; Rea, E.; Battistelli, A.; Colla, G. Yield and quality of leafy lettuce in response to nutrient solution composition and growing season. *J. Food Agric. Environ.* **2009**, *7*, 456–462.
17. Hoagland, D.R.; Arnon, D.I. The water-culture method for growing plants without soil. *Calif. Agric. Exp. Stn. Circ.* **1950**, *347*, 1–32.
18. Neugart, S.; Zietz, M.; Schreiner, M.; Rohn, S.; Kroh, L.W.; Krumbein, A. Structurally different flavonol glycosides and hydroxycinnamic acid derivatives respond differently to moderate UV-B radiation exposure. *Physiol. Plant.* **2012**, *145*, 582–593. [CrossRef] [PubMed]
19. Becker, C.; Kläring, H.P.; Kroh, L.W.; Krumbein, A. Temporary reduction of radiation does not permanently reduce flavonoid glycosides and phenolic acids in red lettuce. *Plant Physiol. Biochem.* **2013**, *72*, 154–160. [CrossRef] [PubMed]
20. Barickman, T.C.; Kopsell, D.A.; Sams, C.E. Selenium influences glucosinolate and isothiocyanates and increases sulfur uptake in Arabidopsis thaliana and rapid-cycling Brassica oleracea. *J. Agric. Food. Chem.* **2013**, *61*, 202–209. [CrossRef] [PubMed]
21. Fu, Y.; Li, H.; Yu, J.; Liu, H.; Cao, Z.; Manukovsky, N.S.; Liu, H. Interaction effects of light intensity and nitrogen concentration on growth, photosynthetic characteristics and quality of lettuce (*Lactuca sativa* L. Var. youmaicai). *Sci. Hortic.* **2017**, *214*, 51–57. [CrossRef]
22. Gaudreau, L.; Charbonneau, J.; Canda, A.; Gv, Q. Photoperiod and Photosynthetic Photon Flux Influence Growth and Quality of Greenhouse-grown Lettuce. *HortScience* **1994**, *29*, 1285–1289.
23. Glenn, E.P. Seasonal effects of radiation and temperature on growth of greenhouse lettuce in a high insolation desert environment. *Sci. Hortic.* **1984**, *22*, 9–21. [CrossRef]
24. Rouphael, Y.; Colla, G. Growth, yield, fruit quality and nutrient uptake of hydroponically cultivated zucchini squash as affected by irrigation systems and growing seasons. *Sci. Hortic.* **2005**, *105*, 177–195. [CrossRef]
25. Bres, W.; Weston, L. A Nutrient Accumulation and Tipburn in NFT-grown Lettuce at Several Potassium and pH Levels. *HortScience* **1992**, *27*, 790–792.
26. Jenni, S.; Truco, M.J.; Michelmore, R.W. Quantitative trait loci associated with tipburn, heat stress-induced physiological disorders, and maturity traits in crisphead lettuce. *Theor. Appl. Genet.* **2013**, *126*, 3065–3079. [CrossRef] [PubMed]
27. Bunning, M.L.; Kendall, P.A.; Stone, M.B.; Stonaker, F.H.; Stushnoff, C. Effects of Seasonal Variation on Sensory Properties and Total Phenolic Content of 5 Lettuce Cultivars. *J. Food Sci.* **2010**, *75*, 156–161. [CrossRef] [PubMed]
28. García-Macías, P.; Ordidge, M.; Vysini, E.; Waroonphan, S.; Battery, N.H.; Gordon, M.H.; Hadley, P.; John, P.; Lovegrove, J.A.; Wagstaffe, A. Changes in the flavonoid and phenolic acid contents and antioxidant activity of red leaf lettuce (Lollo Rosso) due to cultivation under plastic films varying in ultraviolet transparency. *J. Agric. Food Chem.* **2007**, *55*, 10168–10172. [CrossRef] [PubMed]
29. Oh, M.; Carey, E.E.; Rajashekar, C.B. Plant Physiology and Biochemistry Environmental stresses induce health-promoting phytochemicals in lettuce. *Plant Physiol. Biochem.* **2009**, *47*, 578–583. [CrossRef] [PubMed]
30. Barickman, T.C.; Horgan, T.E.; Wheeler, J.R.; Sams, C.E. Elevated Levels of Potassium in Greenhouse-grown Red Romaine Lettuce Impacts Mineral Nutrient and Soluble Sugar Concentrations. *HortScience* **2016**, *51*, 504–509.
31. Hu, Y.; Schmidhalter, U. Drought and salinity: A comparison of their effects on mineral nutrition of plants. *J. Plant Nutr. Soil Sci.* **2005**, *168*, 541–549. [CrossRef]
32. Anderson, T.S.; Martini, M.R.; de Villiers, D.; Timmons, M.B. Growth and Tissue Elemental Composition Response of Butterhead Lettuce (*Lactuca sativa*, cv. Flandria) to Hydroponic Conditions at Different pH and Alkalinity. *Horticulturae* **2017**, *3*, 41. [CrossRef]

9

Evaluation of Two Wild Populations of Hedge Mustard (*Sisymbrium officinale* (L.) Scop.) as a Potential Leafy Vegetable

Marta Guarise [1,*], Gigliola Borgonovo [2], Angela Bassoli [2] and Antonio Ferrante [1]

1 Department of Agriculture and Environmental Science—Production, Landscape, Agroenergy, Università degli Studi di Milano, 20133 Milano, Italy; antonio.ferrante@unimi.it
2 Department of Food, Environmental and Nutritional Science, Università degli Studi di Milano, 20133 Milano, Italy; gigliola.borgonovo@unimi.it (G.B.); angela.bassoli@unimi.it (A.B.)
* Correspondence: marta.guarise@unimi.it.

Abstract: The minimally processed industry is always looking for produce innovation that can satisfy consumer needs. Wild leafy vegetables can be a good source of bioactive compounds and can be attractive for the consumer in term of visual appearance and taste. In this work, *Sisymbrium officinale* (L.) Scop., commonly called hedge mustard, was grown in a greenhouse and evaluated as a potential leafy vegetable. Two wild populations, Milano (MI) and Bergamo (BG), were grown in peat substrate and harvested at the commercial stage for the minimally processing industry. Leaf pigments such as chlorophyll and carotenoids were determined as well as chlorophyll *a* fluorescence parameters. Total sugars, antioxidant compounds such as ascorbic acid, phenolic index, total phenols, anthocyanins, and nitrate were determined at harvest. Significant differences between wild populations were found in April with higher nitrate content in BG, 2865 mg/kg FW than in MI, 1770 mg/kg FW. The nitrate levels of *S. officinale* measured in the present study are significantly lower than the maximum NO_3 level allowed in other fresh leafy vegetables. Ascorbic acid measured in November was higher in MI compared BG with values of 54.4 versus 34.6 mg/100 g FW, respectively. The chlorophyll *a* fluorescence data showed that BG reached optimal leaf functionality faster than MI. Overall results indicated that *Sisymbrium officinale* (L.) Scop. can be suggested as a potential leafy vegetable for the minimally processed industry.

Keywords: *Sisymbrium officinale*; *Brassicaceae*; hedge mustard; leafy vegetables

---

1. Introduction

Minimally processed leafy vegetable production has been evolving in recent years by providing new produce with beneficial effect on human health. There are several wild species that can be considered as potential leafy vegetables. The introduction of new species can be useful for diet enrichment and diversification. Moreover, wild plants can be highly adaptative to different environments.

Sisymbrium officinale (L.) Scop., synonym *Erysimum officinale*, commonly known as hedge mustard in English, *erísimo* in Spanish, *erisimo* or *erba cornacchia* in Italian, and *velar* in French, is a medicinal plant that belongs to the Brassicaceae family. This species could have potential for introduction into the leafy vegetable production for the minimally processed or fresh-cut industry.

S. officinale is a terophyte scapose plant with a reddish-violet erect trunk, that present a lot of trichomes and many branches. Basal leaves are different from the upper ones with a dentate shape. Hedge mustard has a linear racemose inflorescence; each flower has four small (1–2 mm) yellow petals; the fruit is a tiny siliqua, close-fitting to the trunk. Flowering occurs in Spring–Summer, from May to

July–August, depending on the climate. Siliqua pods usually are pubescent, once they reach maturity they release seeds. Seeds are very small, each siliqua can contain from 10 to 20 seeds. *S. officinale* is endemic in the Eurasian continent and widespread in all Italian regions from 0 to 1000 m. above sea level (a.s.l.), and rarely up to 2400 m a.s.l. [1]. This annual or biennial herbaceous plant is described as ruderal, growing on disturbed sites such as field margins and roadsides [2].

Flowers and leaves of hedge mustard are commonly used as a traditional medicinal herb for the treatment of sore throats, coughs, and hoarseness [3–5] under specific indication based upon long-standing use [6] and recent clinical studies [7]. For that reason, *S. officinale* is largely known as "singer's plant" and is used among singers, actors, and professionals who use the voice for working. The therapeutic activity of this plant is attributed to its sulfurated components. Dried flowering aerial parts contain: total glucosinolates (0.63–0.94%), mucilage (13.5–10.9%), total thiols (8.9–10.2%), and total flavonoids (0.50–0.56%). The main glucosinolate in *S. officinale* is glucoputranjivine [8]. It represents 58.3% of total glucosinolates on a fresh weight basis. This percentage declined to 32.5% after autolysis [6].

Brassicaceae is one of the most important botanical families in horticultural production in Mediterranean countries, due to their great diversity expressed both in spontaneous and cultivated species. In Italy, horticultural Brassicaceae are widespread on about 40,000 ha, in particular in the center-southern region [9].

In spite of its long, traditional therapeutic use for treating voice discomfort as dried plants (including leaves, stem, and flowers) for preparing decoctions, tinctures, or propolis, *S. officinale* has barely been investigated for its beneficial proprieties, and there are no data about its possible use and consumption as a fresh leafy vegetable. Its low agronomic requirements allow the cultivation in different Mediterranean environments.

In order to evaluate the possibility of recommending this species as a potential leafy vegetable, the aim of present study was to investigate production of two different wild populations, one collected in Milan and the other one in Bergamo, Italy. Cultivation was performed in pots containing fertilized substrate in a greenhouse. Total chlorophyll content and chlorophyll *a* fluorescence were measured for evaluating photosynthetic activity. The most common quality parameters that are usually considered for leafy vegetable evaluation were determined such as ascorbic acid, carotenoids, phenols, anthocyanins, nitrates, and total sugar. Furthermore, to evaluate the production of leaves, fresh and dry biomass were measured at the baby leaf stage which is usually the developmental stage for leafy vegetables destined to the fresh-cut industry.

2. Materials and Methods

2.1. Plant Material

Seeds of two wild populations of *Sisymbrium officinale* (L.) Scop., or hedge mustard, respectively named MI (Milan) and BG (Bergamo), obtained from controlled seed reproduction at Fondazione Minoprio (Como, Italy) during summer in 2017, were sown separately in polystyrene panels using common horticultural fertilized substrate under controlled conditions in a greenhouse at the Faculty of Agricultural and Food Science of Milan, 16 January 2018, for the first evaluation, and 8 October, for the second evaluation (Supplementary Materials Table S1). Cultivation was performed in the greenhouse of the Agricultural Faculty, which was a single gable covered with glass and provided with a cooling system and supplemental light only for the second growing cycle. The supplemental lighting was provided for 16 h from the 7:00 a.m. to 11:00 p.m. with 400 W/m^2 High Pressure Sodium lamps. The environmental parameters are reported in the Supplementary Materials Figure S1.

Plantlets were transplanted and grown in complete substrate (Vigorplant, Italy) containing the following components: 21% Baltic peat, 22% dark peat, 26% Irish peat, 13% volcanic peat, 18% calibrated peat at a pH of 6.5 in 10 cm diameter plastic pots. The plant density was 16–18 plants/pot or 80–90 plant/m^2.

Harvest was performed at the end of each cultivation cycle, on 3 May, and on 7 November, when the plant reached the commercial baby leaf stage, which corresponded to plants at a 15 cm height with 4–6 fully expanded true leaves. Plants were randomly chosen from each pot and sampled for the analyses. Plants were not supplied with extra-fertilizers in either experimental period and were watered every day to maintain optimal water availability.

2.2. *Non-Destructive Analyses*

Chlorophyll a Fluorescence

For the characterization of the two *S. officinale* wild populations, non-destructive analyses were conducted on fresh leaf tissue. Each week, starting from 18 of April to the first week of May 2018 and 7 November, chlorophyll *a* fluorescence was measured using a hand-portable fluorometer (Handy PEA, Hansatech, Kings Lynn, United Kingdom). Leaves were dark-adapted for 30 minutes using leaf clips. After this time, a rapid pulse of high-intensity light of 3000 μmol m^{-2} s^{-1} (600 W m^{-2}) was administered to the leaf inducing fluorescence. Fluorescence parameters were calculated automatically by the device: variable fluorescence to maximum fluorescence (Fv/Fm). From the fluorescence parameters, JIP analyses were performed to determine the following indices: Performance Index (PI), dissipation energy per active reaction center (DIo/RC), and density of reaction centers (RC/CSm).

2.3. *Destructive Analyses*

To evaluate qualitative characteristics of the two wild populations of *S. officinale*, small samples of fresh leaves, about 1 g for each sample, were sampled one month after transplanting and at the beginning of May for the first cultivation cycle, and on 7 November for the second cycle. Some leaf samples were immediately stored at −20 °C to prevent tissue degradation.

To evaluate the yield at the baby leaf stage, plant fresh weight and dry weight were recorded at the end of biological cycle, in May for the first cultivation cycle and in November for the second cycle.

2.3.1. Chlorophyll and Carotenoids

Chlorophyll and carotenoids were extracted from fresh leaves. Leaf disks of 5 mm diameter (or 20–30 mg) in 5 mL 99.9% methanol as solvent were kept in a dark cold room at 4 °C for 24 h. Quantitative chlorophyll determinations were carried out immediately after extraction. Absorbance readings were measured at 665.2 and 652.4 nm for chlorophyll pigments and 470 nm for total carotenoids. Chlorophylls and carotenoid concentrations were calculated by Lichtenthaler's formula [10].

2.3.2. Phenolic Index, Total Phenols, and Anthocyanins

For the following analyses, fresh leaf tissue (disks of 5 mm diameter, or 20–30 mg) was extracted in 3 mL 1% methanolic HCl. The Phenolic Index of leaf tissue was determined spectrophotometrically by direct measurement of leaf extract absorbance at 320 nm. After overnight incubation the supernatant was read at 320 nm. The values were expressed as ABS_{320nm}/g FW.

Total phenols were determined spectrophotometrically following the Folin-Ciocalteu reagent method [11] using 200 μL of each sample extract 7.8 mL of distilled water, 0.5 mL of Folin-Ciocalteu reagent and 1.5 mL of 20% Na_2CO_3. Samples were extracted for 2 h in the dark and then read at 760 nm. Total phenols were calculated using a standard curve performed with gallic acid.

Anthocyanin content was determined spectrophotometrically. Sample extracts were incubated overnight at 4 °C in darkness. The concentration of cyanidin-3-glucoside equivalents was determined spectrophotometrically at 535 nm using an extinction coefficient (ε) of 29,600 [12].

2.3.3. Ascorbic Acid Determination

Only the November samples were analyzed for ascorbic acid content. For analysis, about 1 g of frozen leaves (frozen at −80 °C) were homogenized in a mortar with 1.3 mL of cold 6% (*w/v*) metaphosphoric acid and centrifuged at 10.000 × *g* at 4 °C. The pellet obtained by centrifugation was washed with 1.06 mL of cold metaphosphoric acid solution and centrifuged again. The supernatants were combined and 6% metaphosphoric acid was added to make a final volume of 3.3 mL.

After filtration through nylon filter, a 10 μL sample aliquot was injected onto an Inertsil ODS-3 GL Science column at 20 °C attached to a Series 200 LC pump. Peaks were converted to concentrations by using the dilution of stock ascorbic acid to construct a standard curve. Chromatographic data were stored and processed with a PerkinElmer TotalChrom 6.3 data Processor (PerkinElmer, Norwalk, CT, USA) [13].

2.3.4. Nitrate Determination

Nitrate concentration was measured by the salicyl sulfuric acid method [14]. About 1 g of fresh leaves was ground in 5 mL of distilled water. The extracts were centrifuged at 4000 rpm for 15 min. After centrifugation, the supernatant was collected for colorimetric determinations. Twenty μL of sample were collected and 80 μL of 5% (*w/v*) salicylic acid in concentrated sulfuric acid were added. After the reaction, 3 mL of NaOH 1.5 N were added. Each sample was cooled, and absorbance was measured at 410 nm. Nitrate concentration was calculated referring to a KNO_3 standard calibration curve.

2.3.5. Total Sugar Determination

To determine total sugar levels, extracts were prepared as above for the determination of nitrate levels. Total sugars were determined using the anthrone assay [15] with slight modification. The anthrone reagent was prepared using 0.1 g of anthrone dissolved in 50 mL of 95% H_2SO_4. The reagent was left 40 min before use; then, 200 μL of extract was added to 1 mL of anthrone, put in ice for 5 min and vortexed. The reaction was heated at 95 °C for 5 min. Samples were cooled and absorbance was read at 620 nm. Total sugar concentration was calculated referring to a glucose standard calibration curve.

2.4. Statistical Analyses

Data from the first cultivation cycle were subjected to two-way ANOVA and differences among means were determined using Tukey's post-test ($P < 0.05$). Data from the second cultivation cycle were analyzed using a t-test ($P < 0.05$). The number of replicate samples used in each analysis or measurement is reported in the legend of the figures or tables.

3. Results

3.1. Total Chlorophylls, Carotenoids, Phenols, and Anthocyanins

The yield at the baby leaf stage, a 15 cm height and 4–6 leaves, for both wild populations ranged from 2.41 g to 4.00 g for the winter-spring season and from 2.26 g to 2.45 g for the autumn season. The dry weight percentage ranged from 10.64 to 10.76 in May and from 8.07 to 7.87 in November. No significant differences were found in FW or DW between values of MI and BG harvested in May and in November (Table 1).

The plants showed different leaf pigment contents at different sampling times. In May a higher chlorophyll content was observed in MI than in April, while no significant differences between MI and BG were found in either cycle (Table 2).

Table 1. Fresh weight and % of dry matter in MI and BG hedge mustard wild populations at the baby leaf stage after two production cycles. Data are expressed as means of five plants ($n = 5$).

Cycle		Wild Population	Fresh Weight (g/plant)		Dry Matter (%)	
I	May	MI BG	2.41 4.00	ns	10.64 10.76	ns
II	November	MI BG	2.26 2.45	ns	8.07 7.87	ns

Data were analyzed using a t-test ($P < 0.05$). ns means no statistical differences.

Table 2. Chlorophyll (a, b) and total carotenoid content in MI and BG hedge mustard wild populations after two production cycles. Data are reported as mg/g FW ($n = 4$).

Cycle		Wild Population	Chl a (mg/g FW)		Chl b (mg/g FW)		Total Carotenoids (mg/g FW)	
I	April	MI BG	0.63 b 0.93 ab		0.19 b 0.37 ab		0.120 0.143	ns
	May	MI BG	1.22 a 0.79 ab		0.42 a 0.27 ab		0.215 0.157	ns
II	November	MI BG	1.56 1.44	ns	0.51 0.47	ns	0.382 0.379	ns

Data of cycle I were subjected to two-way ANOVA and differences among wild populations and dates within a cycle were determined using Tukey's test ($P < 0.05$). Data of cycle II were analysed using a t-test ($P < 0.05$). ns indicates no statistical differences.

In cycle I in April the total phenols and anthocyanins did not differ between populations. In May higher anthocyanins values were observed in BG compared to MI, while no significant differences were observed for total phenol among the two wild populations. Significant differences were found in phenolic index between the two wild populations (MI and BG) only in November with a higher concentration for BG. In November ascorbic acid determination revealed significant differences between MI and BG with a higher concentration for MI (Table 3). Among the antioxidant compounds, ASA and phenolics showed opposite differences between the two *S. officinale* populations cultivated in November, ASA higher and lower phenolic index in MI.

Table 3. Ascorbic acid (ASA), phenolic index, total phenols, and anthocyanin content in MI and BG hedge mustard wild populations ($n = 4$).

Cycle		Wild Population	ASA (mg/100 g FW)		Phenolic Index (ABS_{320nm}/g FW)		Total Phenols		Anthocyanins (mg/100 g FW)	
I	April	MI BG	- -		27.45 26.72	ns	1.13 1.09	ns	32.13 ab 26.38 b	
	May	MI BG	- -		19.05 32.60	ns	0.75 1.35	ns	14.88 c 38.79 a	
II	November	MI BG	54.45 34.57	*	42.51 55.44	*	1.62 2.37	ns	29.53 31.51	ns

Data of cycle I were subjected to two-way ANOVA and differences among wild populations and dates within cycles were determined using Tukey's test ($P < 0.05$). Data of cycle II were analysed using a t-test ($P < 0.05$). An asterisk (*) indicates a significant difference and ns indicates no statistical differences.

3.2. Chlorophyll a Fluorescence Measurements

From the chlorophyll *a* fluorescence data, four parameters were considered: Fv/Fm (maximum quantum yield of PSII), PI (Performance Index), DI0/RC (rate of energy dissipated by PSII per reaction center), and RC/CSm (active RCs per excited cross-section). No significant differences were found between the two wild populations for Fv/Fm and DIo/RC parameters. The Fv/Fm ratio after 26 April in both wild populations showed values higher than 0.83.

PI did not show significant differences between wild populations during cycle I. Significant differences were found between BG measured on 18 April and 2 May. PI index increased from 18 April to 2 May. Then, it decreased by the following measurement on 5 May but remained higher than the initial measurement.

Like the Fv/Fm ratio, the DIo/RC values did not show significant differences between wild populations and measurement times. However, this index declined in both wild populations during cultivation.

Significant differences were found for RC/CSm in MI between 18 April and 2 May and between 26 April and 2 May. In addition, significant differences were found in the BG wild population between 18 April and 26 April (Figure 1).

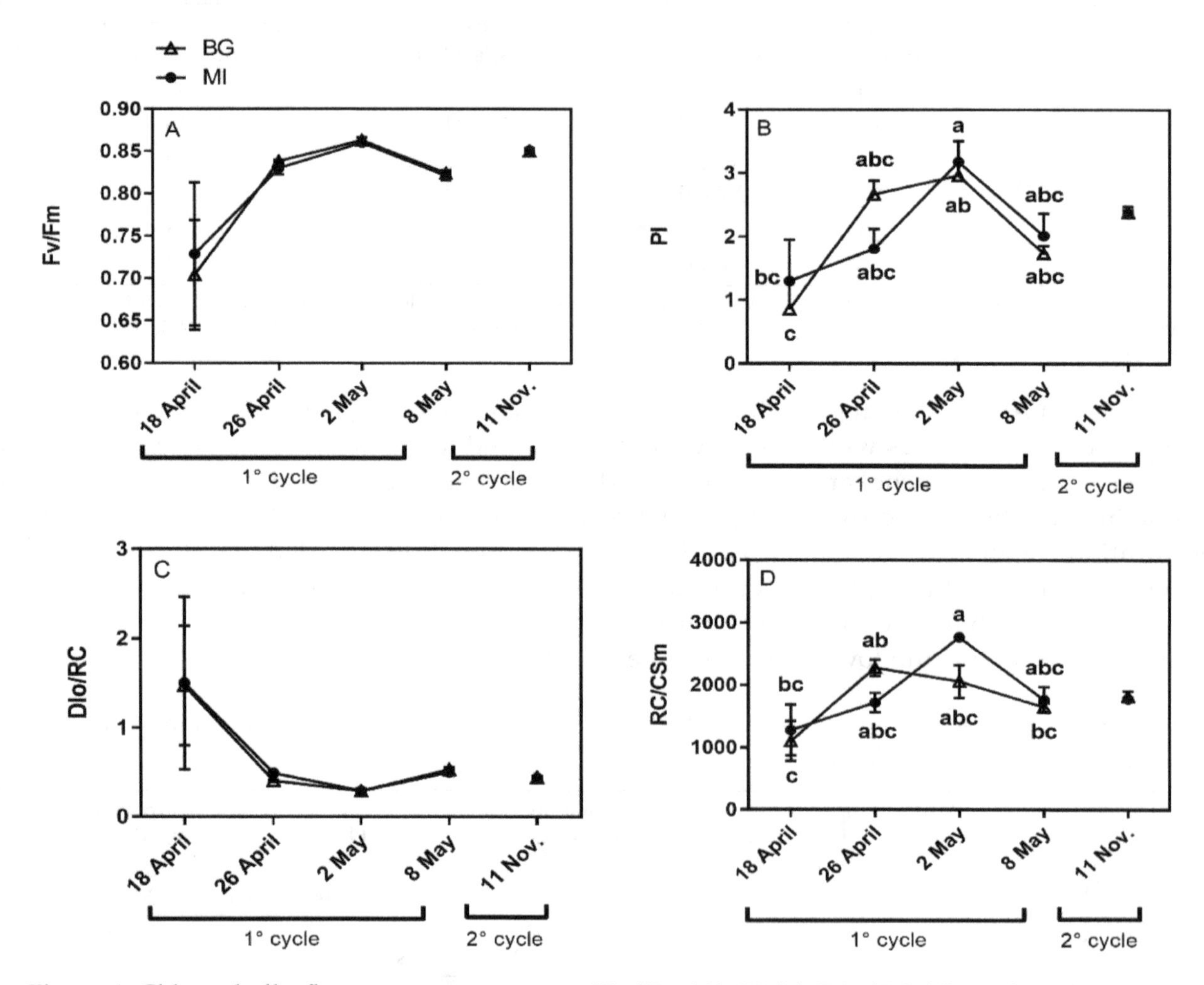

Figure 1. Chlorophyll *a* fluorescence parameters (Fv/Fm (**A**), PI (**B**), DIo/RC (**C**), and RC/CSm (**D**)) in leaves of two hedge mustard wild populations, MI and BG. Data are means with standard errors ($n = 4$ for April and May, $n = 5$ for November). Data of cycle I were subjected to two-way ANOVA and differences among wild populations and dates within cycles were determined using Tukey's test ($P < 0.05$). Data of cycle II were analysed using a *t*-test ($P < 0.05$). Different letters indicate statistical differences, and no letters indicate no significant differences.

3.3. Nitrate and Total Sugars

The quality of hedge mustard grown as baby leaf vegetables was also evaluated in terms of nitrate accumulation and total sugars. The nitrate content was statistically different between wild populations in April, with higher nitrate content in BG, 2865 mg/kg FW than in MI (Figure 2). In May, the two wild populations showed lower and significantly different values compared with those measured in April, nitrate content ranged from 199 to 256 mg/kg, but there was no difference between them. In November, both wild populations showed a nitrate content that ranged from 1756 mg/kg FW to 1683 mg/kg FW, with no significant differences between them (Figure 2).

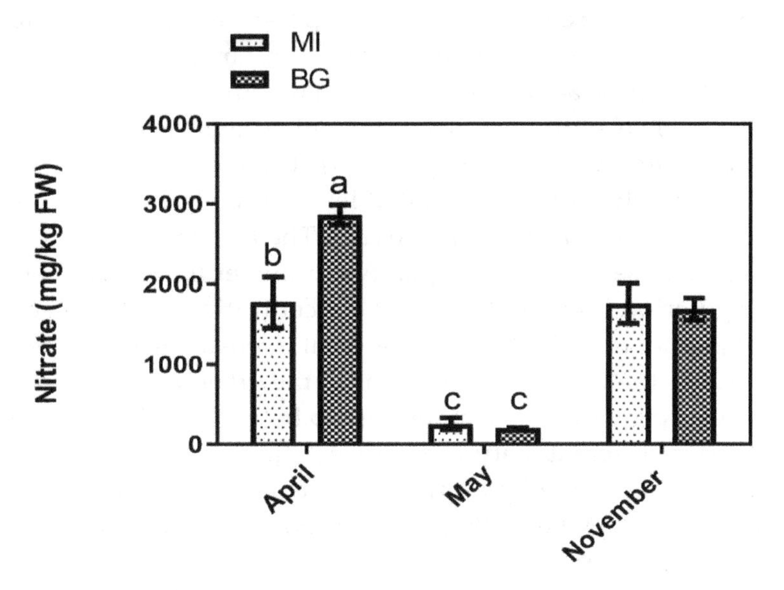

Figure 2. Nitrate content in leaves of two hedge mustard wild populations, BG and MI. Data are means with standard errors (n = 3 for April and May, n = 4 for November). Data of cycle I were subjected to two-way ANOVA and differences among wild populations and dates within cycles were determined using Tukey's test. Data of cycle II were analysed using a t-test ($P < 0.05$). Different letters indicate statistical differences for $P < 0.05$. Population means did not differ in November.

Total sugars were not statistically different between wild populations (MI and BG) or by dates within cycle I (April, May). The total sugar content ranged from to 5.84 mg Glu eq./g FW observed in MI in November, to 3.1 mg Glu eq./g FW in the April (Figure 3).

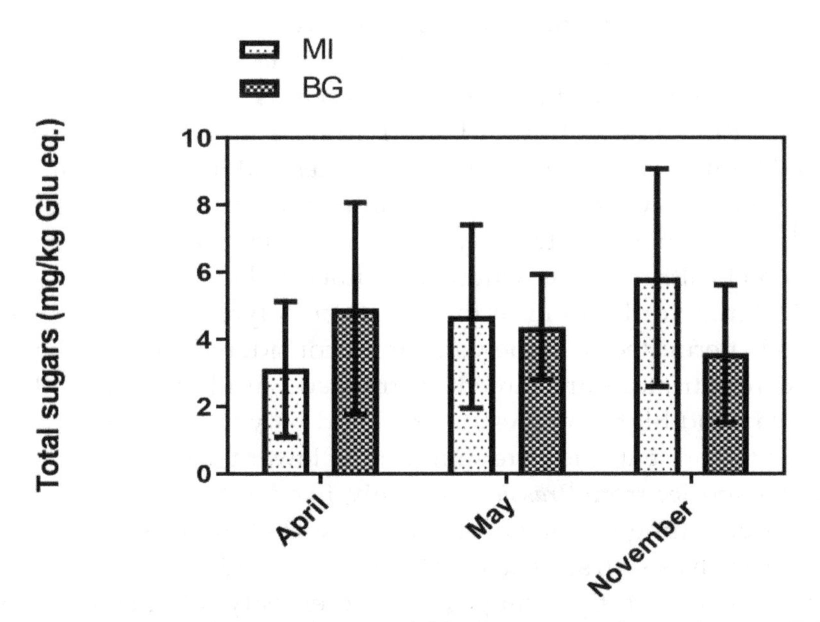

Figure 3. Total sugar content in leaves of two wild hedge mustard populations, BG and MI. Data are means with standard errors (n = 3 for April and May, n = 4 for November). Data of cycle I were subjected to two-way ANOVA. Data of cycle II were analysed using a t-test ($P < 0.05$). There were no significant differences between wild populations or times.

4. Discussion

The Brassicaeae family includes a wide number of species that can be used for vegetable production. Some of them have been described as potential vegetables and sources of antioxidant compounds for the Mediterranean area [16,17]. Hedge mustard is a wild Brassicaceae species widely dispersed and, therefore, it has been evaluated as a potential leafy vegetable for the minimally processed industry. The nutritional components were similar to other leafy vegetables and could provide a good quantity of ascorbic acid that is higher than lettuce, which shows values ranging from 10 to 30 mg/100 g [18]. The leaf pigments observed in hedge mustard were similar to other leafy vegetables such as rocket (*Eruca vesicaria* subsp. *sativa*) [19], lamb's lettuce (*Valerianella locusta*) [20], and lettuce (*Lactuca sativa*) [21]. It is well known that leaf pigments are important parameters because they contribute to leaf color and visual appearance [22]. The leaf color is very important in minimally processed leaf vegetables because it is the first quality parameter that consumers evaluate at purchase.

During the growing period lower chlorophyll concentrations were observed in spring than in autumn. Usually the leaf pigments are higher at lower light intensity; in our experimental conditions the higher values could be due to the low light conditions in the greenhouse during autumn, even though supplementary lighting was provided. This relationship between lower light availability and higher pigments have been found in different leafy vegetables [23].

Chlorophyll *a* fluorescence-derived parameters were used for evaluating the PSII activity of the two *S. officinale* wild populations (MI and BG). The Fv/Fm ratio indicated the maximal efficiency of PSII photochemistry, and it did not significantly change during the experiments. Values of Fv/Fm below 0.83 are usually considered as indicative of stressful conditions in plants [24]. The Fv/Fm ratio increased at the end of April, while the values were slightly lower at the earlier measurement. This result may be due to Fv/Fm increasing with leaf development until reaching the fully expanded stage, when the leaves are fully photosynthetically active. Values above 0.83 during May demonstrated that plants were under optimal growing conditions as was also observed in the second cycle performed in autumn. The higher values of PI in May also indicated higher light use efficiency and better performance of the plants [23]. The BG wild population seemed to have had faster adaptation and reached optimal leaf functionality earlier. In fact, BG also had higher RC/CSm values in April and these results were repeated in the second cycle performed in November

Nitrate plays a crucial role in the nutrition and function of plants and naturally occurs as a compound in the nitrogen cycle. In plants, nitrate levels are higher in leaves, whereas lower levels occur in tubers and seeds. In fact, leaf crops such as spinach (*Spinacea oleracea* L.), lettuce and rocket have high nitrate concentration [25]. Nitrate and nitrite are also commonly used as preservatives in food. Nitrate is non-toxic, but its metabolites and reaction products (nitrite, nitric oxide, and N-nitroso compounds) could be dangerous for human health inducing methaemoglobinaemia or carcinogenesis [26]. In addition to nitrate, leafy vegetables also provide several bioactive compounds with beneficial effects on health, and are widely recommended in the diet.

The European Union, in order to limit the nitrate supply in human nutrition, has defined the maximum nitrate levels permitted in some vegetables considered to have the highest levels of this compound. Nitrate concentrations are directly correlated with light availability. This may explain the lower values in May and higher values in April and November. However, the nitrate levels of *S. officinale* measured in the present study are significantly lower than the maximum NO_3 level allowed in rocket salad, another species from *Brassicacea* family, fixed at 6000 or 7000 mg NO_3/kg FW by the UE Commission [27], depending on the harvest time. Rocket plants grown in different cultivation systems in greenhouses such as soil, substrate or floating, have higher nitrate levels [28,29]. However, the higher ascorbic acid concentration compared to other leafy salad contributes a reduced risk of nitrosamine formation and carcinogenic effects of the nitrate in the diet. In fact, it has been reported that ascorbic acid is a nitrosation inhibitor and could inhibit nitrate reduction [30]. The low nitrate content of *S. officinale* is a good quality trait for possible use of this species as a leafy vegetable. Moreover, under greenhouse cultivation, the level of ascorbic acid is similar or slightly higher than those found in plants grown in the wild [31].

The leaf sugar content is related to photosynthetic activity and biomass production. It is also an important parameter for the storage of the product. Higher sugar content can be potentially associated with higher shelf life, because sugars are used for the basal metabolism and maintaining quality of the product. The total sugar content measured in the present study (maximum level of 5.84 mg/g) is lower if compared to the average sugar content observed in rocket leaves, where the total sugar reached 6.30 mg/g in the first harvest and 7.61 mg/g in second harvest [32]. The total sugars were similar among the two wild populations which also corresponds to biomass production and dry matter percentage.

5. Conclusions

Sysimbrium officinale is a wild species from the *Brassicaceae* family, quite common in all temperate Euroasiatic areas. Our results indicated that this species can be successfully grown in a greenhouse with nutritional components as well as quality parameters such as nitrate, chlorophyll, and sugar content similar to the most common commercial leafy vegetables. It has a good concentration of ascorbic acid, higher than common leafy vegetables. Although these results suggest that *Sysimbrium officinale* can be grown as a leafy vegetable, further investigation will be required for evaluating quality during postharvest storage and handling.

Author Contributions: The authors contributed to the work as follows: conceptualization: A.F. and A.B.; methodology: M.G.; formal analysis: M.G.; investigation: M.G.; resources: G.B.; data curation: M.G.; writing—original draft preparation: M.G., A.F.; writing—review and editing: A.F., A.B.; supervision: A.F.; project administration: A.B.; funding acquisition: A.B.

References

1. Pignatti, S. *Flora d'Italia*; Edizioni Edagricole: Bologna, Italy, 1982; Volume 2, p. 377.
2. Bouwmeester, H.J.; Karseen, C.M. Annual changes in dormancy and germination in seeds of Sisymbriumofficinale (L.) Scop. *New Phytol.* **1993**, *124*, 179–191. [CrossRef]
3. Politi, M.; Braca, A.; Altinier, G.; Sosa, S.; Ndjoko, K.; Wolfender, J.L.; Hostettmann, K.; Jimenez-Barbero, J. Different approaches to study the traditional remedy of "hierba del canto" Sisymbriumofficinale (L.) Scop. *Bol. Latinoam. Caribe Plantasmedicinales Aromàt.* **2008**, *7*, 30–37.
4. Blažević, I.; Radonić, A.; Mastelić, J.; Zekić, M.; Skočibusić, M.; Maravić, A. Hedge Mustard (Sisymbriumofficinale): Chemical diversity of volatiles and their antimicrobial activity. *Chem. Biodivers.* **2010**, *7*, 2023–2034. [CrossRef] [PubMed]
5. Di Sotto, A.; Vitalone, A.; Nicoletti, M.; Piccin, A.; Mazzanti, G. Pharmacological and phytochemicalstudy on *Sisymbriumofficinale* Scop. Extract. *J. Ethnopharmacol.* **2010**, *127*, 731–736. [CrossRef] [PubMed]
6. EMA—European Medicines Agency. *Assessment report on Sisymbriumofficinale (L.) Scop., Herba*; EMA: London, UK, 15 January 2014.
7. Calcinoni, O. Sisymbrium "Singers' Plant" Efficacy in Reducing Perceived Vocal Tract Disability. *J. Otolaryngol. Ent Res.* **2017**, *8*, 00243. [CrossRef]
8. Carnat, A.; Fraisse, D.; Carnat, A.P.; Groubert, A.; Lamaison, J.L. Normalization of hedge mustard, *Sisymbriumofficinale* L. *Ann. Pharm. Fr.* **1998**, *56*, 36–39.
9. Pardossi, A. *Orticoltura. Principi e Pratica*; Edizioni Edagricole: Bologna, Italy, 2018; pp. 261–276.
10. Lichtenthaler, H.K. Chlorophylls and carotenoids: Pigmentsof photosynthetic membranes. *Methods Enzymol.* **1987**, *148*, 350–382.
11. Singleton, V.L.; Orthofer, R.; Lamuela-Raventos, R.M. Analysis of total phenols and other oxidation substrates and antioxidants by means of folin-ciocalteu reagent. *Methods Enzymol.* **1999**, *299*, 152–178.
12. Giusti, M.M.; Rodríguez-Saona, L.E.; Wrolstad, R.E. Molar absorptivity and color characteristics of acylated and non-acylated pelargonidin-based anthocyanins. *J. Agric. Food Chem.* **1999**, *47*, 4631–4637. [CrossRef] [PubMed]

13. Rizzolo, A.; Brambilla, A.; Valsecchi, S.; Eccher-Zerbini, P. Evaluation of sampling and extraction procedures for the analysis of ascorbic acid from pear fruit tissue. *Food Chem.* **2002**, *77*, 257–262. [CrossRef]
14. Cataldo, C.A.; Maroon, M.; Schrader, L.E.; Youngs, V.L. Rapid colorimetric determination of nitrate in plant tissue by titration of salicylic acid. *Commun. Soil Sci. Plant Anal.* **1975**, *6*, 71–80. [CrossRef]
15. Yemm, E.W.; Willis, A.J. The estimation of carbohydrates in plant extracts by anthrone. *Biochem. J.* **1954**, *57*, 508–514. [CrossRef] [PubMed]
16. Branca, F. Studies on some wild Brassicaceae species utilizable as vegetables in the Mediterranean areas. *Plant Genet. Resour. Newsl.* **1995**, *104*, 6–9.
17. Branca, F.; Li, G.; Goyal, S.; Quiros, C.F. Survey of aliphatic glucosinolates in Sicilian wild and cultivated Brassicaceae. *Phytochemistry* **2002**, *59*, 717–724. [CrossRef]
18. Albrecht, J. Ascorbic Acid Content And Retention In Lettuce1. *J. Food Qual.* **1993**, *16*, 311–316. [CrossRef]
19. Žnidarčič, D.; Ban, D.; Šircelj, H. Carotenoid and chlorophyll composition of commonly consumed leafy vegetables in Mediterranean countries. *Food Chem.* **2011**, *129*, 1164–1168. [CrossRef]
20. Ferrante, A.; Maggiore, T. Chlorophyll a fluorescence measurements to evaluate storage time and temperature of Valeriana leafy vegetables. *Postharvest Biol. Technol.* **2007**, *45*, 73–80. [CrossRef]
21. Fallovo, C.; Rouphael, Y.; Rea, E.; Battistelli, A.; Colla, G. Nutrient solution concentration and growing season affect yield and quality of Lactuca sativa L. var. acephala in floating raft culture. *J. Sci. Food Agric.* **2009**, *89*, 1682–1689. [CrossRef]
22. Ferrante, A.; Incrocci, L.; Maggini, R.; Serra, G.; Tognoni, F. Colour changes of fresh-cut leafy vegetables during storage. *J. Food Agric. Environ.* **2004**, *2*, 40–44.
23. Colonna, E.; Rouphael, Y.; Barbieri, G.; De Pascale, S. Nutritional quality of ten leafy vegetables harvested at two light intensities. *Food Chem.* **2016**, *199*, 702–710. [CrossRef]
24. Bulgari, R.; Cola, G.; Ferrante, A.; Franzoni, G.; Mariani, L.; Martinetti, L. Micrometeorological environment in traditional and photovoltaic greenhouses and effects on growth and quality of tomato (*Solanum lycopersicum* L.). *Ital. J. Agrometeorol.* **2015**, *2*, 27–38.
25. Maxwell, K.; Johnson, G.N. Chlorophyll fluorescence—A practical guide. *EXBOTJ* **2000**, *51*, 659–668. [CrossRef]
26. Alberici, A.; Quattrini, E.; Penati, M.; Martinetti, L.; Gallina, P.M.; Ferrante, A.; Schiavi, M. Effect of the Reduction of Nutrient Solution Concentration on Leafy Vegetables Quality Grown in Floating System. *Acta Hortic.* **2008**, *801*, 1167–1176. [CrossRef]
27. EFSA—European Food Safety Authority. *Nitrate in Vegetables. Scientific Opinion of the Panel on Contaminants in the Food Chain*; Question N. EFSA-Q-2006-071; EFSA: Parma, Italy, 2008; Volume 689, pp. 1–79.
28. European Commission. *Maximum Levels for Nitrates in Foodstuffs*; Regulation (EU) N. 1258/2011 Amending Regulation (EC) N. 1881/2006; Official Journal European Union: Brussel, Belgium, 2011; pp. 15–17.
29. Hanafy Ahmed, A.H.; Khalil, M.K.; Farrag, A.M. Nitrate accumulation, growth, yield and chemical composition of Rocket (*Eruca vesicaria* subsp. sativa) plant as affected by NPK fertilization, kinetin and salicylic acid. *Ann. Agric. Sci. Cairo* **2002**, *47*, 1–26.
30. Ferrante, A.; Incrocci, L.; Maggini, R.; Serra, G.; Tognoni, F. Preharvest and postharvest strategies for reducing nitrate content in rocket (*Eruca sativa* L.). *Acta Hortic.* **2003**, *628*, 153–159. [CrossRef]
31. Tannenbaum, S.R.; Wishnok, J.S.; Leaf, C.D. Inhibition of nitrosamine formation by ascorbic acid. *Am. J. Clin. Nutr.* **1991**, *53*, 247S–250S. [CrossRef]
32. Shad, A.A.; Shah, H.U.; Bakht, J. Ethnobotanical assessment and nutritive potential of wild food plants. *J. Anim. Plant Sci.* **2013**, *23*, 92–97.

10

How Water Quality and Quantity Affect Pepper Yield and Postharvest Quality

Elazar Fallik [1,*], Sharon Alkalai-Tuvia [1], Daniel Chalupowicz [1], Merav Zaaroor-Presman [1], Rivka Offenbach [2], Shabtai Cohen [2] and Effi Tripler [2]

1 Agricultural Research Organization, The Volcani Center, Department of Postharvest Science of Fresh Produce, Rishon Leziyyon 7505101, Israel; sharon@volcani.agri.gov.il (S.A.-T.); chalu@volcani.agri.gov.il (D.C.); merav.zaaroor@mail.huji.ac.il (M.Z.-P.)
2 Central and Northern Arava Research and Development, Arava Sapir 8682500, Israel; Rivka@arava.co.il (R.O.); sab@inter.net.il (S.C.); effi@arava.co.il (E.T.)
* Correspondence: efallik@volcani.agri.gov.il.

Abstract: There are gaps in our knowledge of the effects of irrigation water quality and amount on yield and postharvest quality of pepper fruit (*Capsicum annuum* L.). We studied the effects of water quality and quantity treatments on pepper fruits during subsequent simulated storage and shelf-life. Total yield decreased with increasing water salinity, but export-quality yield was not significantly different in fruits irrigated with water of either 1.6 or 2.8 dS/m, but there was a 30–35% reduction in export-quality yield following use of water at 4.5 dS/m. Water quantity hardly affected either total or export-quality yield. Water quality but not quantity significantly affected fruit weight loss after 14 days at 7 °C plus three days at 20 °C; irrigation with water at 2.8 dS/m gave the least weight loss. Fruits were significantly firmer after irrigation with good-quality water than with salty water. The saltier the water, the higher was the sugar content. Vitamin C content was not affected by water quality or quantity, but water quality significantly affected antioxidant (AOX) content. The highest AOX activity was found with commercial quality water, the lowest with salty water. Pepper yield benefited by irrigation with fresh water (1.6 dS/m) and was not affected by water quantity, but post-storage fruit quality was maintained better after use of moderately-saline water (2.8 dS/m). Thus, irrigation water with salinity not exceeding 2.8 dS/m will not impair postharvest quality, although the yield will be reduced at this salinity level.

Keywords: prolonged storage; salinity; shelf-life

---

1. Introduction

The amount of agricultural land destroyed by salt accumulation each year, worldwide, is estimated to be 10 million ha [1]. Furthermore, this destruction rate could be accelerated by: climate change; excessive use of groundwater; increasing use of low-quality water in irrigation; and the massive introduction of irrigation associated with intensive farming. On the other hand, it has been confirmed in many regions that the tendency to increase the efficiency of irrigation water use and to irrigate with low-quality water, because of water scarcity, can lead to accumulation of salts in the soil. It is estimated that by 2050, 50% of the world's arable land will be affected by salinity [2].

During the last decade, salinity and drought were two of the major abiotic stresses in the Arava Valley in the southern part of Israel. This region is predominantly arid and is affected by salinity because of very low rainfall (<30 mm year^{-1}), high evapotranspiration (3000 mm year^{-1}), and groundwater that is mostly saline, with an electrical conductivity (EC) about 2.8 dS/m. Moreover, the amount of water available for irrigation is declining every year; salinity is gradually increasing and there are underground water wells with more than 4 dS/m. Consequently, plant growth and yield can be negatively affected [3]. Azuma et al. [4] reported that the detrimental impact of salinity mainly affects fruits rather than leaves and stems. Thus, high salinity and water scarcity in agricultural soils present the most serious challenges faced by horticultural crops in southern Israel.

The major crop in the Arava Valley during the winter is sweet bell pepper (*Capsicum annuum* L.); about 60% of the sweet bell pepper that is designated for export from Israel is grown in this region during the fall and winter; the growth area is estimated at 2000 ha, with an average yield of about 80–120 ton ha^{-1}. Pepper plants are sensitive to drought stress and moderately sensitive to salt stress [5,6]. Nevertheless, very little is known about the influence of water quantity and quality on pepper fruit quality after harvest and prolonged storage. Therefore, the objective of the present study was to evaluate, for two consecutive years, the effects of water quantity (i.e., irrigation water), and quality (i.e., salinity) on pepper yield and fruit quality after prolonged storage and shelf-life simulation.

2. Materials and Methods

2.1. Plant Materials and Physical Design

The study was performed at Yair experimental station (30°46′45.3″ N; 35°14′31.1″ E) in a 900 m^2 greenhouse, situated in Israel's Central Arava Valley, 130 m below mean sea level. The experiment took place during the growing season 2015/2016, in which sweet red bell-pepper (*Capsicum annuum* L., cv. Cannon) was evaluated for yield, fruit quality and postharvest indicators. The local soil texture is loamy sand, having sand, silt and clay percentages of 83, 8 and 9%, respectively [7]. Two row crops of pepper seedlings were planted on 5 August 2015, in each bed and spaced 0.4 × 0.4 m. The distance between each bed was 1.6 m, which yielded a planting density of 31,250 plants·ha^{-1}. The experiments were equipped with a pressure-compensated drip irrigation system (Netafim Ltd., Hatzerim, Israel), consisting of one lateral for each crop row having an outer diameter of 0.017 m. The integrated drippers were spaced 0.2 m and their discharge was 1.6 L h^{-1}.

Prior to the planting, the greenhouse was enclosed with 25 mesh insect net, with an additional net-shading on the roof which reduced the radiation by 30%. The net shade was removed 6 weeks after the planting, followed by enclosing the greenhouse with translucent plastic (0.12 mm thick, IR—Ginegar Plastic Ltd., Kibbutz Ginegar, Israel), 1 month later. A Spanish trellising method was applied and common cultivation (leaf pruning, side shoots removal, vine-training and canopy-height adjustment) and plant protection practices were used throughout the growing season [8]. Temperature measurements records were downloaded from an adjacent Israeli Meteorological Services (IMS) meteorological weather station.

2.2. Irrigation and Yield

The experimental design was randomized blocks (n = 4), with 20 plants in each replicate. Three irrigation water salinities (EC 1, 2.8 and 4 dS·m^{-1}) and 3 water application levels were applied for each water quality (Table 1). Irrigation application levels were determined based on the long-term (2002–2014) averages of potential evapotranspiration rates of bell peppers in the Arava region. Electrical conductivity of 1 dS·m^{-1} was applied by blending local saline water (EC = 2.8 dS·m^{-1} with desalinated water, while the highest salinity level (EC of 4.5 dS·m^{-1}) was achieved by an equivalent addition of sodium chloride and calcium chloride salts to the local saline water.

Table 1. Irrigation water salinities and their specific application levels, since Day After Planting (DAP). Fertilizer solution in irrigation water contained N as total nitrogen, P as P_2O_5 and K as K_2O.

		Electrical Conductivity of the Irrigation Water ($dS{\cdot}m^{-1}$)										
		1			2.8			4				
		Water Application Levels (% from ET_p)									Fertilizer Application	
DAP	ET_p	70	100	150	100	150	200	100	200	300	N-P-K	N
	($mm{\cdot}d^{-1}$)	Daily Irrigation Water Depths ($mm{\cdot}d^{-1}$)									(%)	($mg{\cdot}L^{-1}$)
0–35	1.3	0.91	1.3	1.95	1.3	1.95	2.6	1.3	2.6	3.9	6-6-6	50
36–51	3.3	2.31	3.3	4.95	3.3	4.95	6.6	3.3	6.6	9.9	6-6-6	50
52–62	2.7	1.89	2.7	4.05	2.7	4.05	5.4	2.7	5.4	8.1	7-3-7	120
63–94	2.5	1.75	2.5	3.75	2.5	3.75	5	2.5	5	7.5	7-3-7	150
95–104	1.7	1.19	1.7	2.55	1.7	2.55	3.4	1.7	3.4	5.1	7-3-7	100
105–114	1.2	0.84	1.2	1.8	1.2	1.8	2.4	1.2	2.4	3.6	7-3-7	100
115–124	1.2	0.84	1.2	1.8	1.2	1.8	2.4	1.2	2.4	3.6	7-3-7	100
125–134	0.8	0.56	0.8	1.2	0.8	1.2	1.6	0.8	1.6	2.4	4-2-6	100
135–144	0.8	0.56	0.8	1.2	0.8	1.2	1.6	0.8	1.6	2.4	4-2-6	100
145–154	1.1	0.77	1.1	1.65	1.1	1.65	2.2	1.1	2.2	3.3	4-2-6	100
155–164	1.3	0.91	1.3	1.95	1.3	1.95	2.6	1.3	2.6	3.9	4-2-6	100
165–194	2	1.4	2	3	2	3	4	2	4	6	4-2-6	100
195–224	3	2.1	3	4.5	3	4.5	6	3	6	9	4-2-6	100
225–243	4	2.8	4	6	4	6	8	4	8	12	4-2-6	100
244–272	5	3.5	5	7.5	5	7.5	10	5	10	15	4-2-6	100

Yield data included the cumulative weight of fruits (total yield) and that of defect-free fruits (export-quality yield), from December through mid-March from all four repetitions per treatment. Results are expressed in ton ha^{-1}.

Petioles were sampled from newly fully-expanded leaves located at the 4th petiole from the apex. Approximately 20–25 petioles were collected at random from each replicate. The samples were taken between 8:00–10:00 am to minimize differences in cell turgidity of plants. The leaflets were stripped, and the petioles placed in a zip-lock bag. One mL of freshly pressed sap was diluted with 50 mL of distilled water. The solution was analyzed for chloride concentration by means of a standard chloridometer instrument.

2.3. *Postharvest Fruit Quality Parameters*

The postharvest quality was determined once monthly, at the end of December, at the beginning of February, and mid-March of each year; there were three harvests per year. Each harvest was collected in four corrugated cartons, each containing 5 kg of export-quality pepper fruits. The fruits were of uniform size of 180–200 g, at 85–90% maturity, with attached calyx and free of defects. Immediately after harvest, fruits were rinsed and brushed in hot water as described by Fallik et al. [9]. Fruit-quality parameters were evaluated immediately after each harvest and at the end of 14 days of storage at 7 °C and relative humidity (RH) of ~95%, followed by 3 days at 20 °C. Weight loss was expressed as percentage loss from the initial weight of 10 fruits. Fruit flexibility was measured by placing the fruit between two horizontal flat plates, the upper of which was loaded with a 2-kg weight, as described by Fallik et al. [9]. A dial fixed to a graduated plate recorded the deformation of the fruit in millimeters. Full deformation was measured 15 s after placing the load on the fruit, the weight was removed, and the residual deformation was measured after a further 15 s. The residual deformation directly indicated fruit elasticity: a fruit with 0–1.5 mm deformation was designated as very firm; with 1.6–3.0 mm deformation as firm; with 3.1–4.5 mm deformation as soft; and with more than 4.6 mm deformation as very soft. Total soluble solids (TSS) were measured in the five fruits that had been tested for firmness, by squeezing juice out of the fruits and recording the readings on an Atago digital refractometer

(Atago, Tokyo, Japan). A fruit was considered decayed if fungal mycelia appeared on the peel or calyx, and decay was expressed as the percentage of decayed fruits in the carton.

The vitamin C content of the bell pepper fruits was determined with the HI3850 Ascorbic Acid Test Kit (Hanna Instruments, Smithfield, RI, USA), which expresses measured quantities as milligrams per 100 g. In accordance with the test kit instructions, 2 g of fresh bell pepper fruit was homogenized with 10 mL of deionized water in a 50-mL vial at high speed for 1 min. The homogenate was passed through filter paper and kept on ice pending mixing of a 1-mL aliquot of homogenate with 49 mL deionized water in a beaker. Then 1 mL of HI3850A-0 reagent and four drops of starch as an indicator were added, and HI3850C-0 reagent was added as 10-mL drops, which were counted until a persistent blue color was developed when the beaker was swirled.

Antioxidant activity (AOX) was measured by using the discoloration method [10] based on 2,2′-azinobis (3-ethylbenzothiazoline-6-sulfonate) ($ABTS^+$) (Sigma-Aldrich, Rehovot, Israel) with slight modification. In the present study, only hydrophilic fractions were isolated from 100 mg of freeze-dried powder by stepwise extraction with acetate buffer, acetone, and hexane, and repeated partitioning of water-soluble and -insoluble portions. Antioxidant activity was evaluated by discoloration of the $ABTS^+$ radical cation. The radical was generated in acetate buffer medium at pH 4.3 to facilitate the activities of the hydrophilic antioxidants. The final reaction mixture contained 150 μmol of $ABTS^+$ and 75 μmol of potassium persulfate ($K_2S_2O_8$) in 249 mL of acetate buffer at pH 4.3. Incubation of the reaction mixture at 45 °C for 1 h was sufficient to generate $ABTS^+$. The resulting stock solution of $ABTS^+$ can be stored for up to 3 days at 4 °C without significant loss of properties. The discoloration test was performed in a 96-well microplate by adding 3 μL of test sample to 300 μL of $ABTS^+$ and comparing the optical density at 734 nm after 15 min of incubation at room temperature, with that of a blank sample. Final results were calculated by comparing the absorbance of the samples with that of the standard (±)-6-hydroxy-2,5,7,8-tetramethylchromane-2-carboxylic acid (Trolox) (Sigma-Aldrich). The antioxidant activity in the samples was determined as Trolox equivalents (TE), according to the formula

$$TE = (A_{sample} - A_{blank})/(A_{standard} - A_{blank}) \times C_{standard}$$

where A is the absorbance at 734 nm and C is the concentration of Trolox (mmol).

The TE antioxidant capacity (TEAC) per unit weight of plant tissue was calculated as follows:

$$TEAC\ (mmol\ TE/mg) = (TE \times V)/(1000 \times M)$$

in which V is the final extract volume and M is the amount of tissue extracted.

The contents of vitamin C and antioxidant activity were measured in 10 fruits taken from each treatment, at each of the three harvests each year.

2.4. Statistical Analysis

The data shown here are the means of two consecutive experiments with three harvests each year; the results were similar. The results were subjected to two-way analysis of variance (ANOVA) with JMP 11 version (SAS, Cary, NC, USA). The means were separated by using the Least Significance Difference (LSD) test at $p < 5\%$. Pairwise correlation analysis was carried out to determine the significance level of the correlation between the parameters of interest.

3. Results

3.1. Yield and Chloride in Petiole

The better the water quality, the higher was the total cumulative yield during the growing season (Figure 1); the total yield decreased as the water salinity increased. The average total yield with water of EC 1.6 dS m^{-1} was about 128-ton ha^{-1}; in water quality of 2.8 and 4.5 dS m^{-1} the average total yield was 115- and 99-ton ha^{-1}, respectively. The export-quality yields were not significantly different at both 1.6 and 2.8 dS m^{-1}, with an average yield of 70- and 65-ton ha^{-1}, respectively. Reductions of 35 and 30% in export-quality yield were observed when plants were irrigated with water at 4.5 dS m^{-1}, compared with those obtained at 1.6 and 2.8 dS m^{-1}, respectively (Figure 1). Water quantity hardly affected either total or export-quality yield, although there was a slight increase in total yield with irrigation at 1.6–1.5 dS m^{-1} and a slight decrease in total yield at 4.5–2.0 or 4.5–3.0 dS m^{-1}.

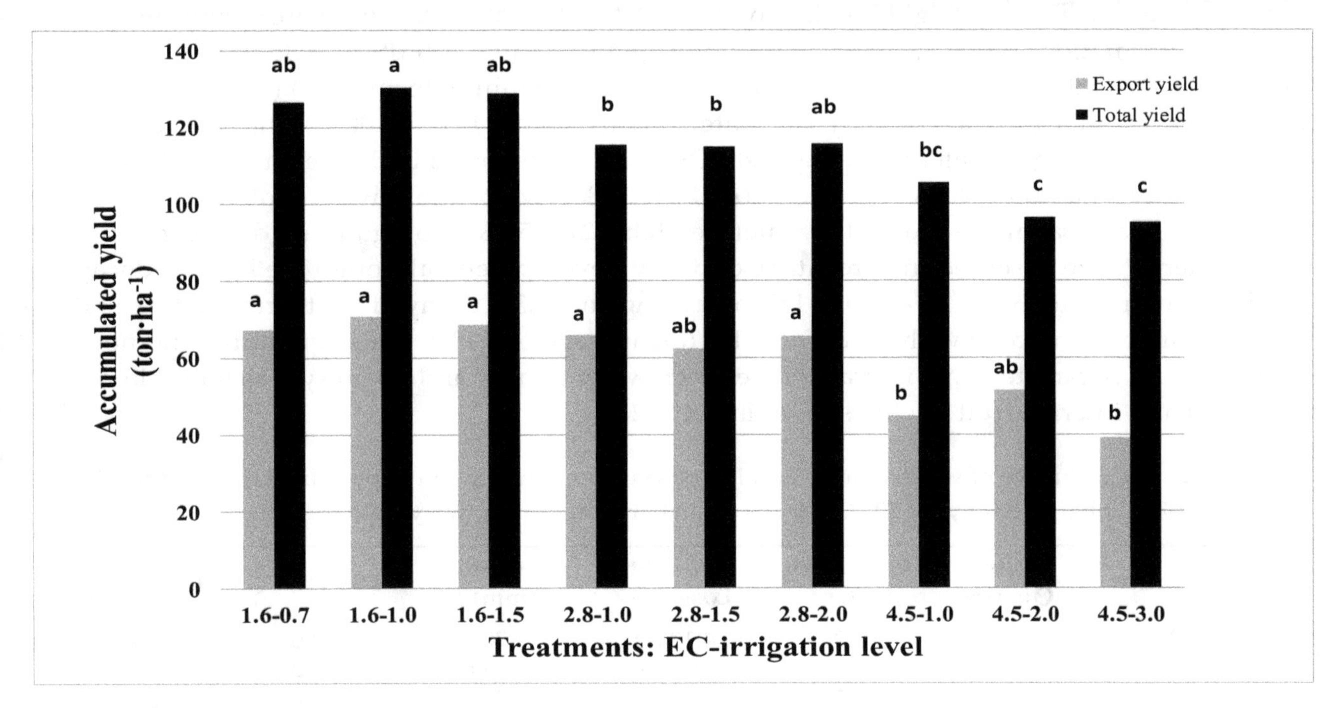

Figure 1. The influence of water quality (electrical conductivity (EC) of the irrigation—1.6, 2.8 and 4.5 dS m^{-1} EC) and water quantity on the cumulative total and export-quality yields of pepper between December and mid-March. Means of columns with the same letter are not significantly different (LSD; $p < 0.05$).

High chloride concentrations were reordered in petioles of peppers treated with low irrigation levels (0.7 and 1). Numerically, at salinities of 1, 2.8 and 4.5 dS·m^{-1}, the chloride concentrations were 150, 159 and 197.5 mg·L^{-1}, respectively. A further increase in irrigation level reduced chloride content in the petioles. However, no differences were observed between the two high water application levels in each salinity treatment (Figure 2).

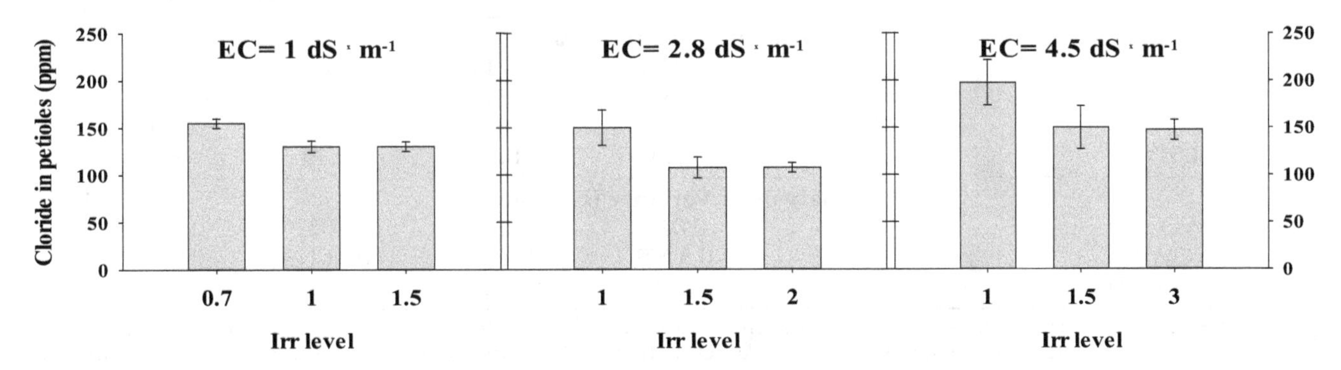

Figure 2. Chloride concentration in the petioles of pepper, treated with combinations of various salinities and irrigation (Irr) levels. Measurements were conducted in December 2015. Error bars indicate standard deviation (n = 4).

3.2. Fruit Quality

After 14 days at 7 °C and an additional three days at 20 °C, no significant differences between the treatments were observed in percentage loss of fruit weight. However, water quality, but not water quantity, affected fruit weight loss significantly (Table 2, F = 0.04). The better the water quality, the higher the weight loss (an average of 4.03%), while fruit harvested from plants irrigated at EC 2.8 dS m^{-1} had the lowest weight loss (an average of 3.55%) (Table 2). Fruit were significantly firmer (2.6 mm deformation) when irrigated with good quality water (1.6 dS m^{-1} EC), while irrigation with very salty water (EC 4.5 dS m^{-1}) gave soft fruits (3.09 mm deformation) (Table 2, F = 0.0058). The TSS was significantly affected by the water quality (Table 1, F = 0.0003); the saltier the water, the higher was the sugar content. The highest TSS content was found in the treatment of 4.5 EC − 3.0 (water salinity − amount of water. See Table 2) (8.72%), while the lowest content was found at 1.6 EC − 1.0 (water salinity − amount of water. See Table 2) (7.53%). No significant differences between the treatments were observed in percentage of decay development, although the highest decay was found in fruit irrigated with 1.6 dS m^{-1} EC (an average of 10.3% decayed fruit) and the lowest decay was found in fruit irrigated with 4.5 dS m^{-1} EC (an average of 7.1%). Water quantities did not affect all fruit quality parameters. No interaction between water quality and quantity was found in relation to external and internal fruit quality shown in Table 2.

Table 2. The influence of water quality and irrigation water amount on pepper fruit quality after 14 days at 7 °C plus three days at 20 °C. Means of six harvests during two years.

Treatment	Water Quality	Amount of Water ^{z}	Weight Loss (%) ^{y}	Flexibility (mm) ^{x}	TSS (%) ^{w}	Decay (%)
1	1.6	0.7	4.13 a ^{v}	2.70 a	7.58 b	14.5 a
2	1.6	1.0	4.05 a	2.58 a	7.53 b	9.3 a
3	1.6	1.5	3.90 a	2.52 a	7.55 b	7.2 a
4	2.8	1.0	3.53 a	2.32 a	7.83 ab	7.5 a
5	2.8	1.5	3.53 a	2.17 a	8.13 ab	6.0 a
6	2.8	2.0	3.58 a	2.43 a	8.12 ab	8.5 a
7	4.5	1.0	3.87 a	3.12 a	8.05 ab	7.0 a
8	4.5	2.0	3.77 a	3.02 a	8.37 ab	7.2 a
9	4.5	3.0	3.62 a	3.13 a	8.72 a	7.2 a
LSD			**0.31**	**0.40**	**0.32**	**5.61**
			Mean of water quality			
	1.6		4.03 a	2.60 ab	7.56 b	10.33 a
	2.8		3.55 b	2.31 b	8.03 a	7.33 a
	4.5		3.75 ab	3.09 a	8.38 a	7.11 a
LSD			**0.18**	**0.23**	**0.19**	**3.24**
			Mean of amount of water			
		Low	3.84 a	2.71 a	7.82 a	9.67 a
		Moderate	3.78 a	2.59 a	8.01 a	7.50 a
		High	3.70 a	2.69 a	8.12 a	7.61 a
LSD			**0.18**	**0.23**	**0.19**	**3.24**
			Analysis of Variance (F-Value)			
WQ ^{u}			0.04 *	0.0058 ***	0.0003 ***	0.54 NS
AOW ^{t}			0.8 NS	0.86 NS	0.31 NS	0.75 NS
WA × AOW			0.97 NS	0.87 NS	0.61 NS	0.83 NS

^{z} From evapo-transpiration; ^{y} Percentage loss from initial weight; ^{x} Deformation as measured in millimeters; ^{w} Percentage of total soluble solids (Brix°); ^{v} Values within each column followed by same letter(s) are not significantly different according to least significance difference test * ($p \leq 0.05$). * $p \leq 0.05$; ** $p \leq 0.01$; *** $p \leq 0.001$; **** $p \leq 0.0001$; NS = non-significant at $p \leq 0.05$; ^{u} Water quality (WQ); ^{t} Amount of water (AOW).

Vitamin C content was not affected by water quality or quantity, although fruits harvested from plants irrigated with water at EC 2.8 dS m^{-1} had the highest average vitamin C content

(130 mg/100 g FW) compared with the other two water qualities (Table 3). Water quality significantly affected AOX content in the fruit after 14 days of storage and marketing simulation. The average AOX activity in fruits harvested from plants irrigated at EC of 2.8 dS m^{-1} was 4.6 μM TE/g FW compared with 4.1 and 4.0 μM TE/g FW in fruits irrigated at EC of 1.6 or 4.5 dS m^{-1}, respectively. The highest AOX activity was found in the 2.8 dS m^{-1} EC-1.5 treatment (4.8 μM TE/g FW), while the lowest activity was found in the 4.5 dS m^{-1} EC-3.0 treatment (3.9 μM TE/g FW). An interaction was found in AOX activity between the water quality and quantity (F = 0.02) (Table 3).

Table 3. Influence of water quality and irrigation water amount on fruit nutritional contents after 14 days at 7 °C plus three days at 20 °C. Means of six harvests over two years.

Treatment	Water Quality	Amount of Water	Vitamin C (mg/100 g FW)	AOX TEAC (μM TE/g FW)
1	1.6	0.7	121 a [z]	4.1 cd
2	1.6	1.0	124 a	4.1 cd
3	1.6	1.5	123 a	4.3 bcd
4	2.8	1.0	124 a	4.4 abc
5	2.8	1.5	133 a	4.8 a
6	2.8	2.0	133 a	4.6 ab
7	4.5	1.0	126 a	4.2 bcd
8	4.5	2.0	119 a	4.0 cd
9	4.5	3.0	118 a	3.9 d
LSD			**10.6**	**0.13**
		Mean of water quality		
	1.6		123 a	4.1 b
	2.8		130 a	4.6 a
	4.5		121 a	4.0 b
LSD			**6.13**	**0.08**
		Mean of water amount		
		Low	123 a	4.2 a
		Middle	125 a	4.3 a
		High	125 a	4.2 a
LSD			**6.13**	**0.08**
		Analysis of Variance (F-Value)		
WQ [y]			0.33 NS	<0.0001 ****
AOW [x]			0.96 NS	0.76 NS
WA × AOW			0.81 NS	0.02 *

[z] Values within each column followed by same letter(s) are not significantly different according to least significance difference test ($p \leq 0.05$). * $p \leq 0.05$; ** $p \leq 0.01$; *** $p \leq 0.001$; **** $p \leq 0.0001$; NS = non-significant at $p \leq 0.05$.
[y] Water quality (WQ). [x] Amount of water (AOW).

In pepper fruit, the correlation coefficient indicated a significantly higher and positive relationship between weight loss and decay development at $p = 0.01$. Weight loss had a significantly high and negative relationship with vitamin C at $p = 0.0001$. Likewise, a negative and significantly higher relationship was also noted between elasticity and decay incidence at $p = 0.01$ (Table 4).

Table 4. Correlation coefficients of weight loss (WL), elasticity (Firm), sugar content (TSS), decay, vitamin C (VC) and antioxidant activity (AOX) in red pepper after 14 days at 7 °C plus three days at 20 °C.

	WL	Firm	TSS	Decay	VC
Firm	−0.065				
TSS	−0.223	0.259			
Decay	0.342 **	−0.310 **	0.193		
VC	−0.471 ****	0.099	−0.014	−0.202	
AOX	−0.101	−0.195	0.076	−0.155	0.059

*, **, ***, and **** = significant at p = 0.05, 0.01, 0.001 and 0.0001 levels, respectively.

4. Discussion

Salinity and water scarcity present crucial problems for many crop species in Mediterranean countries where water resources are the main limiting factor. In these countries, the limited water quantities available to farmers and increasing water salinity impair plant growth and yield, which depend on water quantity and quality, and may vary according to the plant genotype [11,12]. Very little is known about the effect of water quantity on postharvest fruit quality, but the influence of water salinity on fruit yield and quality is well-documented; most vegetable crops have a salinity threshold at $\leq$2.5 dS/m [13]. Pepper plants are categorized as sensitive to moderately sensitive to salinity, although Baath et al. [14] concluded that selected chili pepper cultivars can be irrigated with water of salinity $\leq$3 dS/m, during at least one growing season.

We have found that water quality was more important than water quantity in determining total and export-quality yields: high water quality (1.6 dS m^{-1}) increased yield, whereas high salinity (4.5 dS m^{-1}) significantly decreased it; in both cases water quantity did not affect pepper yields. The decrease in total yield caused by salinity was mainly due to decreases in fruit fresh weight and not to the number of fruits per plant (data not shown). A high export quality fraction from the total yield was found when the salinity increased from 1 to 2.8 dS·m^{-1}. This can be explained by the lower chloride levels measured in EC 2.8 dS·m^{-1}. Previous studies found the same trend. Rameshwaran et al. [15] reported that high salinity reduced pepper yield in two growing seasons and Yasour et al. [16] reported that high water salinity reduced pepper plant biomass and fruit yield in the Arava Valley in Israel. The reduction in total yield at high salinity can be attributed to low water content in the fruit because of poor water uptake at high salt concentration, which affects cell expansion in the growing fruit [17]. It is also possible that the decreased yield and poor fruit quality associated with high salinity are caused by poor photosynthesis, which decreases CO_2 availability as a result of diffusion limitations [18], and by decreased CO_2 conductance in the stomata and mesophyll [19]. Paranychianakis and Chartzoulakis [20] reported that salt accumulation in the root zone caused development of osmotic stress and disrupted cell ion homeostasis, thereby affecting total yield. However, Urrea-Lopez et al. [21] did not find that habanero pepper fruit yield parameters were significantly affected by low photosynthetic activity associated with water salinity, probably because of the fertilizers used in their experiment.

The best fruit quality, as judged by external and internal quality parameters, after prolonged storage and shelf-life simulation, was found at a water of salinity 2.8 dS m^{-1}. Navarro et al. [22] reported that moderately saline water was beneficial when peppers were harvested at the red stage; however, no significant differences in several quality parameters were observed between irrigation with water of 1.6 dS m^{-1} and of 2.8 dS m^{-1}. It might be that plants irrigated with fresh water (1.6 dS m^{-1}) had large canopies, which evaporated more water, thereby increasing canopy humidity, which would increase postharvest decay development because of Botrytis infection (Table 1). At very high salinity (4.5 dS m^{-1}), fruit were softer and more flexible, probably because of severe disturbances in membrane permeability, water channel activity, and stomatal conductance [23]. Salinity increased sugar levels in several crops such as melons, grapes and oranges [24–26]; we have found that the saltier the water, the higher the fruit TSS. The increase in the concentration of these sugars could be due, in part, to a loss of water from the fruit, and/or in part to increased hydrolysis of sucrose, which would yield fructose and glucose, in response to the high osmotic potential in the nutrient solution. The increase in glucose and fructose concentrations could also be associated with an active osmotic adjustment [27]. The increase in sugar level in fruits harvested from high-salinity treatments also could be attributed to the increase in starch biosynthesis in developing fruits, which is believed to increase sink strength [28].

Phytonutrients such as vitamin C or AOX capacity are increasingly important aspects of fruit quality because they are associated with benefits to consumer's health [29], and pepper is considered as one of the healthier fruits [30]. Antioxidant synthesis and accumulation in plants is generally stimulated by biotic or abiotic stress such as salinity; they can protect plant organs from serious oxidative damage to lipids, proteins, and nucleic acids [22]. In the present study, the vitamin C content was not affected by water quality or quantity, but the highest vitamin C concentration was measured in water of EC 2.8 dS m^{-1}. However, in fruits harvested from plants irrigated with water of EC 2.8 dS m^{-1}, AOX was significantly higher. These results may indicate that moderate salt treatment may significantly improve the nutritional benefits of the fruit, with respect to prevention of free-radical-related diseases, as reported by Navarro et al. [22]. On the other hand, Ehret et al. [31] reported that AOX in tomato fruits responded more strongly to light and temperature than to water salinity.

In conclusion, pepper yield was increased by fresh water of good quality (EC 1.6 dS m^{-1}) and not by water quantity, whereas fruit quality after prolonged storage was better maintained in fruits irrigated with moderately saline water, of EC 2.8 dS m^{-1}. Therefore, if the water salinity does not exceed 2.8 dS/m, postharvest quality will not be impaired, although the yield will be reduced at this salinity level. However, if water quality continues to deteriorate and becomes saltier, both pepper yield and postharvest quality will be significantly affected.

Author Contributions: E.F. was the head of the project; he planned the research, analyzed the results and wrote the manuscript with the rest of the team. D.C., S.A.-T. and M.Z.-P. are research engineers in Elazar Fallik's laboratory; they conducted the experiments, evaluated fruit quality, and analyzed the data in both 2016 and 2017. R.O., S.C., and E.T. were in charge of the planning, and growing practices, yield evaluation and harvest.

References

1. Pimentel, D.; Berger, B.; Filiberto, D.; Newton, M.; Wolfe, B.; Karabinakis, E.; Clark, S.; Poon, E.; Abbett, E.; Nandaopal, S. Water resources: Agricultural and environmental issues. *BioScience* **2005**, *54*, 909–918. [CrossRef]
2. Bartels, D.; Sunkar, R. Drought and salt tolerance in plants. *Crit. Rev. Plant Sci.* **2005**, *24*, 23–58. [CrossRef]
3. Singh, S.; Grover, K.; Begna, S.; Angadi, S.; Shukla, M.; Steiner, R.; Auld, D. Physiological response of diverse origin spring safflower genotypes to salinity. *J. Arid Land Stud.* **2014**, *24*, 169–174.
4. Azuma, R.; Ito, N.; Nakayama, N.; Suwa, R.; Nguyen, N.T.; Larrinaga-Mayoral, J.A.; Esaka, M.; Fujiyama, H.; Saneoka, H. Fruits are more sensitive to salinity than leaves and stems in pepper plants (*Capsicum annuum* L.). *Sci. Hortic.* **2010**, *125*, 171–178. [CrossRef]
5. Ben-Gal, A.; Ityel, E.; Dudley, L.; Cohen, S.; Yermiyahu, U.; Presnov, E.; Zigmond, L.; Shani, U. Effect of irrigation water salinity on transpiration and on leaching requirements: A case study for bell peppers. *Agric. Water Manag.* **2008**, *95*, 587–597. [CrossRef]
6. Lee, S.K.D. Hot pepper response to interactive effects of salinity and boron. *Plant Soil Environ.* **2006**, *52*, 227–233.
7. Tripler, E.; Haquin, G.; Koch, J.; Yehuda, Z.; Shani, U. Sustainable agricultural use of natural water sources containing elevated radium activity. *Chemosphere* **2014**, *104*, 205–211. [CrossRef]
8. Suissa, A.; Silverman, D.; Friedman, O.; Tzieli, Y.; Cohen, S.; Ofenbach, R. *Recommendation to Grow Spring Pepper in the Arava*; Ministry of Agriculture and Rural Development, Extension Services, Vegetable Ward: Hanoi, Vietnam, 2017; p. 2. (In Hebrew)
9. Fallik, E.; Grinberg, S.; Alkalai, S.; Yekutieli, O.; Wiseblum, A.; Regev, R.; Beres, H.; Bar-Lev, E. A unique rapid hot water treatment to improve storage quality of sweet pepper. *Postharvest Biol. Technol.* **1999**, *15*, 25–32. [CrossRef]
10. Vinokur, Y.; Rodov, V. Method for determining total (hydrophilic and lipophilic) radical-scavenging activity in the same sample of fresh produce. *Acta Hortic.* **2006**, *709*, 53–60. [CrossRef]
11. Bie, Z.; Ito, T.; Shinohara, Y. Effects of sodium sulfate and sodium bicarbonate on the growth, gas exchange and mineral composition of lettuce. *Sci. Hortic.* **2004**, *99*, 215–224. [CrossRef]
12. Gurmani, A.R.; Khan, S.U.; Ali, A.; Rubab, T.; Schwinghamer, T.; Jilani, G.; Farid, A.; Zhang, J. Salicylic acid and kinetin mediated stimulation of salt tolerance in cucumber (*Cucumis sativus* L.) genotypes varying in salinity tolerance. *Hortic. Environ. Biotechnol.* **2018**, *59*, 461–471. [CrossRef]
13. Machado, R.M.A.; Serralheiro, R.P. Soil salinity: Effect on vegetable crop growth. Management practices to prevent and mitigate soil salinization. *Horticulturae* **2017**, *3*, 30. [CrossRef]
14. Baath, G.S.; Shukla, M.K.; Bosland, P.W.; Steiner, R.L. Irrigation water salinity influences growth stages of *Capsicum annuum*. *Agric. Water Manag.* **2017**, *179*, 246–253. [CrossRef]

15. Rameshwaran, P.; Tepe, A.; Yazar, A.; Ragab, R. The effect of saline irrigation water on the yield of pepper: Experimental and modelling study. *Irrig. Drain.* **2015**, *64*, 41–49. [CrossRef]
16. Yasour, H.; Tamir, G.; Stein, A.; Cohen, S.; Bar-Tal, A.; Ben-Gal, A.; Yermiyahu, U. Does water salinity affect pepper plant response to nitrogen fertigation? *Agric. Water Manag.* **2017**, *191*, 57–66. [CrossRef]
17. Rubio, J.S.; Garcia-Sanchez, F.; Rubio, F.; Martinez, V. Yield, blossom-end rot incidence, and fruit quality in pepper plants under moderate salinity are affected by K^+ and Ca^{2+} fertilization. *Sci. Hortic.* **2009**, *119*, 79–87. [CrossRef]
18. Flexas, J.; Diaz-Espejo, A.; Galmés, J.; Kaldenhoff, R.; Medrano, H.; Ribas-Carbo, M. Rapid variations of mesophyll conductance in response to changes in CO_2 concentration around leaves. *Plant Cell Environ.* **2007**, *30*, 1284–1298. [CrossRef]
19. Ashraf, M.; Harris, P.J.C. Photosynthesis under stressful environments: An overview. *Photosynthetica* **2013**, *51*, 163–190. [CrossRef]
20. Paranychianakis, N.V.; Chartzoulakis, K.S. Irrigation of Mediterranean crops with saline water: From physiology to management practices. *Agric. Ecosyst. Environ.* **2005**, *106*, 171–187. [CrossRef]
21. Urrea-Lopez, R.; Diaz de la Garza, R.I.; Valiente-Banuet, J.I. Effects of substrate salinity and nutrient levels on physiological response, yield, and fruit quality of habanero pepper. *HortScience* **2014**, *49*, 812–818.
22. Navarro, J.M.; Flores, P.; Garrido, C.; Martınez, V. Changes in the contents of antioxidant compounds in pepper fruits at different ripening stages, as affected by salinity. *Food Chem.* **2006**, *96*, 66–73. [CrossRef]
23. Aktas, H.; Abak, K.; Cakmak, I. Genotypic variation in the response of pepper to salinity. *Sci. Hortic.* **2006**, *110*, 260–266. [CrossRef]
24. Botıa, P.; Navarro, J.M.; Cerda, A.; Martınez, V. Yield and fruit quality of two melon cultivars irrigated with saline water at different stages of development. *Eur. J. Agron.* **2005**, *23*, 243–253. [CrossRef]
25. Grieve, A.M.; Prior, L.D.; Bevington, K.B. Long-term effects of saline irrigation water on growth, yield, and fruit quality of 'Valencia' orange trees. *Aust. J. Agric. Res.* **2007**, *58*, 342–348. [CrossRef]
26. Li, X.L.; Wang, C.R.; Li, X.Y.; Yao, Y.X.; Hao, Y.J. Modifications of *Kyoho grape* berry quality under long-term NaCl treatment. *Food Chem.* **2013**, *139*, 931–937. [CrossRef] [PubMed]
27. Sato, S.; Sakaguchi, S.; Furukawa, H.; Ikeda, H. Effects of NaCl application to hydroponic nutrient solution on fruit characteristics of tomato (*Lycopersicon esculentum* Mill). *Sci. Hortic.* **2006**, *109*, 248–253. [CrossRef]
28. Petreikov, M.; Yeselson, L.; Shen, S.; Levin, I.; Schaffer, A.A.; Efrati, A.; Bar, M. Carbohydrate balance and accumulation during development of near-isogenic tomato lines differing in the AGPase-L1 allele. *J. Am. Soc. Hortic. Sci.* **2009**, *134*, 134–140.
29. Laribi, A.I.; Palou, L.; Intrigliolo, D.S.; Nortes, P.A.; Rojas-Argudo, C.; Taberner, V.; Bartual, J.; Pérez-Gago, M.B. Effect of sustained and regulated deficit irrigation on fruit quality of pomegranate cv. "Mollar de Elche" at harvest and during cold storage. *Agric. Water Manag.* **2013**, *125*, 61–70. [CrossRef]
30. Elmann, A.; Garra, A.; Alkalai-Tuvia, S.; Fallik, E. Influence of organic and mineral-based conventional fertilization practices on nutrient levels, anti-proliferative activities and quality of sweet red peppers following cold storage. *Isr. J. Plant Sci.* **2016**, *63*, 51–57. [CrossRef]
31. Ehret, D.L.; Usher, K.; Helmer, T.; Block, G.; Steinke, D.; Fret, B.; Kuang, T.; Diarra, M. Tomato fruit antioxidants in relation to salinity and greenhouse climate. *J. Agric. Food Chem.* **2013**, *61*, 1138–1145. [CrossRef]

11

Automation for Water and Nitrogen Deficit Stress Detection in Soilless Tomato Crops based on Spectral Indices

Angeliki Elvanidi, Nikolaos Katsoulas * and Constantinos Kittas

Department of Agriculture Crop Production and Rural Environment, University of Thessaly, Fytokou Str., 38446 Volos, Greece; elaggeliki@gmail.com (A.E.); ckittas@uth.gr (C.K.)
* Correspondence: nkatsoulas@uth.gr.

Abstract: Water and nitrogen deficit stress are some of the most important growth limiting factors in crop production. Several methods have been used to quantify the impact of water and nitrogen deficit stress on plant physiology. However, by performing machine learning with hyperspectral sensor data, crop physiology management systems are integrated into real artificial intelligence systems, providing richer recommendations and insights into implementing appropriate irrigation and environment control management strategies. In this study, the Classification Tree model was used to group complex hyperspectral datasets in order to provide remote visual results about plant water and nitrogen deficit stress. Soilless tomato crops are grown under varying water and nitrogen regimes. The model that we developed was trained using 75% of the total sample dataset, while the rest (25%) of the data were used to validate the model. The results showed that the combination of *MSAVI*, *mrNDVI*, and *PRI* had the potential to determine water and nitrogen deficit stress with 89.6% and 91.4% classification accuracy values for the training and testing samples, respectively. The results of the current study are promising for developing control strategies for sustainable greenhouse production.

Keywords: remote sensing; hyperspectral; reflectance index; classification tree; machine vision

---

1. Introduction

Nowadays, the need to produce more food with fewer inputs (water, fertilizer, and land, among others) and zero effect on the environment has led to an increase in greenhouse production. However, to cultivate under greenhouse conditions is certainly not an easy task since performance of several farming practices are needed. In this manner, crop yield and quality optimisation will be further improved only if farmers integrate greenhouse control strategies, that is, measurements concerning the dynamic response of plants according to the greenhouse's spatial environment changes.

Within a greenhouse, however, it is a challenge to quantify the spatial impact of biotic or abiotic factors in plant growth. Usually, up to now, environmental patterns under greenhouses have been monitored and managed by sampling at a single position and by considering the indoor microclimate completely homogeneous [1]. This assumption, however, is not valid since an intense heterogeneity that must be taken into account actually occurs, especially in intensive production systems. A sensing system equipped with a multi-sensor platform moving over the canopy is the key to communicate the plant's real state and needs. Obviously, in such moving platforms, the continuous monitoring of the interactions between the microclimate and the physical conditions of the plants is performed using mostly non-contact and non-destructive sensing techniques [2].

Up to now, the methods used for monitoring plant physiology have been quite complex and problematic. Most of these measure plant physiology in a limited spatial scale and, in most cases, they require physical contact with the plants/soil or follow destructive sensing procedures, making their application as a commercial multi-sensor scale rather infeasible [2–4].

The current development of computational hyperspectral machine vision systems allows us to build a real-time plant canopy health, growth, and quality monitoring multisensory platform equipped with remote technologies [5–7]. Hyperspectral machine vision systems allow recording in large scale, the spatial interaction of sunlight with crop canopies and plant leaves providing valuable information about plant growth and health status [6,8]. In this way, changes observed in plant irradiance of visible (*VIS*) and near infrared (*NIR*) spectrum, indicate different types of plant stress. The reflectance variation in *VIS* spectrum, for instance, is recorded to assess a series of several pigments located in the mesophyll area, such as chlorophyll, carotenoids and xanthophyll. The reflectance variation, on the other hand, observed in the *NIR* spectrum, is recorded to obtain information about leaf water content stored in cavities of spongy parenchyma. Additionally, changes in *NIR* spectrum are used to assess the carbon content in different forms (sugar, starch, cellulose and lignin) in mesophyll cells and nutrient compounds (N, P, K) in mesophyll cells and palisade parenchyma [2].

In order to provide more precise information for stress detection, certain parts of the spectrum can be combined to form reflectance indices (*RIs*) [2,7,9,10]. In this way, the spectral differences detected are amplified while the resulting *RIs* are used directly as a metric to quantify different aspects of plant physiology response.

So far, numerous successful case studies related to RIs and their relationship with crop, climate, or soil data from different plants have been performed. The photochemical reflectance index ($PRI = (R_{531} - R_{570})/(R_{531} + R_{570})$) is one of the most widespread indicators estimating rapid changes in de-epoxidation of the xanthophyll cycle [11–14]. Bajwa et al. [15] used the Modified soil-adjusted vegetation index ($MSAVI = 1/2 \times (2 \times (R_{810} + 1) - (\sqrt{(2 \times R_{810} + 1) \times 2 - 8 \times (R_{810} - R_{690})}))$) to predict green biomass ($R^2 > 0.70$). Elvanidi et al. [1,7,13] also showed that Modified red normalised vegetation and simple ration indices ($mrNDVI = (R_{750} - R_{705})/(R_{750} + R_{705} - 2 \times R_{445})$; $mrSRI = (R_{750} - R_{445})/(R_{705} - R_{445})$) were the two most relevant and sensitive indices indicating water deficit stress in tomato plants. Additionally, the Transformed chlorophyll absorption in the reflectance index ($TCARI = 3 \times [(R_{700} - R_{670}) - 0.2 \times (R_{700} - R_{550}) \times (R_{700}/R_{670})]$ was strongly correlated with leaf chlorophyll content variation [13,16].

In addition to reflectance indices, many statistical and machine learning models such as classification tree analysis, support vector machine, artificial neural network, and other classification procedures have been developed to extract optimal information from remotely sensed data [1]. Researchers observed that the classification tree (*CT*) was a very useful model to analyze complex data sets by providing visual results [13,17–19].

Therefore, the object of this work was to develop a model based on the *CT* method to analyse complex *RIs* datasets in order to provide visual assessments of plant water and nitrogen deficit stress. For this reason, *RIs* that indicate different aspects of tomato crop physiology (such as photosynthetic rate (As, $\mu mol\ m^{-1}\ s^{-1}$), chlorophyll_a content (Chl_a, $\mu g\ cm^{-2}$), nitrogen content (N, %) and substrate water content (θ, %)) under a controlled environment, were classified to predict remotely the (i) plant chlorophyll content, (ii) plant water content status and to identify (iii) healthy, water- and nitrogen-deficit stressed plants. The applied objective of this work was to develop a model based on simplified reflectance indices that could be adapted by multisensory platform methodologies to predict future irrigation events.

2. Materials and Methods

2.1. Experimental Set-Up

The experiment was carried out during August of 2014 and April of 2016 in a controlled growth chamber located in Velestino, Central Greece, with a ground area of 28 m^2 (4 m $\times$ 7 m) and height of 3.2 m. Air temperature, relative humidity, light intensity and CO_2 concentration were automatically controlled using a climate control computer (Argos Electronics, Athens, Greece). The light intensity was controlled using 24 high-pressure sodium lamps, 600 W each (MASTER GreenPower 600 W EL 400 V Mogul 1SL, Philips, Eindhoven, The Netherlands), operated in four clusters with six lamps per cluster. The average irradiance when all 24 lamps were used was 240 W m^{-2} (about 350 μmol m^{-1} s^{-1}).

Tomato plants (*Solanum lycopersicum* cv. Elpida, provided by Spyrou SA, Athens, Greece) were grown in slabs filled with perlite (ISOCON Perloflor Hydro 1, ISOCON S.A., Athens, Greece), at different time periods. Two units comprised of two crop lines each (18 plants per line) were used. The precise cultivation practices followed are described in References [1,7,13].

The nutrient solution was supplied via a drip system and was controlled by a time-program irrigation controller (8 irrigation events per day, at 07:00, 10:00, 102:00, 14:00, 16:00, 18:00, 19:30, and 03:30, local time), with set-points for electrical conductivity (*EC*) at 2.4 dS m^{-1} and a pH of 5.6.

In order to quantify the plant physiological response to their environment by crop reflectance characteristics, the plants had about 10 leaves each, were about 1 m in height, had a leaf area index (*LAI*) of about 0.8, and were under the imposition of varying water and nitrogen regimes. A nutrient solution containing from 0 to 100% coverage of plant actual water and nitrogen needs was supplied to the root zone of the plants for several days, varying the chlorophyll_a, nitrogen and substrate water content from 44.62 to 36.04 μg cm^{-2}, 4.77–3.30% and 54.81–35%, respectively. The control plants were irrigated with a nutrient solution with 100% of plant water and nitrogen according to Reference [6] mineral nutrient list (irrigation dose of 120 mL per plant; nutrient solution of 12.9 mmol NO_3 L^{-1} and 1.0 mmol NH_4 L^{-1}; 8 events per day. The concentrations of the rest of the macronutrients in the control treatment were K 7.5 mmol L^{-1}, Ca 4.8 mmol L^{-1}, Mg 2.5 mmol L^{-1}, H_2PO_4 1.5 mmol L^{-1}).

To create a series of plant physiological dataset groups, reflectance measurements (r) along with measurements of plant physiology such as θ (%), *Chl_a* (μg cm^{-2}), and N (%), values were obtained for the same set of plant water and nitrogen characteristics. The measurements were carried out in young and fully developed leaves between the 3rd and 6th branches of three adjacent tomato plants. The resulting correlations between the factors were further presented in References [1,7,13]. In total, 160 groups were performed throughout the period considered, under known conditions of water and nitrogen supply and environmental conditions.

2.2. Measurements

Air temperature (T, °C) and relative humidity (RH, %) were measured using two temperature-humidity sensors (model HD9008TR, Delta Ohm, Caselle di Selvazzano, Italy). Irradiance (Rg,i, W m^{-2}) inside the growth chamber was recorded using a solar pyranometer (model SKS 1110, Skye instruments, Powys, UK). The sensors were calibrated before the experimental period and placed 1.8 m above ground level. The data were automatically recorded in a data logger system (Zeno 3200, Coastal Environmental Systems Inc., Seattle, WA, USA). Substrate volumetric water content (θ, %) was estimated using capacitance sensors (model WCM-control, Rockwool B.V., Roermond, The Netherlands) placed horizontally in the middle (height and width) of the hydroponic slabs. Measurements were performed every 30 s and 10-min average values were recorded.

In plants with known θ values, leaf chlorophyll content measurements were recorded by means of an Opti-Science sensor performing measurements in contact with the leaf (CCM 200, Opti-Science, Hudson, NH, USA). The values recorded by means of the CCM 200 sensor were correlated with *Chl_a* values (μg cm^{-2}) obtained in the lab for the same set of leaves using the Reference [20] protocol. The resulted equation was presented in Reference [13].

The nitrogen content in the plant tissue (leaf dry matter sample of the entire tomato plant) was also analysed in the laboratory with the Total Kjeldahl Nitrogen method (TKN) using [21] protocol. N determination was done with an automatic flow injection analyser system (FIAstar 5000 analyser, Tecator, Foss, Hillerød, Denmark). The impact of varying nitrogen regimes in plant tissue was further examined by Reference [1].

The radiation reflected by the plants were recorded with two spectra sensors: (1) a portable spectroradiometer (model ASD FieldSpec Pro, Analytical Spectral Devices, Boulder, CO, USA) and (2) a hyperspectral camera Imspec V10 (Spectral Imaging Ltd., Oulu, Finland). The spectroradiometer measures the radiation reflected in the range between 350 and 2500 nm, while the camera operates in the visible and near-infrared (*VNIR*) spectrum region between 400 and 1000 nm. The camera system was placed on a moving cart so that images of the vertical canopy axis could be obtained to cover the canopy area of young, fully developed leaves between the 3rd and 6th branches of three adjacent tomato plants. For extra illumination of the target area (70 × 100 cm), four quartz-halogen illuminators (500 W each) were used.

The system calibration procedures of both spectral sensors along with the method concerned the camera's set up, image segmentation, and plant reflectance calculation was done as described in References [1,2,7,13,22]. The basic steps of the experimental setup are described in Figure 1.

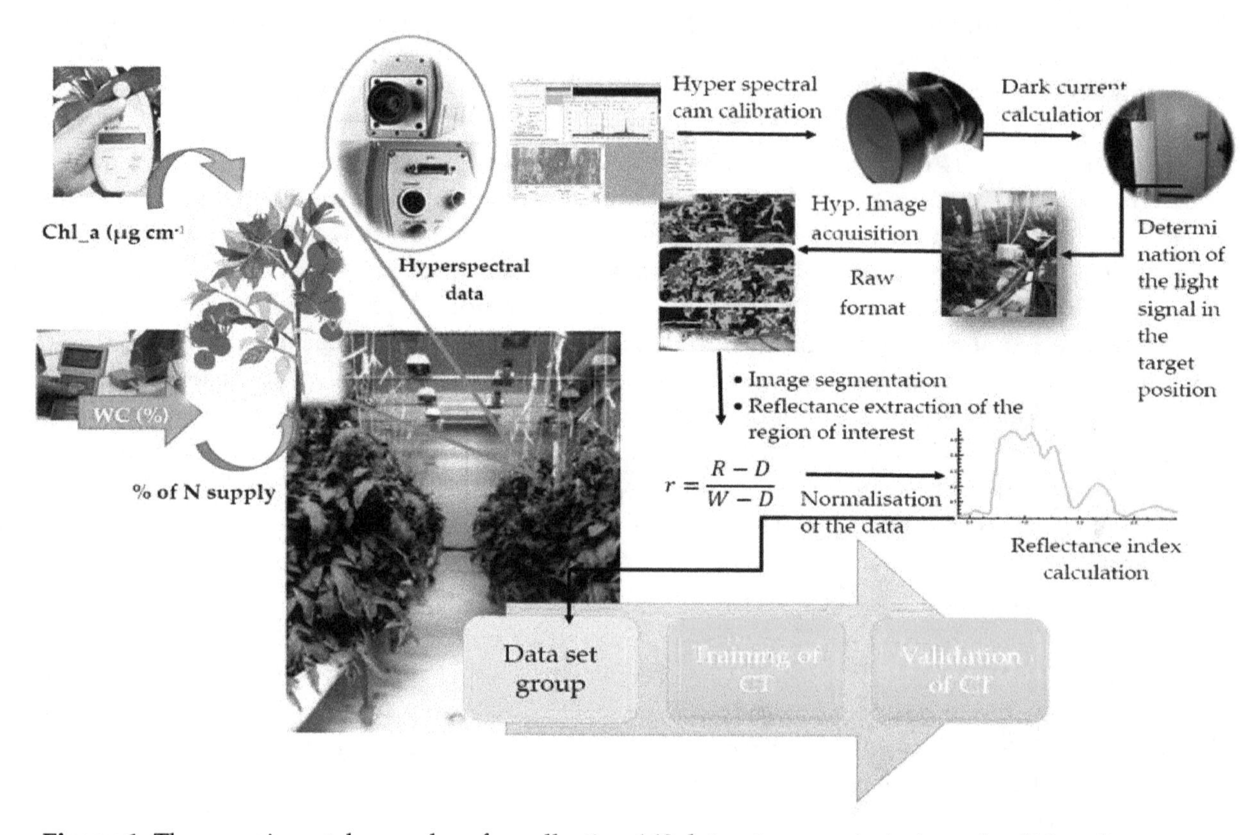

Figure 1. The experimental procedure for collecting 160 dataset groups to train and validate the system.

2.3. Calculations

Based on the available reflectance measurements, the following indices (based on the analysis performed in References [1,7,13]) were calculated and evaluated to train the models:

$$PRI = (R_{531} - R_{570})/(R_{531} + R_{570}) \quad (1)$$

$$TCARI = 3 \times [(R_{700} - R_{670}) - 0.2 \times (R_{700} - R_{550}) \times (R_{700}/R_{670})] \quad (2)$$

$$mrNDVI = (R_{750} - R_{705})/(R_{750} + R_{705} - 2 \times R_{445}) \quad (3)$$

$$mrSRI = (R_{750} - R_{445})/(R_{705} - R_{445}) \quad (4)$$

$$MSAVI = 1/2 \times (2 \times (R_{810} + 1) - (\sqrt{}(2 \times R_{810} + 1) \times 2 - 8 \times (R_{810} - R_{690}))] \quad (5)$$

$$OSAVI = (1 + 0.16) \times (R_{800} - R_{670})/(R_{800} + R_{670} + 0.16) \quad (6)$$

where R is the reflectance value performed in each band expressed in nm that is indicated by the subscript number, *mrSRI* is the Modified red simple ration index and *OSAVI* is the Optimised soil-adjusted vegetation index.

2.4. Statistical Analysis

A classification tree was performed using SPSS (Statistical Package for the Social Sciences, IBM, USA) to create a tree-based prediction model of future irrigation events. Based on *CT* methodology, three hypotheses based on RI prediction rules were used to predict (i) the plant chlorophyll_a content (*1st Hypothesis*); (ii) the substrate water content (*2nd Hypothesis*), and (iii) to identify healthy, water- and nitrogen-deficit stressed plants (*3rd Hypothesis*). Each hypothesis independently consisted of structure trees, in which different *RIs* were involved. The *CTs* developed were based on the classification regression tree (*CRT*) and the chi-squared automatic interaction detection (*CHAID*) method to control the number of *RIs* and the maximum number of levels of growth beneath the root. The p-value was computed each time by applying Bonferroni adjustments. The method followed was done as described in References [13,23].

The models developed were calibrated (using training data) and validated (on a different set of data). A simple-sample validation was performed by using random assignment. A total of 75% of the total (n = 140 data sets) number of datasets were used as the training sample sets and 25% were used as the testing sample sets. The training sets were used to build the classification models, which were subsequently applied to the test set, which consisted of records with unknown class labels. The system tries to decrease the training error by completely fitting all the training examples. Each partition of each hypothesis was marked as the class label. During the 1st and 2nd Hypothesis, the class label was marked as *Chl_a* and θ (the numerical variable) and was expressed in $\mu g\ cm^{-2}$ and %, respectively, to predict the actual chorophyll_a and water status of the plant. To train the model, the measured *Chl_a* and θ values were considered as dependent variables.

During the 3rd Hypothesis, each partition was marked as either C (Control), *WS* (Water Stress), or *NS* (Nitrogen Stress) to answer the question if the crop is under water or nitrogen deficit stress. Thus, it could be considered that C referred to "no stressed plants/no irrigation or fertilization is needed", WS referred to "water stress plants/irrigation is needed", while *NS* referred to "nitrogen stress plants/nitrogen is needed". The model was built according to substrate water and chlorophyll_a content evolution, in which θ values lower than 39% have been defined as *WS*. However, when the θ values were higher than 39%, while the *Chl_a* values were lower than 40 $\mu g\ cm^{-2}$, then the plant was defined as *NS*. These set points were derived from the analyses in References [1,7,13]. The algorithm employed a greedy strategy to grow the decision tree by making a series of locally optimum decisions about which attribute to use for the partitioning data. Each node of the training or testing samples showed the predicted value, which was the mean value for the dependent variable at that node. The mean value along with the measurement variability (standard deviation, $\pm SD$) of the parameters measured are reported. The measure of the tree's predictive accuracy was calculated based on a risk estimation and its standard error, where the proportion of cases were incorrectly classified after adjustment for prior probabilities and misclassification costs. The letter "n" is used to designate the daily sample size of each parameter. The goal of the classification models was to predict the class label of the unknown records. The current methodology followed the steps performed by Morgan [19], Loh [23], Lewis [24] and IBM SPSS Statistics 21 guide [25].

3. Results

3.1. Automation for Plant Chlorophyll_a Content Measuring

During the model analysis of the *1st Hypothesis* where the *Chl_a* status of the tomato is predicted, two independently structured trees with different combinations of *RIs* (*CT1* and *CT2*) were developed (Figures 2 and 3). Both resulting structures consisted of eight nodes (two nodes with at least one child and four nodes without children). For each node, there is a table that provides the number (n) and the percentage (%) of *Chl_a* content cases in each reflectance index category set as a dependent variable.

According to *CT1*, it was calculated that when *TCARI* was $\leq$0.0949, then the *Chl_a* content value was more than 43.61 μg cm^{-2} (Figure 2). Additionally, when the *TCARI* ranged between 0.0949 and 0.1029, the *Chl_a* content varied close to 42.82 μg cm^{-2} ($SD = \pm1$). Since there were no child nodes below it, this was considered as the terminal node.

On the other hand, if the *TCARI* varied between 0.1029 and 0.1386, the *PRI* readings (next best predictor) had to be taken into consideration to identify the plant chlorophyll status and that node three was to be omitted. In this case, when *PRI* readings were $\leq$0.0321, 23% of the sample returned *Chl_a* equal to 42.4 μg cm^{-2} ($SD = \pm0.8$), otherwise ($PRI > 0.03$), 17% of the sample returned *Chl_a* equal to 41.4 μg cm^{-2} ($SD = \pm0.9$). For *TCARI* readings between 0.14 and 0.15 (node 4) and >0.15 (node 5), *Chl_a* was equal to 39.8 μg cm^{-2} ($SD = \pm2.0$) and 37.9 μg cm^{-2} ($SD = \pm1.5$), respectively.

According to *CT2* (Figure 3), it was calculated that when *TCARI* was $\leq$0.0880 and between 0.0880 and 0.0984, then nodes 1 and 2 returned similar *Chl_a* content values to *CT1*. On the other hand, if the *TCARI* varied between 0.0984 and 0.1381, the *mrSRI* readings (next best predictor) had to be taken into consideration to identify the plant chlorophyll status and that node three was to be omitted. In this case, when *mrSRI* readings were $\leq$13.6523, 29% of the sample returned *Chl_a* equal to 42.67 μg cm^{-2} ($SD = \pm0.4$), otherwise ($mrSRI > 13.6523$) 21.5% of the sample returned *Chl_a* equal to 41.73 μg cm^{-2} ($SD = \pm0.9$).

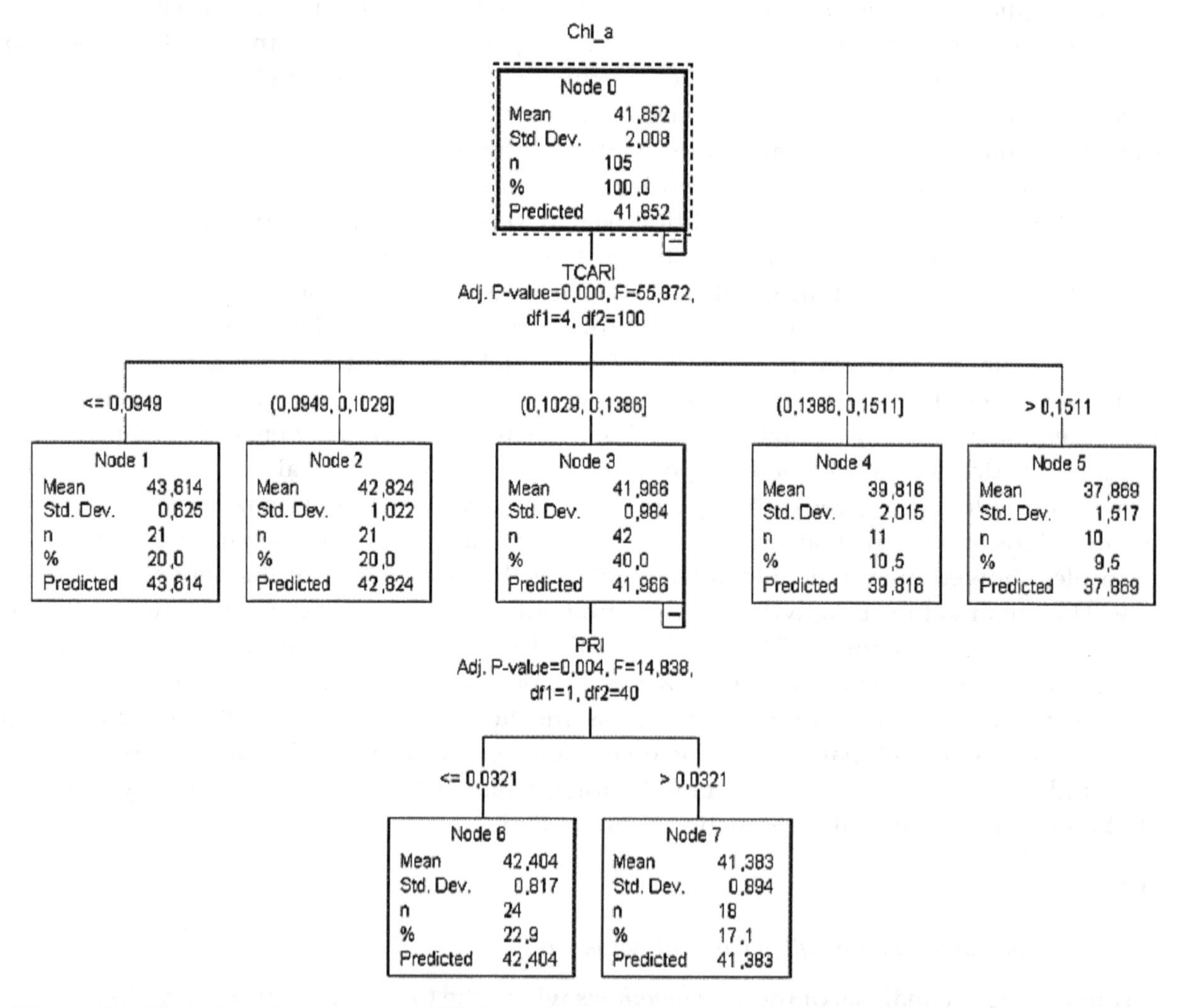

Figure 2. The classification tree (*CT1*) of the training sample for calibration to predict *Chl_a* concentration in the leaf area (μg cm^{-2}).

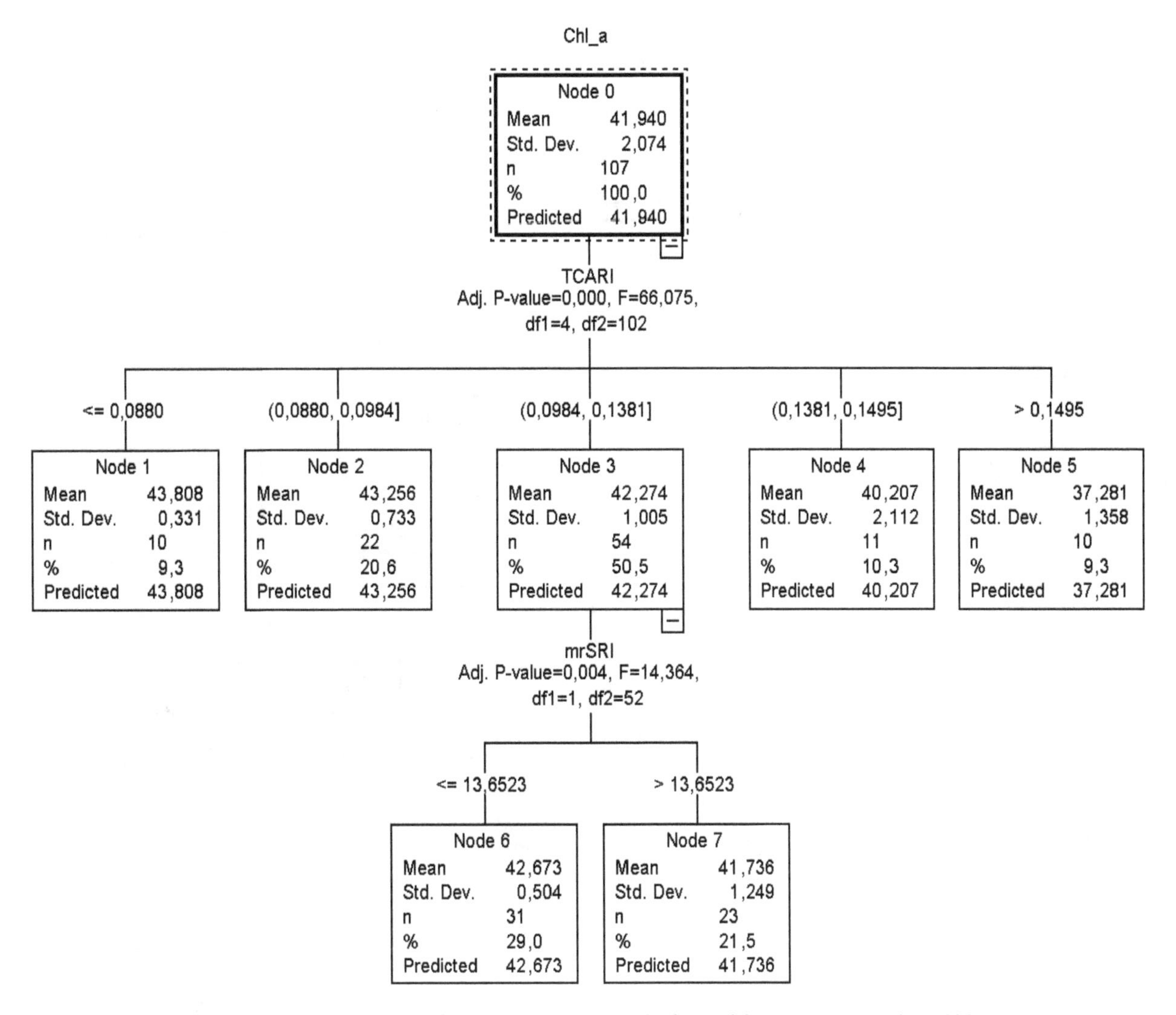

Figure 3. The *CT2* classification tree of the training sample for calibration to predict *Chl_a* concentration in the leaf area (μg cm^{-2}).

Both *CT1* and *CT2* structures did, however, reveal one potential problem with this model: for the plants that had a low level of *Chl_a* content, the standard deviation was high ($>\pm 1.0$), which means that the data were widely spread and more than 20% of the predicted percentage inaccurately classified the *Chl_a* values lower than 39.8 μg cm^{-2}. This was also confirmed by the relevant node of the testing classification trees, where the *SD* was more than ± 2. A comparison of the two structures, however, revealed that both *CT1* and *CT2* had a low estimation risk degree equal to 1.1. The trees resulting from the validation dataset provided an estimation risk degree equal to those of the training dataset (1.5 and 1.7, respectively).

3.2. *Automation for Substrate Water Content Measuring*

In the *2nd Hypothesis*, the tree (*CT3*) developed from both the training and testing samples had eleven nodes (five nodes with at least one child and six nodes without children) (Figure 4). In this tree, the predicted category was the estimation of the substrate water content (varied from 54.81 to 35%).

It was calculated that when *OSAVI* (a starter index) was ≤ 0.70, the tree returned the θ value as > 49.33%. Since there were no child nodes below it, this was considered the terminal node.

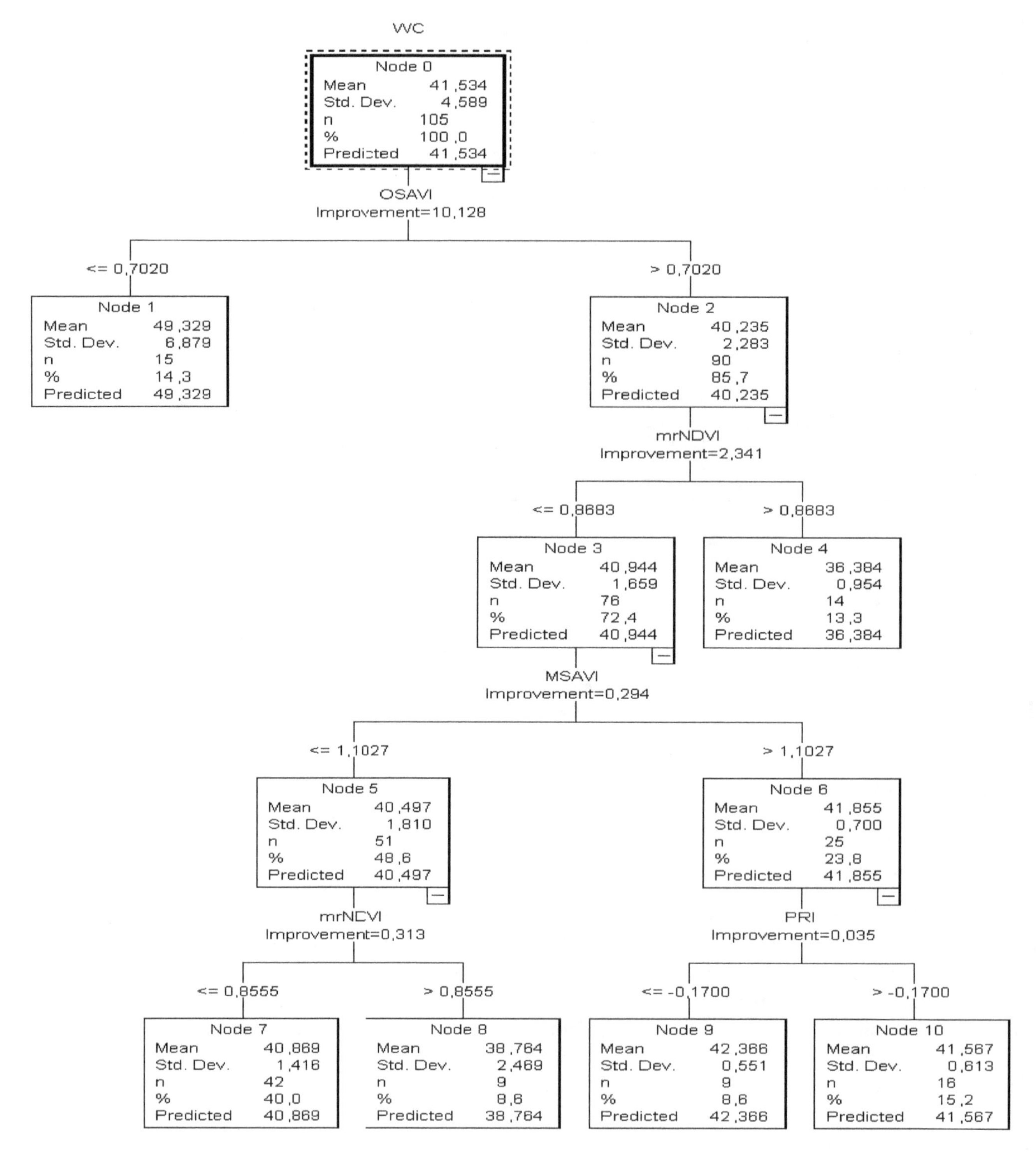

Figure 4. The *CT3* classification tree of the training sample for calibration to predict the plant θ values in the root zone (%).

On the other hand, if the *OSAVI* was higher than 0.7020, the *mrNDVI* readings (next best predictor) had to be taken into consideration to estimate the substrate water status and that node two was to be omitted. In this case, when *mrNDVI* readings were $\leq$0.8683, the sample of the plants returned θ as equal to "36.38%" (13.3%) and the node was considered terminal as no child node developed below. However, by contrast, when *mrNDVI* was less than 0.8683, a combination of *MSAVI* and *PRI* readings had to be taken into account to estimate the substrate water content. This was also confirmed by the relevant node of the testing classification tree. However, the classification tree developed from the training data revealed a high estimation risk degree (7.06), while the risk degree in the tree based on the validation dataset was high as well (5.4). The high values of the risk estimation indicated that the accuracy of this structure was uncertain due to the θ values' estimation and that more water content values should be included. The goal was to reduce the risk values to around 1.

3.3. Automation for Water and Nitrogen Deficit Stress Detection

During the model analysis of the *3rd Hypothesis*, two independently structured trees with different combinations of RIs (*CT4* and *CT5*) were developed (Figures 5 and 6). *CT4* consisted of nine nodes (four nodes with at least one child and five nodes without children) and *CT5* of seven nodes (three nodes with at least one child and four nodes without children). In those trees, the predicted category was the detection of water and nitrogen deficit stress (C or *WS* or *NS*). For each node, the mentioned table provides the number (*n*) and the percentage (%) of the C or *WS* or *NS* cases in each reflectance index category set as a dependent variable.

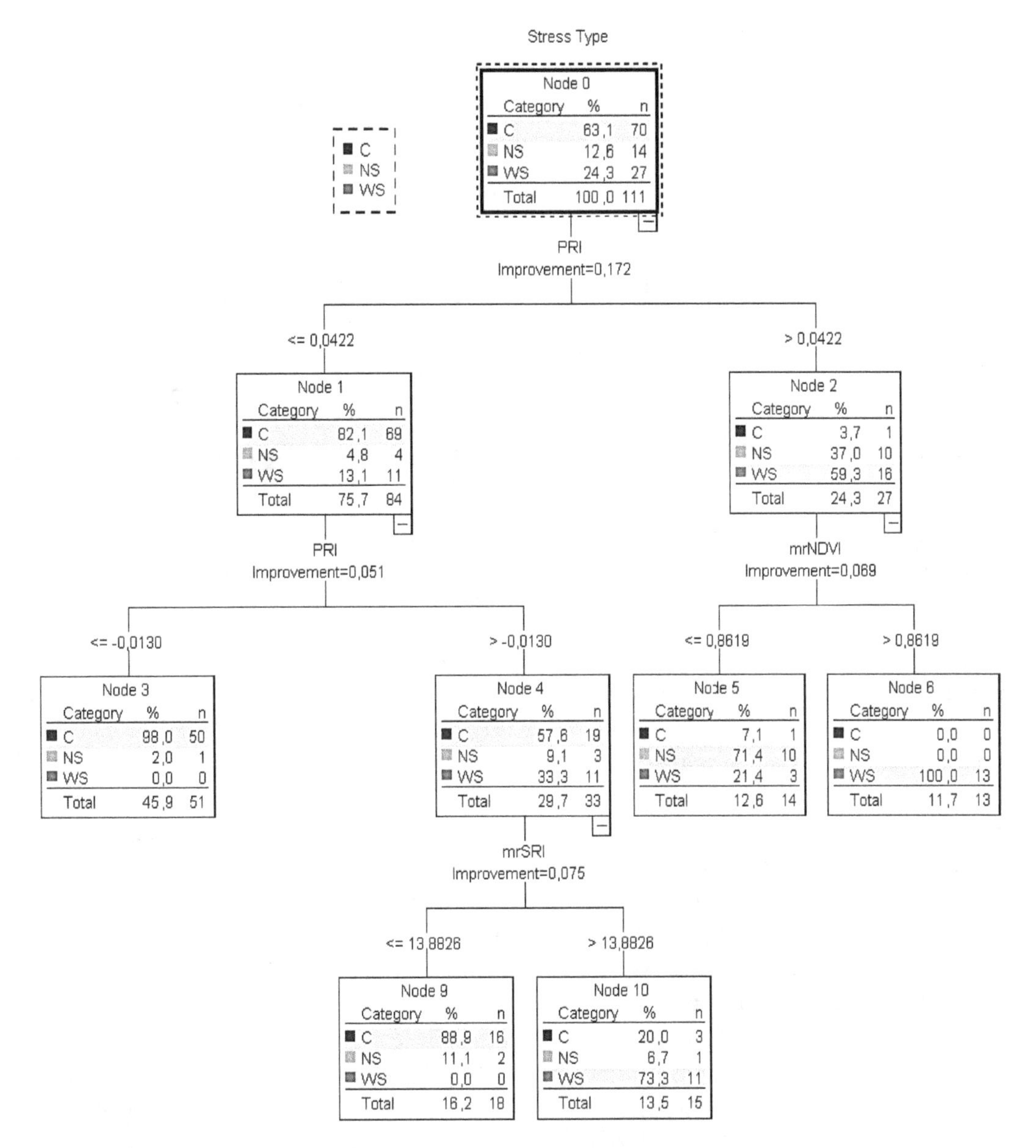

Figure 5. The *CT4* classification tree of the training sample for calibration to identify healthy, water- and nitrogen-deficit stressed plants. *C*: indication of control plants; *WS*: indication of water deficit stress plants; *NS*: indication of nitrogen deficit stress plants.

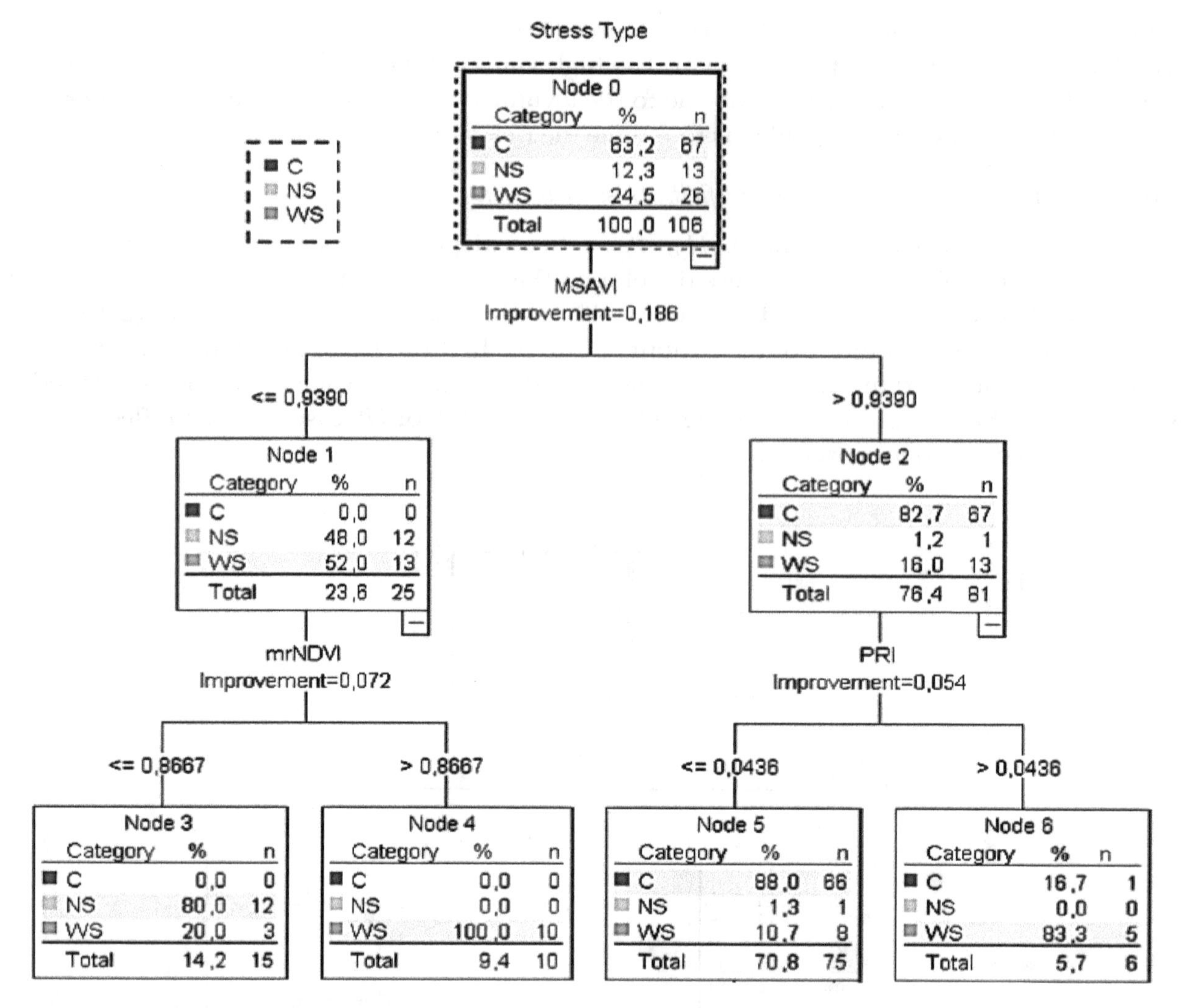

Figure 6. The *CT5* classification tree of the training sample for calibration to identify healthy, water- and nitrogen-deficit stressed plants. *C*: indication of control plants; *WS*: indication of water deficit stress plants; *NS*: indication of nitrogen deficit stress plants.

According to the training *CT4* (Figure 5), it was calculated that there was no water or nitrogen deficit stress when *PRI* was ≤ -0.0130 and *C* was returned (no irrigation is needed). On the other hand, if the *PRI* was between -0.01 and 0.04, the *mrSRI* readings (next best predictor) had to be taken into consideration to identify the water stressed plants and that node 4 was omitted. In this case, when *mrSRI* readings were ≤ 14, *C* was returned (no irrigation is needed). However, when *mrSRI* readings were >14, the majority of the sample (73%) returned *WS*. If the *PRI* was higher than 0.04, the *mrNDVI* readings (next best predictor) had to be taken into account. In this case, when *mrNDVI* readings were ≤ 0.86, *NS* was returned (nitrogen is needed). However, when *mrNDVI* readings were >0.86, the sample returned *WS* (irrigation is needed). However, contrary to node five of the training tree, in the testing tree, it was not clear when the plants were under water or nitrogen deficit stress because the sample of plants returning *NS* (60%) was very close to those returning *WC* (40%).

According to the training *CT5* (Figure 6), it was calculated that when the *MSAVI* was lower than 0.94, the *mrNDVI* readings (next best predictor) had to be taken into consideration. In this case, when the *mrNDVI* readings were ≤ 0.87, the majority of the sample (80%) returned *NS*. On the other hand, when *mrNDVI* was higher than 0.87, the total sample was characterised as water stressed (irrigation is needed). On the other hand, if *MSAVI* readings were higher than 0.94, the *PRI* readings (next best predictor) had to be taken into consideration. In this case, when *PRI* readings were ≤ 0.04 and *C* was returned (no irrigation is needed). However, when *PRI* readings were >0.04, the sample returned *WS* (irrigation is needed). Similar results were confirmed by the relevant node of the testing classification tree.

Table 1 presents the number of cases classified correctly and incorrectly for each category of the dependent variable. The predicted percent of the training sample was 90.1% in *CT4* and 89.6% in *CT5*, indicating that the model classified approximately 90.1% and 89.6%, respectively, of the sample correctly. Although the predicted percent of both *CT4* and *CT5* was similar, the *CT4* result revealed one potential problem with this model: for those plants cultivated under nitrogen stress, *NS* was predicted for 71.4% of them, which means that 28.6% of the stressed plants were inaccurately classified with *C* (non-stressed plants) or *WS* (water stressed plants). On the other hand, only 5.7% of non-stressed plants were inaccurately classified as *WS* or *NS* stressed plants. However, the final estimation risk of the misclassification of the training model was low, with small differences between the two models reflected at 9% and 10%, respectively.

Table 1. The classification accuracy of the training and testing sample for calibration and validation, respectively. *C*: control plants; *WS*: water deficit stress plants; *NS*: nitrogen deficit stress plants.

Classification Model	Sample	Observed	Predicted			
			C	*NS*	*WS*	Percent Correct
CT4	Training *CT4*	*C*	66	1	3	94.3%
		NS	3	10	1	71.4%
		WS	0	3	24	88.9.7%
		Overall Percentage	62.2%	12.6%	25.2%	90.1%
	Validation	*C*	23	2	4	83.3.3%
		NS	0	2	1	75.0%
		WS	1	1	3	62.5%
		Overall Percentage	64.9%	13.5%	21,6%	76.7%
CT5	Training	*C*	61	0	6	91.0%
		NS	1	12	0	92.3%
		WS	1	3	22	84.6%
		Overall Percentage	59.4%	14.2%	26.4%	89.6%
	Validation	*C*	20	0	1	95.2%
		NS	1	3	1	60.0%
		WS	0	0	9	100.0%
		Overall Percentage	60.0%	8.6%	23.5%	91.4%

4. Discussion

Despite many statistical and mathematical models such as principal component analysis [17], artificial neural network, and other classification procedures that have been developed to extract optimal information from remotely sensed data, Yohannes and Hoddinott [26] and Camdeviren et al. [27] believed that the *CT* method was a very useful model for analysing complex datasets by providing visual results. Goel et al. [18] applied a classification tree method to group hyperspectral data in order to identify weed stress and nitrogen status of corn and compared it with artificial neural networks. The advantages of tree-based classification are that it does not require the assumption of a probability distribution, specific interactions can be detected without previous inclusion in the model, non-homogeneity can be taken into account, mixed data types can be used, and dimension reduction of hyperspectral datasets is facilitated [28].

In the current study, in order to determine the water and nitrogen deficit stress severity, reflectance indices were investigated in the *CT* paths. Among the indices, the *CT* model selected (*MSAVI*) as a starter index to predict the water and nitrogen deficit stress. The *CT* analysis revealed that the classification accuracy for the training sample was 89.6% and the testing tree responded to the predicted expectation by approximately 91.4%.

The overall success rate of classification accuracy between the predicted and measured values of the stressed tomato indicated that the combination of *MSAVI*, *mrNDVI*, and *PRI* has the potential to determine water and nitrogen deficit stress and that classification tree algorithms have good potential in the classification of remotely sensed spectral data. Elvanidi et al. [13] also reveal *mrNDVI* and *PRI* in the *CT* path to determine water deficit stress in tomato plants, with classification accuracy values of 84.2% and 78.9% for the training and testing sample, respectively. Additionally, Genc et al. [17] also tested the ability of the classification tree algorithm to assess water stress in corn using hyperspectral reflectance spectra transformed into spectral vegetation indices. Their results demonstrated that water and nitrogen stress in corn was detectable through spectral reflectance analysis.

Generally, the use of machine learning models is not widespread in agriculture since they require a long time-series of datasets. The destructive methods or the complex sensors used in the last decades to quantify plant physiology did not allow for their progress. With the recent integration of a new age of computational intelligent sensors, more and more robust methodologies are being adapted. Up to now, less than 40 articles focused on machine learning models in agriculture. From those, 61% of the articles were related to different aspects of crop management [29].

However, the prediction performed by those models, in most cases, mostly correspond specifically to the area and conditions used in the training data, thereby trying to account for the otherwise invisible variations specific to that land and surroundings [30]. In the open field, for instance, the terrain in which the crop is cultivated affects the process of the crop water demand prediction. With the advent of the new era of computational intelligent sensors that can track various things that were previously not possible, more and more factors can be considered. Nevertheless, this new concept will ensure the development of more robust remote sensing approaches for monitoring plant physiology in order to train a decision support system with the aim of adjusting climate and irrigation control strategies within the greenhouse. Thus, in the future, the widespread usage of machine learning models is expected, allowing for the possibility of integrated and applicable tools. However, improvement in the performance of the decision tree classification approach with increases in the number of data sets further strengthens the belief that, by increasing the amount of data, model performance could probably be further improved.

5. Conclusions

In the current work, the ability to use a classification tree was tested to remotely predict leaf chlorophyll and substrate water content. Additionally, the classification tree was trained to assess different types of tomato stress such as water and nitrogen deficit stress. The model was trained by organizing, in the most effective way, the reflectance values measured by a hyperspectral camera. Among the reflectance indices, the classification tree model selected *TCARI* and *PRI* or *mrSRI* to predict leaf chlorophyll content. To estimate the substrate water content, on the other hand, the process was much more complex since more than four reflectance indices were involved in the procedure. Regarding the model trained to sense the actual plant water and nitrogen status, it was concluded that the combination of *MSAVI*, *mrNDVI*, and *PRI* involved in *CT5* made insufficient provision with a reasonable accuracy. These results are promising for designing a smart decision support system for better managing climate and irrigation control strategies. Additionally, less complicated reflectance sensors recording certain spectral bands could be adapted to monitor plant physiology in real-time in a cost-effective manner. Nevertheless, it has to be noted that the results presented are relevant to the conditions of the measurements and the specific crop studied. Therefore, it is expected that the use of machine learning models will be even more widespread in the near future.

Author Contributions: Conceptualization, methodology and data analysis A.E. and N.K.; Experimental measurements, A.E.; Writing-Original Draft Preparation, A.E.; Writing-Review & Editing, Supervision, N.K. and C.K.

Nomenclature

As	Photosynthesis rate
C	Control: no stressed plants/no irrigation or fertilization is needed
CHAID	Chi-squared automatic interaction detection
Chl_a	Chlorophyll-a content
CT	Classification tree
CRT	Classification regression tree
EC	Electrical conductivity
LAI	Leaf area index
mrNDVI	Modified red edge normalised difference vegetation index
mrSRI	Modified red edge simple ratio index
MSAVI	Modified soil-adjusted vegetation index
n	Number of samples per day per treatment
NIR	Near-infrared region
nm	Nanometre
NS	Nitrogen stress: nitrogen stress plants/nitrogen is needed
OSAVI	Optimisation soil-adjusted vegetation
PRI	Photochemical reflectance index
R or r	Reflectance
R_{445}	Reflectance value in 445 nm band
R_{531}	Reflectance value in 531 nm band
R_{550}	Reflectance value in 550 nm band
R_{570}	Reflectance value in 570 nm band
R_{670}	Reflectance value in 670 nm band
R_{690}	Reflectance value in 690 nm band
R_{700}	Reflectance value in 700 nm band
R_{705}	Reflectance value in 705 nm band
R_{750}	Reflectance value in 750 nm band
R_{800}	Reflectance value in 800 nm band
R_{810}	Reflectance value in 810 nm band
$R_{g,i}$	Irradiance
RH	Relative humidity
RI	Reflectance index
rNDVI	Red edge normalised difference vegetation index
SD	Standard deviation
T	Air temperature
TCARI	Transformed chlorophyll absorption in reflectance index
TKN	Total Kjeldahl nitrogen
VIS	Visible spectrum
VNIR	Visible and near-infrared spectrum region
WS	Water stress: water stress plants/irrigation is needed
θ	Substrate water content

References

1. Elvanidi, A.; Katsoulas, N.; Augoustaki, D.; Loulou, I.; Kittas, C. Crop reflectance measurements for nitrogen deficiency detection in a soilless tomato crop. *Biosyst. Eng.* **2018**, *176*, 1–11. [CrossRef]
2. Katsoulas, N.; Elvanidi, A.; Ferentinos, K.P.; Kacira, M.; Bartzans, T.; Kittas, C. Crop reflectance monitoring as a tool for water stress detection in greenhouses: A review. *Biosyst. Eng.* **2016**, *151*, 374–398. [CrossRef]
3. Ray, S.S.; Das, G.; Singh, J.P.; Panigrahy, S. Evaluation of hyperspectral indices for LAI estimation and discrimination of potato crop under different irrigation treatments. *Int. J. Remote Sens.* **2006**, *27*, 5373–5387. [CrossRef]
4. Alchanatis, V.; Cohen, Y.; Cohen, S.; Moller, M.; Sprinstin, M.; Meron, M. Evaluation of different approaches for estimating and mapping crop water status in cotton with thermal imaging. *Precis. Agric.* **2010**, *11*, 27–41. [CrossRef]

5. Liaghat, S.; Balasundram, S.K. A review: The role of remote sensing in precision agriculture. *Am. J. Agric. Biol. Sci.* **2010**, *5*, 50–55. [CrossRef]
6. Sonneveld, C.; Voogt, W. *Plant Nutrition of Greenhouse Crops*; Springer: New York, NY, USA, 2009.
7. Elvanidi, A.; Katsoulas, N.; Bartzanas, T.; Ferentinos, K.P.; Kittas, C. Crop water status assessment in controlled environment using crop reflectance and temperature measurements. *Precis. Agric.* **2017**, *18*, 332–349. [CrossRef]
8. Story, D.; Kacira, M. Design and implementation of a computer vision-guided greenhouse crop diagnostics system. *Mach. Vis. Appl.* **2015**, *26*, 495–506. [CrossRef]
9. Kim, Y.; Glenn, D.M.; Park, J.; Ngugi, H.K.; Lehman, B.L. Hyperspectral image analysis for plant stress detection. *Am. Soc. Agric. Biol. Eng.* **2010**, 1009114. [CrossRef]
10. Amatya, S.; Manoj, K.; Alva, A.K. Identifying water stress in potatoes using leaf reflectance as an indicator of soil water content. *J. Agric. Eng.* **2014**, *1*, 52–61. [CrossRef]
11. Sarlikioti, V.; Driever, S.M.; Marcellis, L.F.M. Photochemical reflectance index as a mean of monitoring early water Stress. *Ann. Appl. Biol.* **2010**, *157*, 81–89. [CrossRef]
12. Zarco-Tejada, P.J.; González-Dugo, V.; Williams, L.E.; Suárez, L.; Berni, J.A.J.; Goldhamer, D.; Fereres, E.A. PRI-based water stress index combining structural and chlorophyll effects: Assessment using diurnal narrow-band airborne imagery and the CWSI thermal index. *Remote Sens. Environ.* **2013**, *138*, 38–50. [CrossRef]
13. Kacira, M.; Sase, S.; Okushima, L.; Ling, P.P. Plant response-based sensing for control strategies in sustainable greenhouse production. *J. Agric. Meteorol.* **2005**, *61*, 15–22. [CrossRef]
14. Magney, T.S.; Vierling, L.A.; Eitel, J.U.; Huggins, D.R.; Garrity, S.R. Response of high frequency Photochemical Reflectance Index (PRI) measurements to environmental conditions in wheat. *Remote Sens. Environ.* **2016**, *173*, 84–97. [CrossRef]
15. Bajwa, S.G.; Mishra, A.R.; Norman, R.J. Canopy reflectance response to plant nitrogen accumulation in rice. *Precis. Agric.* **2010**, *11*, 488–506. [CrossRef]
16. Lu, S.; Lu, L.; Zhao, W.; Liu, Y.; Wang, Z.; Omasa, K. Comparing vegetation indices for remote chlorophyll measurement of white poplar and Chinese elm leaves with different adaxial and abaxial surfaces. *J. Exp. Bot.* **2015**, *66*, 5625–5637. [CrossRef] [PubMed]
17. Genc, L.; Demirel, K.; Camoglu, G.; Asik, S.; Smith, S. Determination of plant water stress using spectral reflectance measurements in watermelon (Citrullus vulgaris). *Am. Eur. J. Agric. Environ. Sci.* **2011**, *11*, 296–304.
18. Goel, P.K.; Prasher, S.O.; Patel, R.M.; Landry, J.A.; Bonnell, R.B.; Viau, A.A. Classification of hyperspectral data by decision trees and artificial neural networks to identify weed stress and nitrogen status of corn. *Comput. Electron. Agric.* **2003**, *39*, 67–93. [CrossRef]
19. Morgan, J. *Classification and Regression Tree Analysis*; Technical report; University of Boston: Boston, MA, USA, 2014.
20. Lichtenthaler, H.K.; Wellburn, A.R. Determinations of total carotenoids and chlorophylls a and b of leaf extracts in different solvents. *Biochem. Soc. Trans.* **1983**, *11*, 591–592. [CrossRef]
21. Kjeldahl, J. A new method for the determination of nitrogen in organic matter. *Z. Anal. Chem.* **1883**, *22*, 366. [CrossRef]
22. Katsoulas, N.; Elvanidi, A.; Ferentinos, K.P.; Bartzanas, T.; Kittas, C. Calibration methodology of a hyperspectral imaging system for greenhouse plant water stress estimation. In Proceedings of the 6th Balkan Symposium on Vegetables and Potatoes, Zagreb, Croatia, 29 September–2 October 2014; Volume 1142, pp. 119–126.
23. Loh, W.Y. Classification and regression trees. In *Wiley Interdisciplinary Reviews: Data Mining and Knowledge Discovery*; John Wiley & Sons: Hoboken, NJ, USA, 2011; Volume 1, pp. 14–23.
24. Lewis, R. An introduction to classification and regression tree (CART) analysis. In Proceedings of the Annual Meeting of the Society of Academic Emergency Medicine, San Francisco, CA, USA, 22–25 May 2000; pp. 1–14.
25. IBM Corporation. *IBM SPSS Decision Trees 21 Guide*; IBM Corporation: Armonk, NY, USA, 2012.
26. Yohannes, Y.; Hoddinott, J. *Classification and Regression Trees: An Introduction*; International Food Policy Research Institute: Washington, DC, USA, 1999.
27. Çamdeviren, H.; Mendeş, M.; Ozkan, M.M.; Toros, F.; Şaşmaz, T.; ve Oner, S. Determination of depression risk factors in children and adolescents by regression tree methodology. *Acta Med. Okayama* **2005**, *59*, 19–26. [PubMed]

28. Delalieux, S.; Van Aardt, J.; Keulemans, W.; Schrevens, E.; Coppin, P. Detection of biotic stress (Venturia inaequalis) in apple trees using hyperspectral data: Nonparametric statistical approaches and physiological implications. *Eur. J. Agric.* **2007**, *27*, 130–143. [CrossRef]
29. Liakos, K.G.; Busato, P.; Moschou, D.; Pearson, S.; Bochtis, D. Machine learning in agriculture: A review. *Sensors* **2018**, *18*, 2674. [CrossRef] [PubMed]
30. Krupakar, H.; Akshay, J.; Dhivya, G. A review of intelligent practices for irrigation prediction. *arXiv*, 2016; arXiv:1612.02893.

12

Agronomic Management for Enhancing Plant Tolerance to Abiotic Stresses: High and Low Values of Temperature, Light Intensity and Relative Humidity

Antonio Ferrante [1] **and Luigi Mariani** [1,2,*]

[1] Department of Agricultural and Environmental Sciences, Università degli Studi di Milano, via Celoria 2, 20133 Milan, Italy; antonio.ferrante@unimi.it

[2] Lombardy Museum of Agricultural History, via Celoria 2, 20133 Milan, Italy

* Correspondence: luigi.mariani@unimi.it.

Abstract: Abiotic stresses have direct effects on plant growth and development. In agriculture, sub-optimal values of temperature, light intensity, and relative humidity can limit crop yield and reduce product quality. Temperature has a direct effect on whole plant metabolism, and low or high temperatures can reduce growth or induce crop damage. Solar radiation is the primary driver of crop production, but light intensity can also have negative effects, especially if concurrent with water stress and high temperature. Relative humidity also plays an important role by regulating transpiration and water balance of crops. In this review, the main effects of these abiotic stresses on crop performance are reported, and agronomic strategies used to avoid or mitigate the effects of these stresses are discussed.

Keywords: cold; heat; solar radiation; relative humidity; transpiration

---

1. Introduction

Abiotic stress is the result of the action of external environmental factors that affect growth, development, and reproduction of crops. In the first part of this review published last year, we analyzed stressful conditions due to drought, water excess, salinity, and lodging [1]. In this part, high-low values of solar radiation, temperature, and relative humidity will be considered as well as agronomic strategies that can be used for lowering the stressful conditions.

Crop yield is the result of the interaction of multiple factors such as genotype, agronomic management, and environmental conditions. Different genotypes have different yield capabilities, depending on their ability to adapt. Agricultural cropping systems are continuously evolving due to innovation in agronomic tools and to identification of high-performance cultivars coming from traditional or biotechnological genetic improvements. Nevertheless, the major causes of agricultural production losses are due to abiotic stresses such as low water availability, high salinity, high or low temperatures, hypoxia/anoxia, and nutrient deficiency.

Crops exposed to these abiotic stresses respond by activating defense mechanisms. Therefore, in an early stress stage, no visible symptoms are exhibited. The energy used by the crop to counteract or cope with the abiotic stresses is called "fitness cost", and this energy does not contribute to production. While the visual appearance of plants in an early stress stage does not change, the physiology can undergo deep changes, including the accumulation of bioactive compounds able to counteract the stress conditions.

Plants are able to perceive environmental stimuli and to adapt to different environments. However, the degree of tolerance and adaptability to abiotic stresses varies among species and varieties. The global weight of biotic stresses on yield losses was estimated to be 70% by Boyer [2], and 13–94%

by Farooq et al. [3]. A more detailed analysis of yield losses associated with some abiotic stresses was presented by Mariani and Ferrante [1].

In this paper, a review of agronomic strategies aimed to optimize the resilience of crops exposed to abiotic stresses due to sub-optimal values of solar radiation, temperature, and relative humidity is presented. It is, however, necessary to remember that agronomic strategies hereafter presented and discussed can be adopted only if sustainable, not only socially and environmentally but also economically. This is because agriculture is an economic activity that cannot be done without adequate remuneration of the production factors.

2. Solar Radiation as Resource and Limitation for Crops

The sun provides the energy that moves the climate system (ocean and atmosphere circulation), drives the water cycle and feeds the food chains by means of the photosynthetic process. Moreover, an energy balance is in place for which the Earth emits towards space the same quantity of energy received from the sun.

Some basic physical laws rule the energy fluxes at all scales. More specifically the emission of energy by all bodies is ruled by Planck's Law, while the Stefan-Boltzmann Law and Wien's Law state that the total emission and the wavelength of the maximum emission are a function of the temperature of a blackbody, respectively. Moreover, the radiative flux (I) intercepted by a surface is determined by the sine law $I = I_0 * \sin(\alpha)$ where I_0 is the maximum radiative flux intercepted by a surface perpendicular to it and α is the elevation angle (the angle of the sun above the surface) [4]. The cosine law explains, for example, the high quantity of energy received by south-facing slopes at mid-latitudes during spring and autumn when the sun at noon is still low in the sky.

The radiation that comes directly from the sun is defined as beam radiation. The scattered and reflected radiation that arrives at the earth's surface from all directions (reflected from other bodies, molecules, particles, droplets, etc.) is defined as diffuse radiation. The sum of the beam and diffuse components is defined as total (or global) solar radiation (GSR). The fraction of the GSR re-irradiated toward space is named albedo A. Moreover, the surface of the planet emits radiation toward the sky in the far infrared (LR_1) and in its turn, the sky emits radiation in the far infrared towards the surface (LR_2). Consequentially, net radiation for a given surface is expressed with the following equation:

$$R_n = GSR * (1 - A) - LR_1 + LR_2 \quad (1)$$

Solar radiation is a source of energy and information for plants [5]. The relation between plants and light takes place through a series of pigments, which can be classified into four broad categories, namely chlorophylls, carotenoids, anthocyanins and phytochromes. Photosynthesis is the set of two phases, namely luminous and dark phase. In relation to the dark phase of photosynthesis, crops can be classified as C3, C4, and Crassulacean Acid Metabolism (CAM) plants [6].

For C3 plants the whole process of photosynthesis takes place in the mesophyll cells and the first products of photosynthesis catalyzed by Rubisco are two molecules with 3 atoms of carbon (Calvin cycle). In the case of C4 plants, the mesophyll is the site of CO_2 absorption by phosphoenolpyruvate (PEP), a reaction that is catalyzed by the enzyme PEP-carboxylase, which unlike Rubisco, is very efficient—even at low CO_2 levels. From the evolutionary point of view, C4 plants appeared on our planet much later than C3 plants (between 25 and 32 million years ago, according to Osborne and Beerling [7]) and were the product of the adaptation to high temperatures and water shortages.

CAM plants show a nocturnal phase in which, with open stomata, cells of the mesophyll produce OAA starting from PEP, and then malic acid is produced from OAA and is stored in the vacuolar juice that becomes increasingly acidic. During the day, when the stomata are closed, the malic acid returns to the cytoplasm where CO_2 is released from it which enters the Calvin cycle.

The relevance of C3 and C4 plants for global food security is indicated by the fact that about the 70% of the global calories required by humans is fulfilled by three C3 crops (wheat (*Triticum aestivum* L.), rice (*Oryza sativa*) and soybean (*Glycine max* L.) and two C4 crops (maize (*Zea mays*), sorghum (*Sorghum bicolor*)). On the other hand, CAM plants have lower importance and the only species subject to extensive cultivation are the pineapple (*Ananas comosus* (L.) Merr.) and the Indian-fig (*Ficus benghalensis*). For the C3 species, the lower efficiency of the Calvin cycle causes light saturation to occur at values between 30 and 80,000 lux (ex: 30–50,000 lux for grape (*Vitis vinifera* L.), while in the C4 saturation occurs at higher values at 80,000 lux. Therefore, on very bright days, when 100,000 lux is exceeded, the C3 species can exploit only a limited part of the available light energy. As far as light radiation is concerned, the compensation point is defined as the level of radiation at which photosynthesis and respiration reach equal values. Radiation values that coincide with this point are not significantly different between C3 and C4. At the level of general morphology, it is observed that leaves in the shade are wider and thinner (in particular the palisade layer appears less developed than that of leaves in the sun) and often have a greater concentration of chlorophyll in the upper surface. Added to this are a series of characteristic effects of the lack of radiation, i.e., yellowing and abscission of the lower leaves, a lack of branching, excessive elongation of stems and shoots and low or no fertility.

Photosynthesis is a key determinant of yield and quality of crops which in their turn are the pillar to the economic sustainability of farm activity. Therefore, all aspects of agronomic management should focus on crop photosynthesis with the objective of maximizing it even under stressful conditions. This objective can be achieved by selecting the best crops (species and varieties) for the selected environment, combining them in suitable crop successions and optimizing management techniques.

Making good choices of species and varieties are essential to making optimal use of the available solar radiation, avoiding at the same time problems due to radiation excess and water limitation. From this point of view, environments characterized by high temperatures, high values of global solar radiation, and water limitation are in general more suited to C4 plants, because these plants appeared some millions of years ago as an adaptation to savannas and tropical grasslands [8]. An important exception to this general rule is provided by maize, which in the presence of severe water stress shows a decrease in leaf area index (LAI), intercepted photosynthetically active radiation (PAR), and Radiation Use Efficiency (RUE) [9]. These drawbacks are usually associated with reduced competitiveness with weeds and proterandry that creates sterility. Such phenomena are only partially compensated by the increase of water use efficiency (WUE) and, consequently, a relevant reduction in maize biomass production and harvest index is generally observed.

Possible solutions for environments characterized by high temperatures and high levels of solar radiation could be use of C4 species such as sorghum, millet and panicum (Poaceae spp.) and sugarcane (*Saccharum officinarum*) or tropical C3's such as rice, oil palm (*Elaeis guineensis*), common bean (*Phaseolu vulgaris*) and cassava (*Manihot esculenta*). Some mesothermal C3 crops such as wheat and grapevine have shown significant injuries to photosystem II due to high temperatures [10].

Environments characterized by high values of cloud coverage have a greater presence of diffuse radiation that penetrates deeper into the canopy reaching the lower leaf layers. This explains why in these environments there is an increase in the RUE, which partially compensates for the decrease of intercepted PAR.

Obviously, the contrasting effects of the abovementioned factors (temperature, global solar radiation, LAI, intercepted PAR, RUE, soil water content, WUE) could be fully analyzed only through a dynamic crop simulation model [11–13] with a suitable time step (daily or hourly). This could be driven by weather data and run for a number of years sufficient to describe the climate of the selected environment (about 30 years is prescribed by the World Meteorological Organization to describe the climate for a given site).

Agronomic techniques (sowing density, plant nutrition, irrigation, integrated pest management, etc.) can promote the achievement of optimal values of LAI (for crops like maize, rice, sorghum, wheat, tomato (*Solanum lycopersicum*), and soybean the ideal values are 4–6), and should be adopted to maximize the light intercepted by crop canopies. Moreover, plants should be distributed as regularly as possible in the surface unit. From this point of view, an analysis carried out for apple (*Malus domestica*) orchards with the same LAI showed that square plantings (same distance between rows and along rows—1:1 ratio) intercept up to 20% more light than 3:1 system (3 times distance between rows than along rows) for a range of densities between 3000 and 20,000 trees per ha [14]. Other experiments have shown 10% higher yields at a 12% lower tree density than more rectangular single rows [15] (Wertheim, 1985). Moreover, square plantings give better coloration of red fruit than 3:1 plantings due to a more uniform light distribution. Despite these advantages, square plantings are rarely applied on a commercial scale because such plantings need higher capital investment for equipment adapted for over-the-row spraying and transport at harvest that may not compensate for higher yields, particularly with small-scale orchards [14].

Another relevant factor for optimizing light interception is the orientation of rows. In this regard, regardless of any other consideration (e.g., specific guidelines imposed by the effects of slope on mechanization or the presence of strong dominant winds able to break down rows perpendicular to them), in mid latitudes a north–south orientation is optimal for most crops, as it gives the maximum amount of intercepted radiation. In the specific case of grapevines, the east–west orientation of rows is preferable as it guarantees a greater radiation intercepted in coincidence with maximum daily temperatures, which for specific varieties is particularly appreciated in the late phase of sugar accumulation (September–October).

Specific interventions like pruning or choice of species/varieties with a suitable inclination of leaves can be adopted to enhance the penetration of light into the canopy so that each foliar level can ideally receive the same quantity of light. By this point of view, dicotyledon canopies (with almost planar leaves) are farther from this ideal than graminaceous canopies with a suitable inclination of leaves. Furthermore, in the specific case of maize, current cultivars, unlike traditional ones, have fully erect leaves able to enhance penetration of light inside the canopy and homogenize its vertical profile. Light extinction in canopies can be effectively simulated by adopting Lambert-Beer's law as stated by Monsi and Saeki in 1953 [16].

The sunlight intercepted by canopies is a fundamental driver of crop water consumption as expressed by the Priestley-Taylor equation for calculation of reference crop evapotranspiration (ET0) on the basis of net radiation. The sunlight intercepted by a canopy is also directly correlated with water consumption expressed as percentage of reference crop needs. For vineyards with row training systems, Williams and Ayars [17] found that the area shaded by the canopy at noon expressed as the percentage of the surface allotted to a single plant (sh%) multiplied by the empirical coefficient 0.017 gives a reliable estimate of crop coefficient (kc). This latter value can be multiplied by reference crop evapotranspiration ET0 to obtain the maximum evapotranspiration (ETM).

Moreover, solar radiation is a source of information for living beings and a series of morphogenetic effects derive from it. In many plants, initiation of flowering is driven by perception of changes in day length (photoperiodism). Plants adapted to temperate climates (e.g., photoperiodic varieties of wheat) often perceive the lengthening days of spring as a signal to initiate reproduction. Such plants are known as long-day plants, while in short day species (e.g., many varieties of chrysanthemum (*Chrysanthemum indicum*) and poinsettia (*Euphorbia pulcherrima*)) flowering is induced with hours of light below the critical threshold. Moreover photo-indifferent crops (e.g., many fruit tree species) are not sensitive to the length of the day [18].

Light signals such as light quality and length of day are perceived by several types of photoreceptors including phytochromes and cryptochromes. More specifically, phytochromes detect the levels and ratio of red (R) and far-red (FR) light in the environment [17,18]. Phytochromes are blue photosynthetic pigments present in the leaves and sensitive to very low light intensity (even lower than 0.01 W m^{-2}), which explains the effects of full moon light on flowering of some plant species [19,20].

Photoperiodic effects give rise to precautions in moving a plant species from one latitudinal band to another or the agronomic practices useful to induce flowering in flower species in protected cultivations.

2.1. High Light Intensity Stress

Solar radiation has a fundamental importance for crop growth, yield, and quality in agricultural systems. Light intensity and duration cannot be modified in the open field. Therefore, plants must adapt to light stress, while they can be modulated and optimized in a greenhouse. More specifically, crops in the open field during summer must protect themselves from high light intensity which can damage leaves, young shoots or even the fruits. Plants may protect chlorophyll molecules by increasing the biosynthesis and the concentration of carotenoids. These antioxidant compounds act as shields avoiding photo-oxidation of chlorophyll from excessive light intensity (Figure 1). High light intensity also leads to the formation of reactive oxygen species (ROS), which increase photo-damage [21,22]. Leaf damage from high light conditions can be determined by monitoring the lipid peroxidation of leaf cell membranes and the functionality of photosystem II. The most dangerous radiation of the light spectrum that induces severe damage is UV-B radiation (280–320 nm) that has short wavelengths and high energy. The damage from UV-B radiation is especially notable on the vital macromolecules such DNA with negative effects on cellular processes. Light stress reduces the light use efficiency (LUE) and photosynthetic activity and can be observed in both the open field and in protected cultivation.

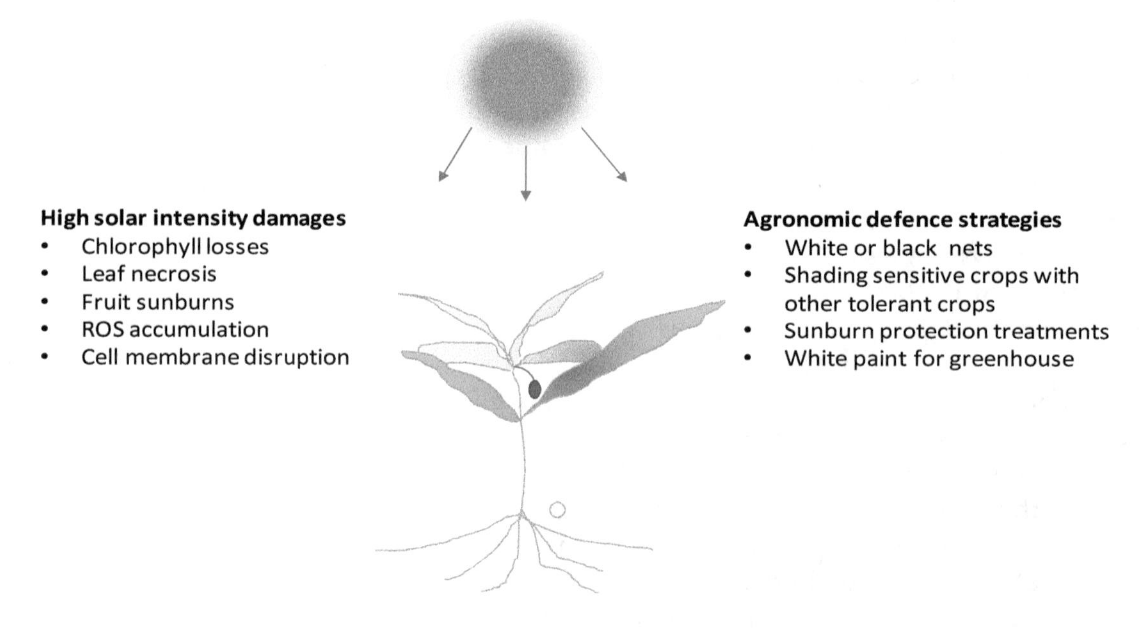

Figure 1. Effects of high light intensity on crops and agronomic strategies that can be adopted for increasing or avoiding crop or produce damage.

In the open field, light stress is particularly severe if associated with high temperature and drought. Agronomic strategies for reducing the negative effects of an excess of light include adequate irrigation systems, with support of sensors able to evaluate soil moisture and crop water requirements. Sufficient water availability can guarantee transpiration and, thus, thermoregulation by the evaporation of water at the leaf level which lowers leaf temperature.

In mixed crops, species with different light stress tolerances and heights can be cultivated together to provide shading of the most sensitive crop species. This strategy can be used to protect young seedlings of sensitive species against damage by strong solar radiation but can give protection against other stress factors like strong winds, low temperatures or salt. Once seedlings are sufficiently developed the crop used to protect them can be eliminated manually or by means of a selective herbicide treatment.

A high light intensity can reduce the yield and quality of young fruit directly exposed to sunlight. This is mostly important for fruit of species whose stomata no longer allow evapotranspiration with the approach of maturation, which makes them unable to thermoregulate fruit temperature. Therefore, fruit temperature shows an increase, especially with dark-colored fruit (e.g., grapevine berries of red varieties after veraison), when temperature in the presence of direct solar radiaton can exceed 45–50 °C. In these cases, damage can be direct (sunburn with cell membranes that lose their integrity) or indirect (slowing or interruption of the biosynthesis of compounds favorable to quality). Temperature of fruit can be measured directly by means of thermocouples or simulated by energy balance models [23].

Recently several products have become commercially available that can be sprayed on plants or fruit for avoiding sun damage. Kaolin containing compounds have been successfully applied to reduce sunburn in pomegranate [24]. Good results can be also obtained using shading nets. In apple trees, transparent and black nets were able to reduce sunburn incidence on fruit [25]. The nets reduced the direct sun light on fruit avoiding the excessive temperature of the exposed tissue and localized physiological disorders. Moreover, in vineyards, canopy management techniques (e.g., winter or green pruning) have protected clusters by covering them with a layer of leaves, which in many cases has enhanced berry quality [26].

In greenhouses during summer, high temperature associated with high light intensity can induce direct damage on plants. Therefore, to avoid the high temperature and light intensity, greenhouses are covered with shading nets or sprayed with white paint. As an alternative, white shading nets can also be placed inside the greenhouse. The aim of these strategies is to avoid excess heat in the greenhouse through the reduction of direct light. Water running along the roof and walls of the greenhouse in special interspaces also has potential application, since water reduces the radiation entering the greenhouse.

2.2. Low Light Intensity Stress

Some plant species are able to grow at low light conditions (termed shade plants) such as under vegetation or in areas with low solar radiation (e.g. valley bottoms, entering of caves). Most agricultural crops need high light intensity and are classified as sun plants. Shade plants have low light compensation and saturation points, while sun plants in contrast have high light compensation and saturation points. However, most plants can adapt to a range of low light conditions. The adaptation induces physiological and morphological changes in plants exposed to such conditions. In general, low light intensities induce stem elongation to overcome the shade conditions. Leaves of shaded plants increase their size and reduce their thickness and have a higher chlorophyll concentration. At a physiological level, the plants lower their light compensation point for balancing the reduced photosynthetic activity.

Low light conditions present in greenhouses during winter can be overcome by supplemental lighting using fluorescent lights, metal halide lights, high-pressure sodium lamps (HPS), or light emitting diodes (LEDs). The outputs of these lamps ideally must match the crop light utilization spectra. If the lamps have a higher emission in the regions of the leaf absorbance spectra, the LUE as well as the yield and quality are higher. LEDs can provide precise outputs and emission spectra (wavelengths) and can be readily adjusted based on the species requirements. Few studies have been performed on the effects of light intensity and quality on plant morphological and developmental processes [27]. Plants sense and translate environmental light signals, which interfere directly or indirectly with metabolism [28]. Light can modify auxin transport and gibberellin (GA) biosynthesis [29,30]. These changes can affect many aspects of plant development, including seed germination, stem elongation, and floral initiation. Light quality can modify the profile of carbon and nitrogen metabolites and those of organic acids and aromatic amino acids [27,31]. An alteration of the organic acids level, in particular of α-ketoglutarate, can affect amino acid biosynthesis, plant growth, and crop yield [27]. Supplemental lighting is often very expensive, and its application is only used for the most lucrative crops such as tomato or rose.

3. Thermal Resources and Limitations for Crops

3.1. Temperature and Agriculture

Typically, in agriculture we refer to surface air temperature, which is the temperature measured with a thermometer placed about 1.50–1.80 m above a soil with a regularly mowed lawn and protected by an anti-radiation screen that ensures a suitable circulation of air around the sensor. This measure is used to estimate thermal resources and limitations for cultivated and spontaneous plants.

The temperature that would be most useful to know is that of the organs of plants. The leaf temperature is mitigated by the transpiration process, which transfers to the atmosphere 2450 J g^{-1} via transpired water. Consequently, the leaf temperature of well-watered plants is close to that of the air. However, when soil water is insufficient, plants close their leaf stomata, and the leaf temperature rises to levels significantly higher than the air.

Another important variable is the temperature of the soil layer explored by roots. It can be measured at different depths, although usually for agricultural purposes it is only gauged at 10 cm in depth because the most superficial soil layer is also the only one that shows a significant daily cyclicity.

3.2. Temperature and Some Physical Presuppositions

Temperature values and variability in space and time obey physical presuppositions that are briefly summarized here. Hot and cold air behave as non-miscible fluids and cold air, being denser, flows downward along the sides of hills and mountains and gathers in valleys, depressions and concavities of the ground. Moreover, during the day, the sun warms the ground and this warms the air layers over it while, during the night, the ground cools radiating energy toward space and cools the layers of air over it. The strength of the ground radiative cooling towards space is directly proportional to the fraction of sky visible from it (sky view factor (SVF)), and it is at a maximum for a flat plain without surrounding obstacles (mountains, trees, walls, buildings, etc.) or for the tops of hills and mountains. These latter locations have a higher SVF than the bottom of the valleys, which also cool indirectly due to the cooler dense air that descends from higher elevations along the slopes with cold air accumulation in the lowlands (cold lake effect).

Above the cold lake there is a milder area called a thermal belt, which is an area of thermal optimum. Historically, villages of the European Alps and Italian Apennines were often built in the thermal belt mainly because the need for winter heating is lower.

The most favorable exposure, from the perspective of thermal resources, is the south face, while the north face is the least favored. In an intermediate position with respect to these two extremes are the sides facing west and east, the latter being thermally more favored because the eastward-side is the first to receive the sun in the morning, when light energy must warm the surfaces that are cold after the night and are often dew-covered (informally described as "the sun works on the cold") before the air is heated. When the sun illuminates the west facing slopes in the afternoon, it heats already warm slopes.

Maximum and minimum daily air temperatures are mainly determined by:

1. The energy balance of the underlying surface;
2. The short distance transport by downslope and upslope air movement;
3. The advection of air masses from more distant areas; for example, in Europe, the advection of subtropical air masses from the intertropics, Arctic air masses from the polar region and polar continental air masses from Siberia;

4. The rising of air from the surface resulting in cooling (convection);
5. The compressional effect typical of dynamic anticyclones where air masses are animated by a vertical descending motion of high pressure.

In addition, soil covered by vegetation heats up less than a bare soil during the day and cools less during the night, and soil covered with a sufficiently thick snow layer shows a constant temperature near the surface close to 0 °C, which also maintains some microbial activity.

3.3. Quantitative Approach to Temperature Resources and Limitations for Plants

Temperature is a primary driver for plant growth and development (G&D). Therefore, the objective of the grower is to optimize thermal resources and minimize limitations, to meet the business objectives (i.e., quantity and quality of crop production). Given specific values of daily or hourly temperature, to discriminate between resources and limitations, adequate G&D happen only in a specific thermal range delimited by the two reference values, the lower cardinal (LC) and the upper cardinal (UC) temperatures [32]. Furthermore, in this primary thermal range it is possible to detect a sub-optimal range (delimited by lower optimal (LO) and upper optimal (UO) temperatures) where G&D happens without thermal limitation. Below the LC and above the UC values, there are mortality thresholds for low (CL) and high (CH) temperatures [25].

Values of cardinal temperatures for various species are shown in Table 1. However, it should be noted that cardinal temperatures for each species and variety may vary with phenological stage. For wheat, the values of LC, LO, UO, and UC are 3.5, 20, 24, and 33 °C, respectively, from sowing to emergence; −1.5, 4, 6, and 15.7 °C for vernalization; 1.5, 9, 12, and >20 °C for terminal spikelet; and 9.5, 19, 23, 31 °C for anthesis and 9.2, 19, 22, 35.4 °C for caryopsis development until ripening. On the other hand, LC, LO, UO and UC for some general processes of wheat are −1, 21.5, 22.5, 24 °C, respectively, for leaf initiation; 3, 20, 21, >21 °C for shoot growth; and 2, 12.5, 20, >25 °C for root growth [27,33].

If T_m is the daily average temperature, daily thermal resources useful for G&D can be expressed as growing degree days (GDD) or normal heat hours (NHH). GDD are calculated by the following equation:

$$GDD = (T_m - LC) \tag{2}$$

where GDD = 0 for $T_m < LC$. Moreover, to evaluate the depleting effect of high temperatures on G&D, the following alternative truncated conditions are adopted: If $T_m > UC$ then $T_m = UC$ or $T_m = UC - (T_m - UC)$.

A more physiologically-based way to simulate thermal resources, taking into account the effect of sub-optimal or supra-optimal temperatures, is provided by the NHH method that converts hourly average temperatures to a normal value in the range 0–1 by means of a suitable response curve like that shown in Figure 2, growing from 0 to 1 between LC and LO, equal to 1 between LO and UO and linearly decreasing from 1 to 0 between UO and UC [34]. This method enables quantification of thermal limitation by low temperatures, when normal hours are useless because they are sub-optimal (LHH), and by high temperatures, when normal hours are useless because they are supra-optimal (HHH), respectively expressed as a complement to 1 for the NHH value.

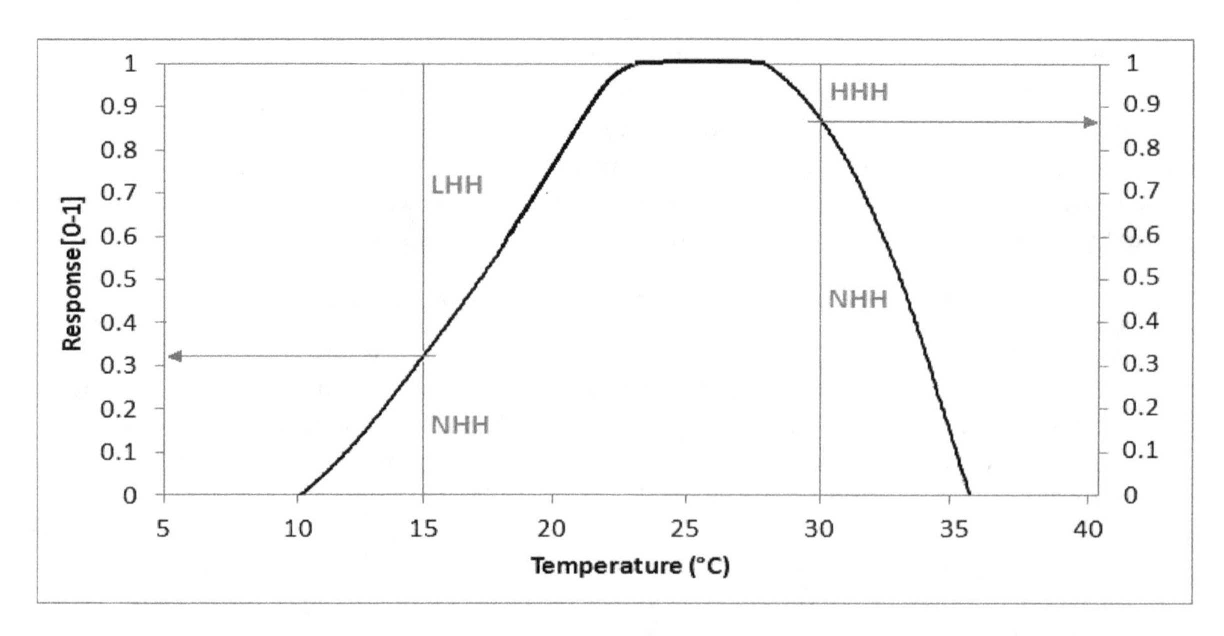

Figure 2. A response curve derived from grapevine which allows translating an hour spent at a given temperature into a fraction of normal heat hours (NHH) (range 0–1) while the complement to 1 represents normal hours which are useless because they are sub-optimal (LHH) or supra-optimal (HHH). For example, an hour spent at 5 °C gives 0 NHH and 1 LHH, an hour spent at 15 °C gives 0.32 NHH and 0.68 LHH and finally an hour spent at 30 °C is 0.87 NHH and 0.13 HHH.

NHH, LHH and HHH values for a given mean hourly temperature (ATH) can be obtained by means of the following simple algorithm (statements are given in Pascal pseudo-language code).

```
if (ATH ≤ LC) or (ATH ≥ UC) then NHH = 0 else
if (ATH ≥ LO) and (ATH ≤ UO) then NHH = 1 else
if (ATH > LC) and (ATH < LO) then
begin
m: = 1/(LO − LC);
q: = 1 − LO/(LO − LC)
NHH = m * ATH + q
end else
if (val1 > UO) and (val1 < UC) then
begin
m: = −1/(UC − UO)
q: = 1 + UO/(UC − UO)
NHH = m * ATH + q
end
if ATH < LO then LHH = 1 − NHH
if ATH > UO then HHH = 1 − NHH
```

This algorithm is based on a theoretical response curve with a trapezoidal shape. The real curve is the result of the whole set of physiological processes ruled by temperature, which means that the availability of experimental curves could be important for enhancing the proposed approach.

GDD or NHH cumulated from the sowing date to a given day can be used to simulate the phenological stage (flowering, fruit-set, ripening and so on), cumulative biomass, or some derived indexes such as the LAI [35]. The scatterplot in Figure 3 shows the relation between values of NHH and GDD without truncation calculated for 202 Italian stations in 2017. The shape of the cloud of points are the consequence of the less physiological approach based on GDD subject to an overestimation growing with the increase of thermal resources. The scatterplot in Figure 4 also from 2017 shows the relation between NHH and Global Solar Radiation for the same stations analyzed in Figure 3. The poor correlation is a result of high altitude stations characterized by high values of radiative resources in coincidence with low or null NHH.

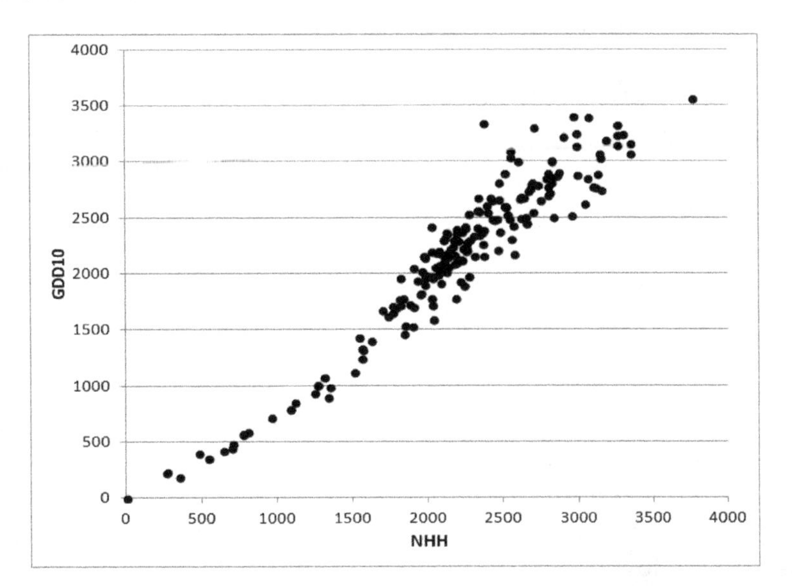

Figure 3. The scatterplot shows the relationship between NHH [h] and growing degree days (GDD) [°C] calculated for a minimum cardinal of 10 °C without truncation. Data are from 2017 for 202 stations in Italy located at elevations between 0 and 3488 m and latitudes between 35.498° N and 46.943° N (our elaborations on NOAA-GSOD data). NHH was calculated with the algorithm based on a response curve with a trapezoidal shape described in the text.

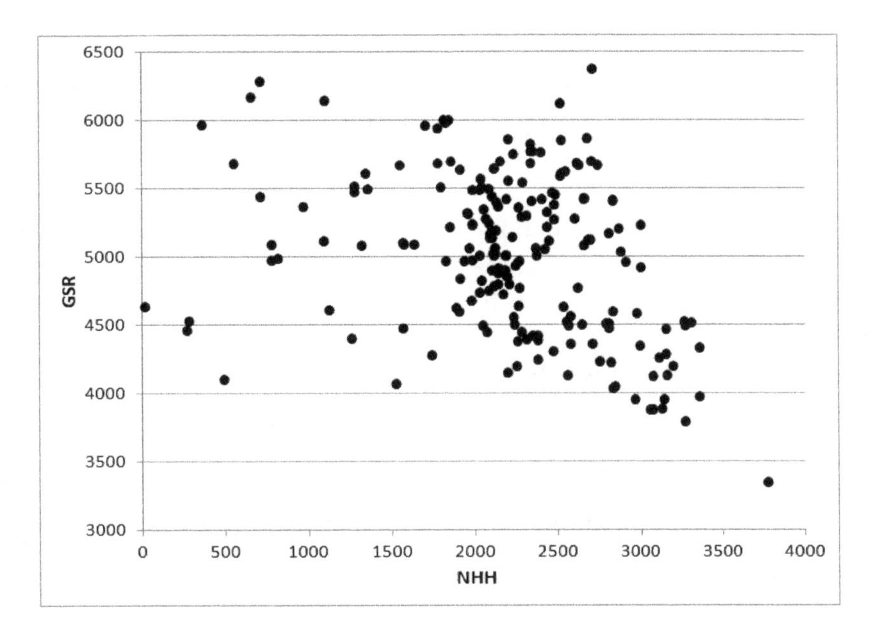

Figure 4. The scatterplot shows the correlation between values of NHH [h] and Global Solar Radiation [MJ m^{-2}]. Data are from 2017 for 202 stations in Italy located at elevations between 0 and 3488 m and latitudes between 35.498° N and 46.943° N (our elaborations on NOAA-GSOD data). NHH was calculated with the algorithm based on a response curve with a trapezoidal shape described in the text.

Temperature is also an important source of information for plants because it indicates the season, allowing optimization of their phenological rhythm in relation to thermal, radiative, and pluviometric features. An example of this is the vernalization process to which several species of medium-high latitudes are subjected. Exposure to low temperatures for a certain period of time, is essential for spring vegetative recovery and the development of vegetative and reproductive organs. Chilling is expressed as chill units (CU), and calculated in the same way as NHH, but by adopting an LC = 0 °C, LO = UO = 7 °C, and UC = 14 °C. CU can be obtained with the same algorithm previously described for NHH calculation. In many species vernalization and photoperiod cooperate to induce flowering in periods suitable for thermal and water resources [36].

Particular attention should be paid to temperatures below the critical threshold CL. By this, frost tolerant species when in full vegetative rest (tissues hardened which means low water content and high concentrations of cellular soluble solids) are able to withstand very low temperatures, such as critical low temperatures of −9/−10 °C for olive (*Olea europaea* L.), −15/−18 °C for grapevine and −20/−22 °C for wheat. It is more difficult to establish the upper critical threshold CH because heat stress from high temperatures depends primarily on canopy temperature rather than on air temperature. So, the interplay between various environmental variables (soil water content, relative humidity, wind speed and global solar radiation) and the stomatal response of plants affects leaf temperatures which can vary substantially for the surrounding air temperature. In general, as stated by De Boeck et al. [37], leaves tend to heat up when stomatal conductance is low (drought conditions) and this effect is strengthened by high solar radiation and high relative humidity, while high wind speeds brings the leaf temperature closer to the air temperature, which can imply either cooling or warming (i.e., abating or reinforcing heat stress) depending on other prevailing conditions. Furthermore, the effect of wind is more relevant with small leaves which have a reduced boundary layer and oscillate easily, enhancing the exchanges of heat with air. For this reason, plants affected by heat stress tend to develop more small leaves than plants growing in environments without stress.

Higher temperatures change root architecture, acting on primary root elongation rate and the overall shape of the root system giving a shallower and broader root distribution and a general increase in lateral root branching as stated by Gray and Brady [38]. The same authors noted that reproductive growth was altered by heat stress for rice which showed an optimum at 33 °C for vegetative growth, while grain formation and yield were negatively affected by temperatures above 25 °C. Moreover temperatures above 33 °C reduced viability of pollen which reached zero for temperatures of 40 °C with a similar phenomenon for sorghum (optimum at 26–34 °C for vegetative growth and at 25–28 °C for reproductive growth) and for Arabidopsis (*Arabidopsis thaliana*), where the abortion of the whole inflorescence was observed at a temperature of 36 °C. Gray and Brady [39] emphasized the cellular response to temperature stress which includes altered organization of organelles, cytoskeleton, and membrane structure. To maintain membrane stability and normal cellular functions in the presence of heat stress, plants synthesize heat shock proteins (HSPs), molecular chaperones that prevent protein misfolding or aggregation, as well as other co-chaperones, hormones, and other protective molecules. Expression of HSPs is induced by heat-stress transcription factors (HSFs) that bind to heat shock elements in the promoters of HSPs. There are many steps of regulation allowing dynamic control of the heat stress response, as the HSFs themselves can be post-transcriptionally modified. In addition to the constitutive role that HSPs play in heat stress responses across cell types, these proteins can acquire specialized functions that regulate developmental responses of organs to environmental stress.

3.4. Agronomic Approach to Thermal Resources and Limitations

The choice of agronomic solutions should be based on a sufficiently detailed territorial agroclimatic analysis, and finalized to quantify thermal resources and limitations for different environments. In such an analysis, it is appropriate to consider the traditional agronomic choices typical of the selected territory because in many cases they are the consequence of a secular adaptation that has discarded fewer effective solutions. In this sense, for example, the Italian alpine territories have a traditional use for forestry and pasture for the slopes exposed to the north, while those exposed to the south are primarily used for fruiting or viticulture which are concentrated in the thermal belt, as in great valleys such as the Valtellina, Valle d'Aosta, and Adige River Valleys, which are located between 250 and 750 m above sea level (asl). Similarly, the Apennine territories have a typical summer crops of maize, sunflower (*Helianthus annuus*), and tobacco (*Nicotiana tabacum*) in the valley bottoms, most exposed to winter and spring frost, while grapevines are located on the lower slopes, at intermediate risk of frost, and finally olive, most sensitive to frost, is traditionally located in the thermal belt which in the Apennines is often between 150 and 350 m asl.

Agroclimatic analysis must be founded on meteorological data of at least 20–30 years close to the present time to provide information representative of the current climate. In the light of resources and limitations (not only the average but also the absolute minimum and maximum and the probability of yearly values lower or higher than specific thresholds), agronomic practices useful for optimizing resources and minimizing limitations can be employed.

Almost the entire set of agronomic practices (soil cultivation, pruning, weeding, fertilization, irrigation, and so on) influence temperatures by interfering with the terms of the energy balance of the field. Thus, only the most important decisions for optimizing thermal resources and minimizing thermal limitations in a given field/territory will be reported. They can be subdivided into strategic decisions (that involve farm activity for many years) and tactical actions (that involve the farm activity only for the current year). Obviously, all these decisions should be founded on both technical and economic evaluation.

The first fundamental decision is the right choice of species and varieties best suited to the planting area, and this should be done after evaluating the length of the growing season and the mean and extreme dates of the beginning of the main phenological stages (e.g., dates of bud-break and ripening must be compatible with the dates of the last killing frost), and the compatibility with the temperature resources (chill units for vernalization, NHH or GDD needed to close the crop cycle), and limitations (LTL, HTL).

The prevention of cold damage should be founded on the uniformity of field morphology to prevent cold air storage which may occur due to the presence of depressions and irregularities in the field slope. It is also important to have suitable openings in barriers (drywalls, compact rows of trees or bushes, etc.) that limit the drainage of cold air enhancing the risk of frost. The adoption of active frost protection systems (low volume sprinkler irrigation over or under-canopy, stoves, candles, etc.) is a relevant decision that should be taken with a careful evaluation of the installation and handling costs and staffing requirements related to system operability.

For poorly drained fields, where a relevant limitation is water excess during winter that slows the spring warming of soils, the hydraulic arrangement of the terrain with ditches or subsurface drains is of paramount importance.

The choice of suitable training systems is important, both for the prevention of cold and heat damage. In fact, the height of the canopies of grapevines and fruit trees is directly related to the risk of frost damage because the coldest layers are closest to the ground. Furthermore, fruit damage from heat should be avoided by adopting training systems that protect fruit with a layer of leaves mitigating thermal extremes.

The risk of high temperatures should be evaluated during planning of harvest activities to prevent loss of quantity and quality of products. In this regard, consideration of options may include harvest machinery, work organization and machinery utilization. For example, harvesting grapes at night, use of refrigerated bins for transport of harvested clusters, and a quick crushing soon after harvesting are solutions sometimes adopted for hot climate viticulture to prevent unwanted fermentation.

Among the tactical decisions, making a rational choice regarding the sowing or transplanting periods for annual herbaceous species and of the planting times for perennial crops (trees, shrubs and herbaceous species) is also important. This decision should be founded on weather forecast information derived from good quality climatological data.

It should not be forgotten that, in conditions of water scarcity, plants close their stomata and the cooling of the foliar tissue are entrusted to the air. In this case, it is useful to plant the crops in rows parallel to the dominant winds to favor the penetration of air into the canopies.

The appropriate use of irrigation is also important in order to limit the negative effects of high temperatures on field crops. The mitigating effect of irrigation on maximum temperatures at a territorial scale was described by Lobell and Bonfils [39] who, working on a long time series of air temperatures for California (1934–2002), highlighted the strong mitigating effect of irrigation that was quantified in a 5 °C reduction in maximum temperature of the entire surface of the State had it been irrigated.

A melon (*Cucumis melo* L.) crop is transplanted early in spring when air and soil temperatures reach 14 °C in order to harvest the fruit early to get the highest price on the market. Lower temperatures can induce chilling injury and delay the recovery and the production of the plants [40]. The agronomic strategy to reduce low temperature damage is to use mulches or rowcovers [41]. Mulches with black or white plastic films can reduce thermal excursion. A black film can increase the soil temperature and increase G&D shortening the production cycle. Rowcovers can increase the soil and air temperature around the transplanted plants.

In summer, high temperatures can accelerate G&D of the plants too much, inducing a fast ripening of fruits with a shortened growing cycle. Such fast growth usually reduces the quality of fruis since the photosynthetic period is short and the sugar loading in fruit is reduced, resulting in low quality fruit. The best strategy is to avoid this problem is to identify the best genotype, cultivar or variety, since there are differences in sugar accumulation during ripening related to sucrose phosphate synthase and acid invertase activities [42]. In the open field, another agronomic strategy to reduce the negative effect of high temperature is to improve thermoregulation by keeping the soil water content high, if there is adequate water availability.

Thermal stress conditions can be detected by measurements carried out with thermal infrared cameras managed by hand or installed on drones or satellites. Symptoms of thermal stress can also occur in the presence of pests and diseases that alter stomatal functionality which is essential for plant thermal regulation.

Table 1. Cardinal temperatures for some crops with the sources of information used.

Common Name	Scientific Name	Minimum Cardinal	Optimal Range	Maximum Cardinal	Reference
Alfalfa	*Medicago sativa* L.	8	24–26	36	[43]
Asparagus	*Asparagus officnalis* L.	4	18–22	28	[44]
Banana	*Musa* seppe. L.	12	25–30	40	[43]
Barley	*Hordeum vulgare* L.	2	18–28	34	[45]
Bean	*Phaseolus vulgaris* L.	10	24–30	36	[45]
Carrot	*Daucus carota* L.	3	16–22	28	[44]
Cotton	*Gossypium hirsutum* L.	14	25–30	38	[33]
Durum wheat	*Triticum durum* L.	2	18–26	32	[44]
Flax	*Linum usitastissimum* L.	2	18–24	30	[44]
Grapevine	*Vitis vinifera* L.	7–10	22–28	36	[26]
Lemon	*Citrus limon* L.	13	23–30	35	[43]
Maize	*Zea mays* L.	8	22–30	35	[44]
Melon	*Cucumis melo* L.	15	25–35	40	[44]
Oat	*Avena sativa* L.	2	18–26	32	[44]
Okra	*Abelmoschus esculentus* L.	16	25–35	40	[44]
Olive	*Olea europaea* L.	10	22–28	38	[43]
Onion	*Allium cepa* L.	2	20–28	34	[43]
Sweet potato	*Hipomea batata* L.	15	25–33	38	[44]
Peanuts	*Arachis ipogea* L.	11	23–30	40	[33]
Pineapple	*Ananas comosus* L.	15	22–30	35	[43]
Pea	*Pisum sativum* L.	4	15–20	30	[43]
Tomato	*Solanum lycopersicum* L.	12	22–26	35	[33]
Potato	*Solanum tubrerosum* L.	4	14–23	33	[46]
Rapeseed	*Brassica napus oleifera* L.	5	15–20	30	[47]
Rice	*Oryza sativa* L.	12	25–32	38	[33]
Rye	*Secale cereale* L.	2	20–26	31	[44]
Safflower	*Carthamus tinctorius* L.	10	18–28	35	[43]
Soft wheat	*Triticum aestivum* L.	2	18–26	32	[45]
Sorghum	*Sorghum bicolor* L.	12	24–30	36	[43]
Soybean	*Glycine max* L.	10	20–28	34	[43]
Strawberry	*Fragaria X ananassa* L:	4	15–20	28	[44]
Sugar Beet	*Beta vulgaris* L.	2	18–24	30	[47]
Sugarcane	*Saccharum officinarum* L.	15	22–30	35	[43]
Sunflower	*Helianthus annuus* L.	7	18–25	35	[43]
Tobacco	*Nicotiana tabacum* L.	15	22–30	38	[43]
Watermelon	*Citrullus lanatus* L.	12	22–30	35	[43]

4. Relative Humidity and Effects on Crops

Plants are insensitive to the absolute atmospheric water content expressed for example as mixing ratio or absolute humidity [4,48], while they are quite sensitive to relative humidity (RH) which at a given temperature is the water content of the atmosphere expressed as a percentage of the saturated water content, which is a constant at a given temperature. RH is an important environmental variable for crop productivity, because it regulates the transpiration rate at the leaf level and can influence the water balance in crops. A high RH limits transpiration and reduces growth and nutrient assimilation. A low RH increases water flux through plants and increases transpiration with severe problems in species with a reduced ability to regulate stomatal aperture.

4.1. Space and Time Variability of Relative Humidity

The RH in a plant canopy is the result of a balance between humidity received from soil evaporation (which is enhanced when the soil surface is well-watered by rainfall, irrigation or by lifting from water tables) and plant transpiration (which in the absence of soil water limitation is a function only of the atmosphere and canopy features). Wind is crucial for removing water from canopy layers, including breezes and local winds induced by lack of thermal homogeneity among land surfaces, sea, lakes, forests, swamps, cropping areas, etc. During the day lack of homogeneity is induced by differential solar heating of surfaces with different characteristics that trigger the establishment of stationary convective cells in the planetary boundary layer (first 1000–1500 m asl), while at night it is induced by differential radiative cooling with production of cold air pools that drain along the relief. Another important determinant of RH is foehn, a katabatic wind typical of areas located downwind of mountain chains [49]. Foehns cause substantial drops in RH producing stress conditions for crops mainly due to the substantial increase of water demand of the atmosphere. Also relevant to RH are other macroscale and circulation patterns typical of different tropical and mid latitude contexts; these are beyond the scope of this review but can be appreciated in meteorological treatises [50]. In the specific case of dynamic anticyclone weather patterns, which for Italy accounts for 50–60% of the total days of the year, the daily pattern of RH shows a regular behavior with a nighttime minimum in coincidence with low values of air temperature and a daily maximum in coincidence with high values of air temperature. Nevertheless, while AT shows a behavior more similar to a sinusoid with minimum at sunrise and maximum 3–4 h before sunset [51,52], RH shows two abrupt changes with an abrupt decrease in coincidence with the beginning of convective vertical exchanges between a canopy layer and upper layers triggered by solar heating in the morning and an abrupt increase with the fade out of vertical exchanges (disruption of convective cells) in coincidence with the evening decrease of solar radiation before sunset.

4.2. Effects of Low/High Relative Humidity on Crops

Low RH increases evapotranspiration, enhancing water needs of rainfed and irrigated crops and, consequently, the risk of water stress conditions due to the lack of easily accessible water in the area explored by roots. On the other hand, high values of RH reduce the quantity of transpired water [51], which reduces soil water stress. Nevertheless, a high RH in the presence of high values of solar radiation can create problems of thermal excess [53], because plants control the temperature of their tissues by transpiring water, which changes its state from liquid to vapor and removes 2450 J g^{-1} from plant tissues in the evaporated water.

It must also be considered that many organic substances like starch are hygroscopic, which means that grains in presence of high relative humidity are re-hydrated. Multiple cycles of drying/re-hydration enhance the cracking phenomenon in rice with a significant reduction in commercial value of the product [54].

RH plays an important role in plant nutritional status because plants exposed to high RH show two contrasting effects on growth. On the one hand, plants might show increased growth due to higher stomatal opening, leading to increased uptake of CO_2. On the other hand they might show reduced growth due to a reduced transpiration volume leading to a lower translocation of nutrients [55].

High values of RH are favorable to parasitic fungi with a direct effect on their growth and development and an indirect one on the family of Peronosporaceae that needs the presence of condensed water on plant organs (leaves and shoots) for the activity of zoospores which is an essential part of their cycle. Consequently, many agronomic practices have as main or secondary objectives the avoidance of a long persistence of excessively low or high values of RH inside the canopy. Among these practices are:

1. The choice of areas not excessively humid, avoiding for example valley bottoms or basins with low air movement;
2. The adoption of suitable training systems for grapevine and other fruit crops;
3. The adoption of suitable plant distances along and between rows;
4. The execution of suitable tillage practices that enhance air circulation inside the canopy (e.g., weed management, cutting of grass in orchards and vineyards, leaf removal, winter or green pruning, etc.).

Dew and hoarfrost are the result of water condensation in the presence of a saturated atmosphere on objects (plants, rocks, and others) which have a temperatures below the dew point or the 0 °C threshold. The contribution of dew or hoarfrost to satisfaction of the water needs of crops is generally low because the dew accumulating in a single night at mid latitudes is often below 0.2 mm [56,57] while the mean daily water need for a reference crop in summer is about 4–6 mm. Dew contribution may be important in areas located close to large sources of humidity like seas or lakes and with high daily thermal ranges where water from dew can exceed 0.5–1 mm which means 0.5–1 L m^{-2}.

5. Future Prospective and Conclusions

Understanding the sources of abiotic stresses, how plants respond to them for improving tolerance, and the use of specific agronomy strategies for stress alleviation is essential. With reference to the scheme of the production model in Figure 2 of Mariani and Ferrante [1], the present analysis of resources and limitations related to light radiation, temperature, and RH has been performed. This analysis follows the analysis of abiotic factors carried out in Mariani and Ferrante [1].

It is important to consider that crop management needs an overall view of the atmospheric variables and the meteorological events that occur in a given year. For example, in the case of the temperate rice cropping areas (about the 35% of the total rice world area), it is well known that:

1. Low values of global solar radiation and suboptimal/supra-optimal temperatures reduce photosynthesis, while low daily thermal ranges reduced the translocation of biomass from leaves to storage organs:
2. Water for paddy rice submersion can be insufficient in years with low levels of water resources;
3. Long rainy periods are serious obstacles to the rational management of weeds, pests, and pathogens, preventing timely spraying of pesticides, fungicides, and herbicides;
4. Low temperatures during floral differentiation enhance rice male sterility with drastic reduction of yields. The critical threshold for rice sterility is 12 °C and farmers generally counteract this problem by increasing the water level in rice ponds;
5. Long periods of high temperatures (mean daily values above the 26 °C threshold) enhance rice cracking and rice chalking, phenomena that reduce significantly the quality of the final product and that are also strictly related to varieties and nutrient management;
6. Meteorological variables like temperature, relative humidity, and rainfall have a strong influence on rice pests, weeds, and pathogens.

Diseases impose a series of strategic and tactical choices suitable for preventing and controlling them, ranging from the choice of varieties and sowing periods to the strategies of water, pest, and weed management. These strategies have evolved over the centuries, and have been refined based on the available technologies and the experience acquired by the farmers, in response to the constant interaction with the climate which characterizes the area under examination. For the main European rice cropping area, located in the north of Italy (where rice has been cultivated since the 15th century), an historical document of great value for appreciating this aspect is provided by the correspondence between Camillo Benso, Count of Cavour, and his partner Giacinto Corio [58,59] who managed the three large farms of Leri, Torrone, Montarucco with a total area of more than 400 hectares. From these letters emerge the profound agronomic culture that characterizes not only Giacinto Corio, but also Camillo Benso who was then prime minister of the kingdom of Piedmont and the main father of the unification of Italy. Today, this agronomic culture derived from constant interaction of the farmer with atmospheric phenomena is a key element of global food security, and the task of science is to make it more rational and based on qualitative data and interpretative and predictive models. The mitigation of stressful conditions can be achieved using appropriate agronomic management practices that have to be chosen harmonizing all production factors in a specific area with well known pedo-climatic conditions.

In this two part review, we have analyzed some relevant abiotic stress factors (drought, hypoxia, and lodging in the first part [1], and temperature, light intensity, and relative humidity in this second part). We have described the stress factors, and the agronomic strategies for mitigation and adaptation to them. From this we can derive some suggestions to guide future approaches to such abiotic stresses.

Firstly, it is necessary to improve our knowledge of the main stress factors in each production area, including their average incidence, and their variability in space and time. These analyses are now feasible by the availability of monitoring tools for agroclimatic conditions, soils, and crops. For the future a more systematic approach to these analyses is needed. It is of paramount importance that a more general adoption of techniques for timely monitoring of crop performance and pedoclimatic conditions (e.g., agrometeorological stations and remote sensing techniques based on satellites and drones) that allow defining the mesoscale and microscale variability of the various stress factors be established. In addition, the availability of time series of meteorological, phytopathological and productive data is also essential. These series at the regional level should be available to farmers and technicians by authorities. Farmers should take care of collection and archiving of time series within their farm.

Another essential factor to promote is the knowledge of the ability of different species and varieties to adapt to stress factors to be able to counteract the stress with a suitable varietal choice and crop succession. Genetic improvement of crops must be accompanied by the adoption of suitable and innovative cultivation techniques (mechanization, training systems for orchards and vineyards, hydraulic-agrarian arrangements, tillage, irrigation, fertilization, weeding, pest control, etc.).

The choice of the optimal times for the execution of the different field activities (times for soil cultivation, seeding periods, and harvesting) is another crucial factor to limit the negative effects of stress.

The adoption of precision farming techniques is useful for dominating the microscale variability of the different stress factors. Moreover, the adoption of conservative agriculture practices allows maintaining the levels of the resources within the limits of acceptability for different crops (e.g., dry farming).

Use of computerized decision support systems (DSS), and extension activities, are essential for engaging farmers in the process of innovation in genetics and crop production techniques. This aspect is more important considering that at the world level agriculture is carried out by 590 million farms with characteristics that are extremely diversified by size, number of workers, levels of mechanization, and openness to the market.

It is also important to maintain a central role for agronomy seen as the science of cultivation. This discipline should maintain its own centrality as a science that guides the producer to optimize their results in terms of quantity and quality of the products obtained.

A further element worthy of comment is the interaction between agronomic techniques and genetic innovation aimed at the improvement of the resistance to abiotic stresses. A complementary relationship should generally be desirable because every new variety needs to be embedded in an agronomic context favorable to the full expression of its potential resistance to abiotic stresses. Furthermore, agronomic techniques can be a realistic way to mitigate the negative effects of abiotic stresses while strategic actions of genetic improvement are carried out and the new varieties are available to farmers. In conclusion, we hope that this analysis, obviously somewhat brief given the vastness of the topic, can help those who for different purposes are concerned with the determinants of agricultural production.

Author Contributions: Temperature and relative humidity, L.M.; solar Radiation A.F. and L.M.; writing A.F. and L.M.

References

1. Mariani, L.; Ferrante, A. Agronomic Management for Enhancing Plant Tolerance to Abiotic Stresses—Drought, Salinity, Hypoxia, and Lodging. *Horticulturae* **2017**, *3*, 52. [CrossRef]
2. Boyer, J.S. Plant productivity and environment. *Science* **1982**, *218*, 443–448. [CrossRef] [PubMed]
3. Farooq, M.; Wahid, A.; Kobayashi, N.; Fujita, D.; Basra, S.M.A. Plant drought stress: Effects, mechanisms and management. *Agron. Sustain. Dev.* **2009**, *29*, 185–212. [CrossRef]
4. Stull, R. *Practical Meteorology: An Algebra-Based Survey of Atmospheric Science*; Version 1.02b; University of British Columbia: Vancouver, BC, Canada, 2017; p. 940, ISBN 978-0-88865-283-6. Available online: https://www.eoas.ubc.ca/books/Practical_Meteorology/ (accessed on 15 July 2018).
5. Long, S.P.; Bernacchi, C.J. Gas exchange measurements, what can they tell us about the underlying limitations to photosynthesis? Procedures and sources of error. *J. Exp. Bot.* **2003**, *54*, 2393–2401. [CrossRef] [PubMed]
6. Hagemann, M.; Weber, A.; Eisenhut, M. Photorespiration: Origins and metabolic integration in interacting compartments. *J. Exp. Bot.* **2016**, *67*, 2915–2918. [CrossRef]
7. Osborne, C.P.; Beerling, D.J. Nature's green revolution: The remarkable evolutionary rise of C4 plants. *Philos. Trans. R. Soc. Lond. B Biol. Sci.* **2006**, *361*, 173–194. [CrossRef] [PubMed]
8. Osborne, C.P.; Freckleton, R.P. Ecological selection pressures for C4 photosynthesis in the grasses. *Proc. R. Soc. Lond. B Biol. Sci.* **2009**, *276*, 1753–1760. [CrossRef] [PubMed]
9. Greaves, G.E.; Wang, Y.M. The effect of water stress on radiation interception, radiation use efficiency and water use efficiency of maize in a tropical climate. *Turk. J. Field Crops* **2017**, *22*, 114–125. [CrossRef]
10. Kadir, S.; Von Weihe, M. Photochemical efficiency and recovery of photosystem ii in grapes after exposure to sudden and gradual heat stress. *J. Am. Soc. Hortic. Sci.* **2007**, *132*, 764–769.
11. Parent, B.; Tardieu, F. Can current crop models be used in the phenotyping era for predicting the genetic variability of yield of plants subjected to drought or high temperature? *J. Exp. Bot.* **2014**, *65*, 6179–6189. [CrossRef] [PubMed]
12. Bouman, B.A.M.; van Keulen, H.; van Laar, H.H.; Rabbinge, R. The 'School of de Wit' crop growth simulation models: A pedigree and historical overview. *Agric. Syst.* **1996**, *52*, 171–198. [CrossRef]
13. Avnish Kumar, B. *Crop Growth Simulation Modeling, in Modelling and Simulation of Diffusive Processes*; Springer: Berlin, Germany, 2014; pp. 315–332.
14. Wagenmakers, P.S. *Light Relations in Orchard Systems*; Wageningen University: Wageningen, The Netherlands, 1995; ISBN 90-5485-340-9. Available online: http://edepot.wur.nl/205150 (accessed on 15 July 2018).
15. Wertheim, S.J. Productivity and fruit quality of apple in single-row and fullfield plantings. *Sci. Hortic.* **1985**, *26*, 191–208. [CrossRef]
16. Hirose, T. Development of the Monsi-Saeki Theory on Canopy Structure and Function. *Ann. Bot.* **2005**, *95*, 483–494. [CrossRef] [PubMed]
17. Williams, L.E.; Ayars, J.E. Grapevine Water Use and the Crop Coefficient are Linear Functions of the Shaded Area Measured beneath the Canopy. *Agric. For. Meteorol.* **2005**, *132*, 201–211. [CrossRef]
18. Woods, D.P.; Ream, T.S.; Minevich, G.; Hobert, O.; Amasino, R.M. Phytochrome C Is an Essential Light Receptor for Photoperiodic Flowering in the Temperate Grass Brachypodium distachyon. *Genetics* **2014**, *198*, 397–408. [CrossRef] [PubMed]

19. Quail, P.H. Phytochrome photosensory signaling networks. *Nat. Rev. Mol. Cell Biol.* **2002**, *3*, 85–93. [CrossRef] [PubMed]
20. Pham, V.N.; Kathare, P.K.; Huq, E. Phytochromes and phytochrome interacting factors. *Plant Physiol.* **2018**, *176*, 1025–1038. [CrossRef] [PubMed]
21. Foyer, C.H.; Lelandais, M.; Kunert, K.J. Photooxidative stress in plants. *Physiol. Plant.* **1994**, *92*, 696–717. [CrossRef]
22. Trivellini, A.; Cocetta, G.; Francini, A.; Ferrante, A. Reactive oxygen species production and detoxification during leaf senescence. In *Reactive Oxygen Species and Antioxidant Systems in Plants: Role and Regulation under Abiotic Stress*; Khan, M.I.R., Khan, N., Eds.; Springer Nature Singapore Ltd.: Singapore, 2017; pp. 115–128.
23. Cola, G.; Failla, O.; Mariani, L. BerryTone, a simulation model for the daily course of grape berry temperature. *Agric. For. Meteorol.* **2009**, *149*, 1215–1228. [CrossRef]
24. Iglesias, I.; Alegre, S. The effect of anti-hail nets on fruit protection, radiation, temperature, quality and profitability of 'Mondial Gala'apples. *J. Appl. Hortic.* **2006**, *8*, 91–100.
25. Melgarejo, P.; Martınez, J.J.; Hernández, F.; Martınez-Font, R.; Barrows, P.; Erez, A. Kaolin treatment to reduce pomegranate sunburn. *Sci. Hortic.* **2004**, *100*, 349–353. [CrossRef]
26. Mariani, L.; Failla, O. *Clima e Viticoltura, Capitolo 2 del Testo Progressi in Viticoltura, a Cura di M. Boselli*; EDISES Universitaria: Napoli, Italy, 2016; pp. 19–38.
27. Annunziata, M.G.; Apelt, F.; Carillo, P.; Krause, U.; Feil, R.; Mengin, V.; Lunn, J.E. Getting back to nature: A reality check for experiments in controlled environments. *J. Exp. Bot.* **2017**, *68*, 4463–4477. [CrossRef] [PubMed]
28. Zhao, X.Y.; Yu, X.H.; Liu, X.M.; Lin, C.T. Light regulation of gibberellins metabolism in seedling development. *J. Integr. Plant Biol.* **2007**, *49*, 21–27. [CrossRef]
29. Grebe, M. Out of the shade and into the light. *Nat. Cell Biol.* **2011**, *13*, 347. [CrossRef] [PubMed]
30. Halliday, K.J.; Fankhauser, C. Phytochrome-hormonal signalling networks. *New Phytol.* **2003**, *157*, 449–463. [CrossRef]
31. Woodrow, P.; Ciarmiello, L.F.; Annunziata, M.G.; Pacifico, S.; Iannuzzi, F.; Mirto, A.; Carillo, P. Durum wheat seedling responses to simultaneous high light and salinity involve a fine reconfiguration of amino acids and carbohydrate metabolism. *Physiol. Plant.* **2017**, *159*, 290–312. [CrossRef] [PubMed]
32. Larcher, W. *Physiological Plant Ecology*, 3rd ed.; Springer: Berlin, Germany, 1995; p. 506.
33. Luo, Q. Temperature thresholds and crop production: A review. *Clim. Chang.* **2011**, *109*, 583–598. [CrossRef]
34. Mariani, L. *Agronomia*; CUSL: Milano, Italian, 2014; p. 344. (In Italian)
35. Mariani, L. Carbon plants nutrition and global food security. *Eur. Phys. J. Plus* **2017**, *132*, 69. [CrossRef]
36. Körner, C.; Basler, D. Phenology under Global Warming. *Science* **2012**, *327*, 1461–1462. [CrossRef] [PubMed]
37. De Boeck, H.J.; Van De Velde, H.; De Groote, T.; Nijs, I. Ideas and perspectives: Heat stress: More than hot air. *Biogeosciences* **2016**, *13*, 5821–5825. [CrossRef]
38. Gray, S.B.; Brady, S.M. Plant developmental responses to climate change. *Dev. Biol.* **2016**, *419*, 64–77. [CrossRef] [PubMed]
39. Lobell, D.B.; Bonfils, C. The Effect of Irrigation on Regional Temperatures: A Spatial and Temporal Analysis of Trends in California, 1934–2002. *J. Clim.* **2008**, *21*, 2063–2071. [CrossRef]
40. Korkmaz, A.; Dufault, R.J. Developmental consequences of cold temperature stress at transplanting on seedling and field growth and yield. II. Muskmelon. *J. Am. Soc. Hortic. Sci.* **2001**, *126*, 410–413.
41. Jenni, S.; Stewart, K.A.; Cloutier, D.C.; Bourgeois, G. Chilling injury and yield of muskmelon grown with plastic mulches, rowcovers, and thermal water tubes. *HortScience* **1998**, *33*, 215–221.
42. Hubbard, N.L.; Huber, S.C.; Pharr, D.M. Sucrose phosphate synthase and acid invertase as determinants of sucrose concentration in developing muskmelon (*Cucumis melo* L.) fruits. *Plant Physiol.* **1989**, *91*, 1527–1534. [CrossRef] [PubMed]
43. Schultink, G.; Amaral, N.; Mokma, D. *Users Guide to the CRIES Agro-Economic Information System Yield Model*; Michigan State University: East Lansing, MI, USA, 1987; p. 125.
44. USDA (United States Department of Agriculture). *Technical Bulletin*; USDA: Washington, DC, USA, 1977; pp. 1516–1525.
45. Weikay, Y.; Hunt, L.A. An Equation for Modelling the Temperature Response of Plants using only the Cardinal Temperatures. *Ann. Bot.* **1999**, *84*, 607–614.
46. Rykaczewska, K. The Impact of High Temperature during Growing Season on Potato Cultivars with different response to Environmental Stresses. *Am. J. Plant Sci.* **2013**, *4*, 2386–2393. [CrossRef]

47. North Dakota Agricultural Weather Network. *Sunflower Development and Growing Degree Days (GDD)*; Crop and Pest Report 2016; North Dakota State University: Fargo, ND, USA; Available online: https://ndawn.ndsu.nodak.edu/help-sunflower-growing-degree-days.html (accessed on 15 July 2018).
48. Mc Intosh, D.H.; Thom, A.S. *Essentials of Meteorology*; Wikeham Publications: London, UK, 1972; p. 239.
49. Ambrosetti, P.; Mariani, L.; Scioli, P. Climatology of north foehn in Canton Ticino and Western Lombardy. *Riv. Ital. Agrometeorol.* **2005**, 2, 24–30.
50. Barry, R.G.; Chorley, R.J. *Atmosphere, Weather and Climate*; Routledge: London, UK, 2009; p. 499.
51. Allen, R.G.; Pereira, L.S.; Raes, D.; Smith, M. *Crop Evapotranspiration-Guidelines for Computing Crop Water Requirements*; FAO Irrigation and Drainage Paper 56; Food and Agriculture Organization of the United Nations: Rome, Italy; Available online: http://www.fao.org/docrep/X0490E/X0490E00.htm (accessed on 15 July 2018).
52. Parton, W.J.; Logan, J.A. A model for diurnal variation in soil and air temperature. *Agric. Meteorol.* **1981**, *23*, 205–216. [CrossRef]
53. Hatfield, J.L.; Prueger, J.H. Temperature extremes: Effect on plant growth and development. *Weather Clim. Extremes* **2015**, *10*, 4–10. [CrossRef]
54. Lu, R.; Siebenmorgen, T.J. Modeling rice field moisture content during the harvest season-part 1-model development. *Trans. ASAE* **1994**, *37*, 545–551. [CrossRef]
55. Roriz, M.; Carvalho, S.M.P.; Vasconcelos, M.W. High relative air humidity influences mineral accumulation and growth in iron deficient soybean plants. *Front. Plant Sci.* **2014**, *5*, 726. [CrossRef] [PubMed]
56. Beysens, D.; Muselli, M.; Nikolayev, V.; Narhe, R.; Milimouk, I. Measurement and modelling of dew in island, coastal and alpine areas. *Atmos. Res.* **2005**, *73*, 1–22. [CrossRef]
57. Kabela, E.D.; Hornbuckle, B.K.; Cosh, M.H.; Anderson, M.C.; Gleason, M.L. Dew frequency, duration, amount, and distribution in corn and soybean during SMEX05. *Agric. For. Meteorol.* **2009**, *149*, 11–24. [CrossRef]
58. Fondazione Camillo Cavour. *Lettere di Giacinto Corio a Camillo Cavour (1843–1855)*; Fondazione Camillo Cavour: Santena, Italy, 1980; p. 474. (In Italian)
59. Visconti, E. *Cavour Agricoltore, Lettere Inedite di Camillo Cavour a Giacinto Corio Precedute da un Saggio di Ezio Visconti, G*; Barbera editore: Firenze, Italy, 1913; p. 390.

13

Assessing Quantitative Criteria for Characterization of Quality Categories for Grafted Watermelon Seedlings

Filippos Bantis [1], Athanasios Koukounaras [1,*], Anastasios Siomos [1], Georgios Menexes [1], Christodoulos Dangitsis [2] and Damianos Kintzonidis [2]

1 School of Agriculture, Aristotle University of Thessaloniki, 54124 Thessaloniki, Greece; fbanths@gmail.com (F.B.); siomos@agro.auth.gr (A.S.); gmenexes@agro.auth.gr (G.M.)
2 Agris S.A., Kleidi, 59300 Imathia, Greece; cdaggitsis@agris.gr (C.D.); damianos@agris.gr (D.K.)
* Correspondence: thankou@agro.auth.gr.

Abstract: Vegetable grafting is a practice employed worldwide since it helps prevent biotic and abiotic disorders, and watermelon is one of the most important species grafted. The objective of this study was to set critical limits for the characterization of quality categories for grafted watermelon seedlings. Specifically, watermelon (scion) seedlings were grafted onto squash (rootstock) seedlings, moved into a healing chamber for 7 days, and then transferred into a greenhouse for seven more days. At 7 and 14 days after grafting, experienced personnel assessed grafted seedling quality by categorizing them. The categories derived were Optimum and Acceptable for both time intervals, plus Not acceptable at 14 days after grafting. Optimum seedlings showed greater leaf area, and shoot and root fresh and dry weights at both time intervals. Moreover, they had greater stem diameter, root-to-shoot ratio, shoot dry weight-to-length ratio and Dickson's quality index compared to the other category at 14 days after grafting. Therefore, Optimum seedlings would likely develop into marketable plants of high quality, with better establishment in the field. Not acceptable seedlings showed considerably inferior development, while Acceptable seedlings were between the other categories, but were still marketable.

Keywords: *Citrullus lanatus*; vegetable grafting; optimal production; marketable seedlings; quality indicators

---

1. Introduction

The use of grafting for vegetable seedlings is a well-established practice worldwide [1–3]. The important advantage of using grafted seedlings to prevent significant crop loss due to biotic [3,4] and abiotic [5,6] factors (soil-borne diseases, salinity, low temperatures, etc.), as well as the reduction of the use of agrochemical products, provide grafting as an environmentally-friendly practice [3]. Applications of grafting mainly focus on species of the Cucurbitaceae and Solanaceae families, particularly watermelon and tomato, respectively [2,7]. In Greece the use of grafted watermelon seedlings is almost 99% of growers using low tunnel protected cultivation in order to achieve early yield.

To enjoy the advantages of grafting, use of high quality grafted seedlings is a prerequisite. As a result, a rapid development and expansion of a vegetable nursery industry is in progress. Successful grafting requires good connection between the rootstock and the scion for healthy uniform growth and development of the grafted plants [8]. The most important stage for seedling evaluation for the grower is at the time of purchase and transplanting to avoid possibly negative results during subsequent cultivation. However, the definition of what constitutes a high-quality seedling is a very complicated issue [8].

Generally, high-quality seedlings could be defined as plants uniform in size and traits, proper size or height with a thick healthy stem with large thick leaves, a well-developed root system, good root-to-shoot ratio, and a good ratio of shoot dry weight divided by shoot length [3,8]. However, most of the above characteristics could be suitable for some species such as tomatoes, peppers, and eggplants, but could be adapted for the cucurbits [3]. Moreover, most seedling quality parameters have been qualitatively determined by experienced personnel. Therefore, it is critical to define quantitative parameters that could be easily and objectively applied by anyone in this chain (industry, grower, etc.).

Therefore, the aim of this study was to set critical limits for objective measurements of grafted watermelon seedling quality categories as well as to suggest the most accurate and convenient among them for application by the industry and growers. Evaluations were conducted at 7 and 14 days after grafting. Grafted watermelon seedlings exit the healing chamber at 7 days after grafting and therefore quality assessment at that time is essential for possible later research. By 14 days, seedlings are considered "final product" and therefore quality evaluation at that time is valuable to assess the product marketability.

2. Materials and Methods

2.1. Plant Material

The experiments were conducted in the facilities of Agris S.A. in Kleidi, Imathia, Greece. All measurements were executed at Aristotle University of Thessaloniki, Greece. During the experiment, standard commercial practices were applied.

Watermelon (scion—*Citrullus lanatus*) "Celine" (HM. Clause SA, Portes-Les-Valence, France) and squash (rootstock—*Cucurbita moschata*) "TZ-148" (HM. Clause SA, Portes-Les-Valence, France) were used for the production of grafted seedlings. Watermelon seeds were sown in plastic 171-cell plug trays, while squash seeds were sown in plastic 128-cell plug trays (both types: 67 × 33 cm, G.K. Rizakos S.A., Lamia, Greece). Both plug tray types were filled with a 5:1:2 mixture of peat, perlite, and vermiculite.

2.2. Germination, Grafting, Healing, and Acclimatization

Following planting, the plug trays of scions and rootstocks were moved into a growth chamber (25 °C, 95–98% relative humidity (RH)) until germination. Watermelon and squash germinated after 72 and 48 h, respectively, and afterwards they were moved to a glass greenhouse for 9 (scion) and 10 (rootstock) days at a 21.5 °C minimum night temperature for both species, and 100 ± 10 μmol m^{-2} s^{-1} photosynthetic photon flux density (PPFD) emitted by high-pressure sodium (HPS) lamps (MASTER GreenPower 600 W, 400 V E40, Philips Lighting, Eindhoven, The Netherlands) with an 18 h photoperiod only for watermelon. The natural light photoperiod during the experiment was 12.5 h from sunrise to sunset. Supplemental lighting is commonly practiced for the production of watermelon seedlings in order to achieve high quality product. On the other hand squash seedlings have adequate development under natural light conditions and no supplemental lighting is employed.

Grafting was performed with the "splice grafting" technique, 12 days after sowing. Using a razor blade, the scion was diagonally cut just below the cotyledons while the rootstock was diagonally cut on the cotyledon level leaving only one cotyledon. The rootstock was also cut just above ground level, which is a commonly practiced technique for cucurbit rootstocks in order to achieve increased grafting efficiency [9]. Afterwards, the grafted seedlings were placed in polystyrene 72-cell plug trays (50 × 30 cm, G.K. Rizakos S.A., Lamia, Greece) filled with a 3:1:1 mixture of peat, perlite, and vermiculite. Grafting was performed by experienced personnel to minimize critical errors.

Following grafting, healing and acclimatization of grafted seedlings was achieved during 7 days in a growth chamber at 25 °C, recirculating air, 45 μmol m^{-2} s^{-1} PPFD emitted by fluorescent tubes (Fluora 58W, Osram, GmbH, Munich, Germany) for an 18 h photoperiod, and RH of 98% for days 1–4, 93% for day 5, and 89% for days 6 and 7. RH was high at the beginning of healing in order to prevent leaf dehydration and it was gradually decreased in order for the seedlings to get acclimated to lower RH conditions. The growth conditions were monitored using a climate control system (Priva SA, De Lier, The Netherlands).

After 7 days in the healing chamber the grafted seedlings were placed in a glass greenhouse (21.5 °C minimum night temperature, 60 ± 10 μmol m^{-2} s^{-1} PPFD emitted by HPS lamps for an 18 h photoperiod). The high RH was applied to prevent leaf dehydration due to water loss.

2.3. *Quality Categorizing and Measurements*

Seedling quality categorizing, sampling and measurements were conducted at two times, 7 and 14 days after grafting. Experienced personnel categorized quality. The quality categories are listed in Table 1. Critical parameters for categorizing seedling quality were true leaf and cotyledon area, cotyledon color, and root system development (personal communication with Agris S.A., Kleidi, Imathia, Greece). In total, 50 seedlings per category were sampled and the number of samples was equally distributed throughout the three periods of production

Table 1. Seedling quality categories derived seven and 14 days after grafting of watermelon seedlings.

Days after Grafting	Quality Categories	Marketable
7	Optimum	Yes
	Acceptable	Yes
14	Optimum	Yes
	Acceptable	Yes
	Not Acceptable	No

A digital caliper (Powerfix, Milomex, Pulloxhill, UK) was used to measure shoot height, stem diameter (about 1 cm above the substrate surface), and thickness of true leaves, scion cotyledons or rootstock cotyledons. Leaf area of true leaves, scion cotyledons or rootstock cotyledons were measured using a leaf area meter (LI-3000C, LI-COR Biosciences, Lincoln, NE, USA). Fresh and dry weights of shoots (stem and leaves) and roots were determined. Dry weights were obtained after three days of drying in an oven. Moreover, root-to-shoot (R/S) dry weight ratio, shoot dry weight-to-length (DW/L) ratio, and Dickson's quality index (DQI) were estimated. DQI was calculated as follows [10]:

$$\text{Quality index} = \frac{\text{Seedling total dry weight (g)}}{\frac{\text{Height (mm)}}{\text{Stem diameter (mm)}} + \frac{\text{Shoot dry weight(g)}}{\text{Root dry weight (g)}}} \tag{1}$$

Relative chlorophyll content was measured using a portable chlorophyll meter (CCM-200 plus, Opti-Sciences, USA). Maximum quantum yield of primary photochemistry, variable to maximal fluorescence of dark-adapted leaves (F_v/F_m), was measured with a fluorometer (Pocket-PEA, Hansatech Instruments, Norflock, UK). Finally, the color of true leaves, scion cotyledons, and rootstock cotyledons was characterized using the colorimetric coordinates lightness (L*), Hue (h°), Chroma (C*), a*/b* (a*: red/green coordinate; b*: yellow/blue coordinate), and Hue (h°) obtained from a digital colorimeter (CR-400 Chroma Meter, Konica Minolta Inc., Tokyo, Japan) according to McGuire. [11].

2.4. *Statistical Analysis*

Statistical analysis was performed using IBM SPSS software (SPSS 23.0, IBM Corp., Armonk, NY, USA). Data measured at 7 days after grafting were analyzed using a t-test ($P \leq 0.05$), since at that time point the seedlings were grouped into only two quality categories (Optimum and Acceptable). Data measured at 14 days after grafting were analyzed within the methodological frame of one-way analysis of variance (ANOVA), since at that time point the seedlings were grouped at three quality categories (Optimum, Acceptable and not Acceptable). In this case, mean comparisons were conducted using the Scott-Knott procedure [12], at a significance level of $\alpha = 0.05$, using the StatsDirect v.2.8.0. statistical software (StatsDirect, Ltd., Grantchester, Cambridge, UK). The choice of the Scott-Knott method was based on its important and unique characteristic that does not present any overlapping in its grouping results. The above is critical in order to obtain quality indices that segregate the different quality categories and, therefore, overlapping results between groups must not occur.

3. Results

Seven days after grafting, shoots of Optimum seedlings were significantly longer than Acceptable ones (Table 2). However, stem diameter, shoot DW/L ratio, DQI and thickness of leaves and cotyledons (both scion and rootstock) did not show any differences between Optimum and Acceptable seedlings. Relative chlorophyll content of true leaves and cotyledons, as well as F_v/F_m were also similar in the two quality categories tested (Table 2). Colorimetric parameters of true leaves also did not show differences between the two categories, while scion and rootstock cotyledons showed significant differences in parameters such as lightness (L*), Hue (h°), a*/b* (a*: red/green coordinate; b*: yellow/blue coordinate), and h° (Table 3). Moreover, Optimum seedlings developed significantly greater true leaves and scion cotyledons compared to the Acceptable ones, while no differences were observed in rootstock cotyledons (Figure 1A). Similarly, fresh and dry weight production of shoots and roots were significantly greater for Optimum compared to Acceptable characterized seedlings (Figure 2A,B). However, R/S ratio did not exhibit any significant differences between the different categories (Figure 2C).

Table 2. Morphological and developmental parameters of grafted watermelon seedlings from two quality categories 7 days after grafting.

Parameters		Quality Categories	
		Optimum	Acceptable
Height (mm)		57.82 ± 1.14 a [y]	47.38 ± 1.03 b
Stem diameter (mm)		4.34 ± 0.04 a	4.23 ± 0.05 a
DW/L [z]		0.004 ± <0.001 a	0.004 ± <0.001 a
DQI		0.012 ± <0.001 a	0.012 ± 0.001 a
F_v/F_m		0.82 ± <0.01 a	0.82 ± <0.01 a
Thickness (mm)	True leaf	0.62 ± 0.02 a	0.59 ± 0.02 a
	Scion cot.	0.69 ± 0.01 a	0.69 ± 0.01 a
	Roots. cot.	1.20 ± 0.03 a	1.15 ± 0.03 a
Relative chl. content	True leaf	27.17 ± 1.26 a	29.14 ± 1.04 a
	Scion cot.	44.25 ± 1.38 a	43.37 ± 1.22 a
	Roots. cot.	56.50 ± 1.88 a	57.16 ± 2.31 a

[z] shoot dry weight-to-length ratio; DQI: Dickson's quality index; F_v/F_m: maximum quantum yield of primary photochemistry of a dark-adapted leaf; [y] Mean values (±SE) (n = 50), within a row, followed by different letters are significantly different by t-test ($P \leq 0.05$).

Table 3. Colorimetric parameters of grafted watermelon seedlings from two quality categories 7 days after grafting.

Plant Tissue	Colorimetric Parameters	Quality Categories	
		Optimum	Acceptable
True leaves	L* [z]	41.75 ± 1.39 a [y]	42.30 ± 1.96 a
	C*	24.76 ± 4.08 a	24.18 ± 3.69 a
	h°	129.05 ± 1.64 a	128.32 ± 1.62 a
	a*/b*	−0.81 ± 0.05 a	−0.79 ± 0.05 a
Scion cotyledons	L*	42.85 ± 1.53 b	44.12 ± 2.11 a
	C*	20.42 ± 2.31 a	20.61 ± 3.15 a
	h°	128.05 ± 1.21 a	126.54 ± 1.98 b
	a*/b*	−0.78 ± 0.03 b	−0.74 ± 0.05 a
Rootstock cotyledons	L*	39.32 ± 1.87 a	39.87 ± 2.00 a
	C*	19.66 ± 1.79 a	19.69 ± 2.54 a
	h°	130.07 ± 1.11 a	129.26 ± 1.68 b
	a*/b*	−0.84 ± 0.03 a	−0.82 ± 0.05 a

[z] lightness; C*: chroma; h°: hue angle; a*: red/green coordinate; b*: yellow/blue coordinate; [y] Mean values (±SE) (n = 50), within a row, followed by different letters are significantly different by t-test ($P \leq 0.05$).

Quite similar results were obtained between the quality categories at 14 days after grafting. Specifically, shoot height was significantly greater for Optimum seedlings compared to Acceptable ones, while shoot DW/L ratio was significantly greater for the two marketable categories compared to Not Acceptable seedlings. Stem diameter and DQI were greater for Optimum seedlings compared to the other categories. Nevertheless, thickness of leaves and cotyledons (both scion and rootstock) were similar for all categories. Relative chlorophyll content and F_v/F_m were not different between the three quality categories, similar to the 7 day measurements (Table 4). However, differences between the three categories were observed in colour, particularly parameters such as h° and a*/b* for the true leaf, and C*, L*, h°, and a*/b* for rootstock cotyledons (Table 5). At 7 days after grafting, Optimum scion cotyledons had a darker (lower L*) and greener (lower a*/b*) color compared to Acceptable seedlings. At 14 days after grafting, scion cotyledons did not show any color differences but rootstock cotyledons categorized as Optimum had darker (lower L*) and greener (lower a*/b*) color compared to the rest of the quality categories. True leaves of Optimum seedlings had significantly greater area compared to the other categories, but scion and rootstock leaf area were not different (Figure 1B). Moreover, fresh and dry biomass of shoots and roots, as well as R/S ratio were significantly greater for Optimum seedlings compared to the other categories (Figure 2D–F).

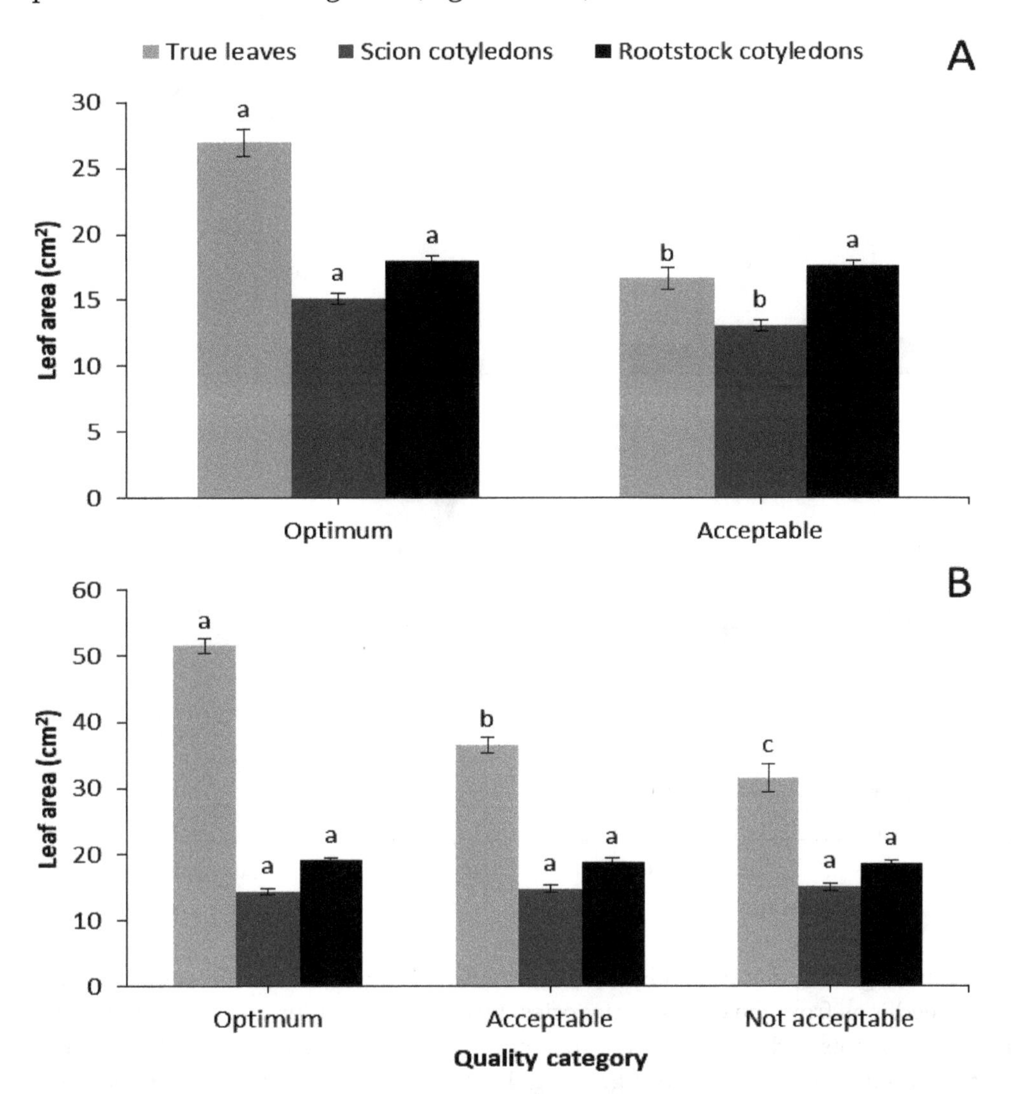

Figure 1. Leaf area of true leaves, scion cotyledons and rootstock cotyledons of grafted watermelon seedlings from the quality categories derived 7 days (**A**) or 14 days (**B**) after grafting. Each data point is a mean value ± standard error (SE) of the mean (n = 50). Bars of the same color (same tissue type) across categories with different letters are significantly different ($P \leq 0.05$) according to the results of the t-test (**A**) or the Scott-Knott method (**B**).

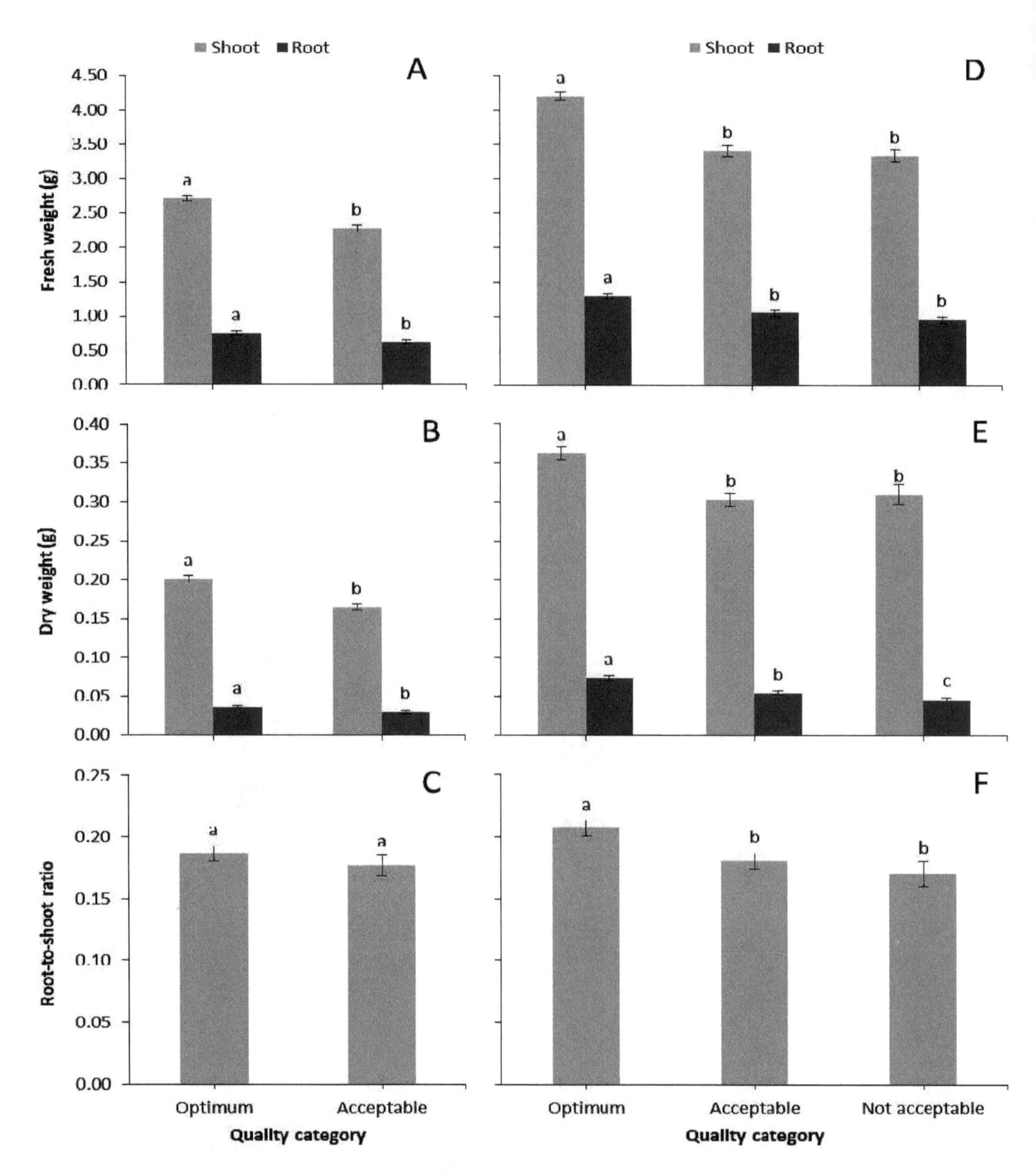

Figure 2. Fresh weight (**A**) and dry weight (**B**) of shoots and roots, and root-to-shoot ratio (**C**) of grafted watermelon seedlings from the quality categories derived 7 days after grafting. Fresh weight (**D**) and dry weight (**E**) of shoots and roots, and root-to-shoot ratio (**F**) of grafted watermelon seedlings from the quality categories derived 14 days after grafting. Each data point is a mean value of 50 observations. Error bars correspond to the standard error (SE) of the mean. Bars of the same colour (same tissue type) across categories followed by different letters are significantly different ($P \leq 0.05$) according to the results of the t-test (**A–C**) or the Scott-Knott method (**D–F**).

Table 4. Morphological and developmental parameters of grafted watermelon seedlings from two quality categories 14 days after grafting.

Parameters		Quality Categories		
		Optimum	Acceptable	Not Acceptable
Height (mm)		59.10 ± 1.19 a [y]	52.00 ± 1.20 b	56.80 ± 1.59 a
Stem diameter (mm)		4.65 ± 0.05 a	4.49 ± 0.07 b	4.35 ± 0.05 c
DW/L [z]		0.006 ± <0.001 a	0.006 ± <0.001 a	0.005 ± <0.001 b
DQI		0.025 ± 0.001 a	0.021 ± 0.001 b	0.018 ± 0.001 c
F_v/F_m		0.84 ± <0.01 a	0.84 ± <0.01 a	0.84 ± <0.01 a
Thickness (mm)	True leaf	0.69 ± 0.03 a	0.62 ± 0.02 a	0.64 ± 0.02 a
	Scion cot.	0.72 ± 0.01 a	0.73 ± 0.01 a	0.74 ± 0.01 a
	Roots. cot.	1.17 ± 0.03 a	1.12 ± 0.02a	1.18 ± 0.03 a
Relative chl. content	True leaf	32.60 ± 1.13 a	31.91 ± 1.40 a	31.76 ± 1.17 a
	Scion cot.	27.56 ± 1.10 a	30.32 ± 1.12 a	29.21 ± 1.48 a
	Roots. cot.	51.11 ± 1.98 a	46.51 ± 2.45 a	50.16 ± 2.98 a

[z] shoot dry weight-to-length ratio; DQI: Dickson's quality index; F_v/F_m: maximum quantum yield of primary photochemistry of a dark-adapted leaf; [y] Mean values (±SE) (n = 50), within a row, followed by different letters are significantly different ($P \leq 0.05$) according to the results of the Scott-Knott method.

Table 5. Colorimetric parameters of grafted watermelon seedlings from two quality categories derived 14 days after grafting.

Plant Tissue	Colorimetric Parameters	Quality Categories		
		Optimum	Acceptable	Not Acceptable
True leaves	L* [z]	40.00 ± 0.46 a [y]	41.26 ± 0.49 a	41.30 ± 0.53 a
	C*	19.39 ± 0.51 a	20.73 ± 0.53 a	21.23 ± 0.67 a
	h°	129.40 ± 0.42 b	129.40 ± 0.42 b	128.56 ± 0.48 b
	a*/b*	−0.86 ± 0.01 b	−0.83 ± 0.01 a	−0.80 ± 0.01 a
Scion cotyledons	L*	44.70 ± 0.47 a	43.87 ± 0.42 a	44.89 ± 0.39 a
	C*	23.29 ± 0.69 a	23.34 ± 0.55 a	23.74 ± 0.51 a
	h°	126.54 ± 0.41 a	126.35 ± 0.37 a	126.20 ± 0.30 a
	a*/b*	−0.74 ± 0.01 as	−0.74 ± 0.01 a	−0.73 ± 0.01 a
Rootstock cotyledons	L*	40.52 ± 0.43 b	41.92 ± 0.43 a	42.53 ± 0.59 a
	C*	19.47 ± 0.43 b	20.92 ± 0.46 a	22.10 ± 0.71 a
	h°	129.46 ± 0.29 a	128.07 ± 0.33 b	127.82 ± 0.40 b
	a*/b*	−0.82 ± 0.01 b	−0.79 ± 0.01 a	−0.78 ± 0.01 a

[z] lighting; C*: chroma; h°: hue angle; a*: red/green coordinate; b*: yellow/blue coordinate; [y] Mean values (±SE) (n = 50), within a row, followed by different letters are significantly different ($P \leq 0.05$) according to the results of the Scott-Knott method.

4. Discussion

Seedling quality is one of the major concerns among farmers, and grafted watermelon seedlings are mainly produced by professional nurseries instead of individual farmers. Many factors influence quality evaluation and, therefore, it is difficult to define and categorize seedlings of different qualities. Seedlings of high-quality should have uniformity in terms of size and traits [3].

During healing, grafted seedlings remained in an environmentally controlled growth chamber where microclimate was almost identical for all seedlings. Nevertheless, two quality categories derived 7 days after grafting (i.e., after healing). The seedlings from both categories showed promising potential to develop into marketable plants of high quality. However, the days between the exit from the healing chamber and planting by the grower (i.e., between day 6 and day 14 after grafting) are crucial for maintaining high seedling quality. Many seedlings suffer during the period of acclimatization which might lead to quality deterioration [3]. In our study, three quality categories were developed at 14 days after grafting: two marketable categories (Optimum and Acceptable) and one non-marketable category (Not acceptable).

After 7 days, shoot height was highly distinguishable between the two quality categories, however, it was not confirmed after 14 days. Therefore, shoot height alone cannot be used as an efficient index of seedling quality before going to market. On the other hand, stem diameter was greater for Optimum seedlings at 14 days after grafting. Even though the differences were very slight and could not easily be detected by eye, this parameter is reliable for distinguishing the quality of grafted seedlings before going to market [13–15]. Color differences were not visually detectable, but colorimetry revealed slight distinctions at both times. However, relative chlorophyll content was similar at all measurement dates and quality categories. Additionally, no differences were detected between the quality categories in parameters such as leaf or cotyledon thickness and F_v/F_m, both seven and 14 days after grafting.

Seven days after grafting, one of the greatest morphological parameters that distinguished the Optimum and Acceptable seedlings was the area of true leaves, which leads to greater absorption of incident light in the first very important days of seedling development. Therefore, leaf area is a valuable indicator not only between marketable and not marketable seedlings, but also between the different quality categories of grafted watermelon.

As discussed above, Optimum seedlings were defined by faster leaf development, i.e., larger photosynthetic area in a shorter amount of time. Subsequently, this quality category contained seedlings with greater fresh and dry biomass production compared to the other categories, both 7 and 14 days after grafting, proving that these parameters can be used as index of marketable seedlings. Seven days is considered a short amount of time for grafted seedlings to develop a vigorous root system, especially when the original roots were completely removed, as in our case. However, the quality categories were also distinguished by root biomass.

R/S ratio is a parameter related to the possibility of successful seedling establishment in the field which depends on the proper allocation of biomass between the above and below ground parts. Fourteen days after grafting, R/S ratio was greater for Optimum seedlings which developed a vigorous root system. Shoot DW/L ratio, which is a good indicator of seedling quality [3] revealed that Optimum plants were of higher value. Since this parameter was comprised of shoot dry weight and length (which is similar for the quality categories), biomass accumulation is decisive for the production of high quality seedlings at the nursery. DQI is commonly used for the evaluation of forest or fruit tree seedlings, but recently it has also been employed for assessing horticultural species such as cucumber, muskmelon and tomato [16–18]. Even though the parameter incorporates a number of destructive measurements, it is a useful indicator of seedling quality and plantation performance [19] since its values was correlated with the quality categories in our study.

Optimum seedlings have a better chance of developing into high quality marketable plants, with better establishment in the field, since they excelled in almost all tested parameters, including the essential leaf area and root dry weight. This superiority was also highlighted by shoot DW/L ratio and DQI. Acceptable seedlings were on the border between the other two categories. Parameters, such as stem diameter, shoot DW/L ratio, and DQI, were valuable for identifying them as marketable or not. Not acceptable seedlings after 14 days had inferior development with smaller leaves and considerably weaker root systems compared to the marketable seedlings. Their lower chance of successful establishment and slower development do not favor these seedlings as marketable.

It is concluded that leaf and cotyledon area of scion, stem diameter, shoot and root dry weights as well as shoot DW/L and DQI are good indicators for categorizing grafted watermelon seedlings. Specifically, seedlings of the highest quality must have a leaf area of about 25 cm^2 and 50 cm^2 at 7 and 14 days after grafting, respectively. Scion (watermelon cv. Celine) cotyledons must be fully expanded (about 15 cm^2) 7 days after grafting, while rootstock cotyledons were not good indicators of seedling quality. Shoot height was a weak quality indicator 14 days after grafting, since Not acceptable values were similar to Optimum seedlings. However, stem diameter was a good quality index even though values between the categories were very close. Moreover, shoot and root dry weights, as well as shoot

DW/L ratio and DQI, proved valuable indicators of grafted watermelon seedling quality. The benefits of grafting is associated with the use of high quality watermelon seedlings, and their categorization could help both the nursery industry and growers.

Author Contributions: Conceptualization, methodology, and data analysis: F.B., A.K., A.S., and G.M.; experimental measurements: F.B., C.D. and D.K.; writing—original draft preparation: F.B. and A.K.; writing—review and editing: F.B., A.K, A.S., and G.M., supervision and project administration: A.K.

References

1. Lee, J.-M.; Bang, H.-J.; Ham, H.-S. Grafting of vegetables. *J. Jpn. Soc. Hort. Sci.* **1998**, *67*, 1098–1104. [CrossRef]
2. Davis, A.R.; Perkins-Veazie, P.; Hassell, R.; Levi, A.; King, S.R.; Zhang, X. Grafting Effects on Vegetable Quality. *HortScience* **2008**, *43*, 1670–1672. [CrossRef]
3. Lee, J.-M.; Kubota, C.; Tsao, S.J.; Bie, Z.; Hoyos Echevarria, P.; Morra, L.; Odag, M. Current status of vegetable grafting: Diffusion, grafting techniques, automation. *Sci. Hortic.* **2010**, *127*, 93–105. [CrossRef]
4. Louws, F.J.; Rivard, C.L.; Kubota, C. Grafting fruiting vegetables to manage soilborne pathogens, foliar pathogens, arthropods and weeds. *Sci. Hortic.* **2010**, *127*, 127–146. [CrossRef]
5. Savvas, D.; Colla, G.; Rouphael, Y.; Schwarz, D. Amelioration of heavy metal and nutrient stress in fruit vegetables by grafting. *Sci. Hortic.* **2010**, *127*, 156–161. [CrossRef]
6. Schwarz, D.; Rouphael, Y.; Colla, G.; Venema, J.H. Grafting as a tool to improve tolerance of vegetables to abiotic stresses: Thermal stress, water stress and organic pollutants. *Sci. Hortic.* **2010**, *127*, 162–171. [CrossRef]
7. Kyriacou, M.C.; Rouphael, Y.; Colla, G.; Zrenner, R.; Schwarz, D. Vegetable Grafting: The Implications of a Growing Agronomic Imperative for Vegetable Fruit Quality and Nutritive Value. *Front. Plant Sci.* **2017**, *8*, 741. [CrossRef] [PubMed]
8. Lee, S.G. Production of high quality vegetable seedling grafts. *Acta Hortic.* **2007**, *759*, 169–174. [CrossRef]
9. Lee, J.M.; Oda, M. Grafting of herbaceous vegetable and ornamental crops. In *Horticultural Review*; Janick, J., Ed.; John Wiley & Sons: New York, NY, USA, 2003; pp. 61–124.
10. Dickson, A.; Leaf, A.L.; Hosner, J.F. Quality appraisal of white spruce and white pine seedling stock in nurseries. *For. Chron.* **1960**, *36*, 10–13. [CrossRef]
11. McGuire, R.G. Reporting of objective color measurements. *HortScience* **1992**, *27*, 1254–1255. [CrossRef]
12. Scott, A.J.; Knott, M. A Cluster analysis method for grouping means in the analysis of variance. *Biometrics* **1974**, *30*, 507–512. [CrossRef]
13. Mattsson, A. Predicting field performance using seedling quality assessment. *New For.* **1996**, *13*, 223–248.
14. Rawat, J.S.; Singh, T.P. Seedling indices of four tree species in nursery and their correlations with field growth in Tamil Nadu, India. *Agrofor. Syst.* **2000**, *49*, 289–300. [CrossRef]
15. Davis, A.S.; Jacobs, D.F. Quantifying root system quality of nursery seedlings and relationship to outplanting performance. *New For.* **2005**, *30*, 295–311. [CrossRef]
16. Guisolfi, L.P.; Lo Monaco, P.A.V.; Haddade, I.R.; Krause, M.R.; Meneghelli, L.A.M.; Almeida, K.M. Production of cucumber seedlings in alternative substrates with different compositions of agricultural residues. *Rev. Caatinga* **2018**, *31*, 791–797. [CrossRef]
17. Vendruscolo, E.P.; Campos, L.F.C.; Nascimento, L.M.; Seleguini, A. Quality of muskmelon seedlings treated with thiamine in pre-sowing and nutritional supplementation. *Sci. Agrar.* **2018**, *19*, 164–171.
18. Costa, E.; Leal, P.A.M.; Benett, C.G.S.; Benett, K.S.S.; Salamene, L.C.P. Production of tomato seedlings using different substrates and trays in three protected environments. *Eng. Agríc.* **2012**, *32*, 822–830. [CrossRef]
19. Bayala, J.; Dianda, M.; Wilson, J.; Ouedraogo, S.J.; Sanon, K. Predicting field performance of five irrigated tree species using seedling quality assessment in Burkina Faso, West Africa. *New For.* **2009**, *38*, 309–322. [CrossRef]

14

Monitoring Dormancy Transition in Almond [*Prunus Dulcis* (Miller) Webb] during Cold and Warm Mediterranean Seasons through the Analysis of a *DAM* (*Dormancy-Associated MADS-Box*) Gene

Ángela S. Prudencio, Federico Dicenta and Pedro Martínez-Gómez *

Department of Plant Breeding, CEBAS-CSIC (Centro de Edafología y Biología Aplicada del Segura-Consejo Superior de Investigaciones Científicas), PO Box 164, 30100 Espinardo, Murcia, Spain; asanchez@cebas.csic.es (A.S.P.); fdicenta@cebas.csic.es (F.D.)
* Correspondence: pmartinez@cebas.csic.es.

Abstract: For fruit tree (*Prunus*) species, flower bud dormancy completion determines the quality of bud break and the flowering time. In the present climate change and global warming context, the relationship between dormancy and flowering processes is a fundamental goal in molecular biology of these species. In almond [*P. dulcis* (Miller) Webb], flowering time is a trait of great interest in the development of new cultivars adapted to different climatic areas. Late flowering is related to a long dormancy period due to high chilling requirements of the cultivar. It is considered a quantitative and highly heritable character but a dominant gene (*Late bloom*, *Lb*) was also described. A major QTL (quantitative trait loci) in the linkage group (LG) 4 was associated with *Lb*, together with other three QTLs in LG1 and LG7. In addition, *DAM* (*Dormancy-Associated MADS-Box*) genes located in LG1 have been largely described as a gene family involved in bud dormancy in different *Prunus* species including peach [*P. persica* (L.) Batsch] and Japanese apricot (*P. mume* Sieb. et Zucc.). In this work, a *DAM* transcript was cloned and its expression was analysed by qPCR (quantitative Polymerase Chain Reaction) in almond flower buds during the dormancy release. For this purpose two almond cultivars ('Desmayo Largueta' and 'Penta') with different chilling requirements and flowering time were used, and the study was performed along two years. The complete coding sequence, designated *PdDAM6* (*Prunus dulcis DAM6*), was subjected to a phylogenetic analysis with homologous sequences from other *Prunus* species. Finally, expression dynamics analysed by using qPCR showed a continuous decrease in transcript levels for both cultivars and years during the period analysed. Monitoring almond flower bud dormancy through *DAM* expression should be used to improve almond production in different climate conditions.

Keywords: flowering; breeding; chilling requirements; qPCR; transcription; cloning

---

1. Introduction

During autumn temperate fruit trees (*Prunus*) activate a survival strategy called endodormancy, to protect against unfavourable chill conditions. Trees cease growth and form structures called buds in order to protect meristems from unfavourable environmental conditions, including low temperature and desiccation [1]. Chill accumulation allows the progression from flower bud endodorrmancy stage to flower bud ecodormancy which is regulated by heat accumulation [2]. Flowering time in almond [*Prunus dulcis* (Miller) Webb] is mainly dependent on chilling requirements to overcome this endodormancy stage [3]. These chilling requirements are considered a cultivar-dependent trait, correlated with species or cultivar origin [4,5].

Warm winter temperatures affect cold accumulation and if chilling requirement is not fully satisfied such a condition could lead to irregular and insufficient flowering with a loss of production [4,6]. Due to its economic importance, dormancy release is being studied in different species, although knowledge is still scarce and no common mechanism has been described. Thus, expression analysis of candidate genes may be a useful tool for the interannual monitoring of endodormancy progression within the flower bud. This is especially interesting for commercial fruit tree cultivars displaying a wide range of flowering and ripening time phenotypes, as in case of almond [6]. Adaptation to climatic conditions largely depends on an adequate flowering time, and it is one of the most important agronomic traits in almond breeding programs, as it determines whether the pollination period will occur in favourable climatic conditions [4]. In this context, the development of new extra-late flowering cultivars to avoid the spring frosts is one of the main objectives of almond breeding programs [7].

Late flowering is considered a quantitative and highly heritable character but a major gene (*Late bloom*, *Lb*) was also described. A QTL (quantitative trait loci) explaining 57% of the observed variance in the Linkage Group (LG) 4 was associated with *Lb*, together with other three QTLs (explaining 20, 12, and 8% of the variance) in LG1 and LG7 [8,9]. In addition, bud endodormancy has a set of genetic controls which may be characterized through examination of gene expression in bud tissues over time. Prior studies showed the *Dormancy Associated MADS-Box* (*DAM*) gene family is a group of transcription factors that regulated peach [*P. persica* (L.) Batsch] dormancy [10]. This gene family was discovered in the *Evergrowing* (*evg*) mutant of peach. The mutation consisted of a deletion in the *EVG* (*EVERGROWING*) locus affecting up to four genes which prevents terminal buds from entering the endodormancy stage [11]. The map-based cloning analyses of *EVG* locus revealed that it included six tandemly arrayed genes [11,12]. Moreover, *DAM5* expression was analysed in different cultivars of peach [13] and *PpDAM6* was postulated as one of the main factors involved in the regulation of dormancy in different *Prunus* species including peach [14–16] and Japanese apricot (*P. mume* Sieb. et Zucc.) [17,18].

In this work, a candidate *DAM* transcript was cloned and expression from endodormancy to ecodormancy stages in two almond cultivars with different chilling requirements and flowering time: 'Desmayo Largueta' and 'Penta' was determined.

2. Materials and Methods

2.1. Plant Material

'Desmayo Largueta' is a traditional Spanish almond cultivar with very low chilling requirements and extra-early flowering time, and 'Penta', a cultivar released from the Almond Breeding Program of CEBAS-CSIC (Murcia, South-East Spain) with high chilling requirements and extra-late flowering time, were used [6]. The plant material consisted of flower buds sampled weekly between stages A (dormancy phase) and B (after dormancy release) referenced to the phenological stages described by Felipe [19] (Figure 1).

2.2. Chilling Requirements Evaluation

Experiments for the evaluation of chilling and heat requirements were conducted in the experimental field of CEBAS-CSIC, in Murcia (South-East Spain), during two seasons of study: 2015–2016 and 2016–2017. Temperatures were recorded hourly with a data logger (HOBO® UX100-003 Temp/Relative Humidity, Madrid, Spain) from November to February during both seasons. Three branches (40 cm in length and 5 mm in diameter) were collected weekly from the same tree in the field, and placed in a growth chamber in controlled conditions (25 ± 1 °C, RH 40 ± 3.5% during a 16 h light photoperiod and 20 ± 1 °C, RH 60% during the dark period). Almond branches were placed in the growth chamber, in a 5% sucrose solution and 1% aluminium sulphate, making a fresh cut in the base of the branches. After 10 days, the development state of the flower buds was recorded.

The date of dormancy breakage was established when, after 10 days in the growth chamber, 50% of the flower buds were in the B-C state [20]. The calculation of chilling accumulation in field conditions was calculated as the chill contributions in the field necessary for breaking of dormancy (transition from stage A to stage B, see Figure 1) in chill units (CUs) according to the model described by Richardson et al. [16] with an initial date for chilling accumulation when consistent chilling accumulation occurred and temperatures producing a negative effect (chilling negation) were rare [21]. These CUs were calculated as hours below 7 °C. In addition, chilling accumulation was calculated in chill portions (CPs) according to the dynamic model [22,23] with an initial chilling accumulation of 0. The model is based on the assumption that dormancy completion may be estimated as a dynamic two-stage process controlling an accumulated bud break factor. The model is "dynamic" in the sense that relatively high temperatures, typically 19 °C and above, effectively negate earlier chilling; alternatively, moderate temperatures, typically around 13–14 °C effectively enhance moderate earlier chilling temperatures.

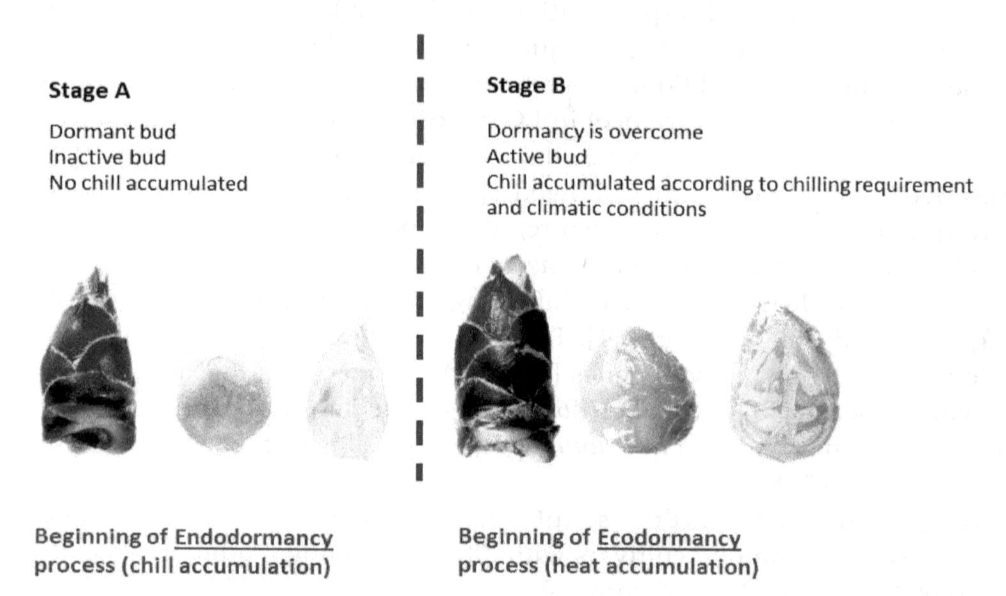

Figure 1. Plant material assayed from the almond cultivar 'Desmayo Largueta'. Flower buds in the dormant stage (A) and after dormancy release (B).

2.3. cDNA Isolation and Cloning

Almond samples assayed include flower buds in state A (completely dormant bud), state B (when 40–50% of the chilling requirement of each cultivar are satisfied) and state B (when the flower bud has broken its dormancy) (Figure 1). Total RNA was extracted from almond flower buds [24] and treated with DNAseI (Ambion). cDNA was synthetized using SSIII Reverse Transcriptase (ThermoFisher Scientific, Waltham, MA, USA). The full-length cDNA was isolated from cDNA of flower buds of 'Desmayo Largueta' and 'Penta' cultivars using 3′-RACE strategy and specific primers from *Prunus persica* available sequences in databases. High-fidelity PCR (Polymerase Chain Reaction) was performed using KOD (from Archaeon Thermococcus kodakaraensis) Hot Start DNA polymerase (Novagen, Berlin, Germany) and the product was cloned into *E. coli* using Zero Blunt Topo PCR Cloning Kit (Life Technologies, Carlsbad, CA, USA) for sequencing.

2.4. Phylogenetic Analysis

A BLAST (Basic Local Alignment Search Tool) search was performed with the full-length PdDAM6 cDNA (*Prunus Dulcis DORMANCY-ASSOCIATED MADS-BOX 6*), in order to confirm the identity of the sequence and to collect homologous proteins from the Prunus genus with a high percentage identity. A pPutative PdDAM6 protein sequence was obtained using the ExPASY translate tool (http://web.expasy.org/translate/). A phylogenetic tree was created using Philogeny.fr (http://phylogeny.lirmm.fr/phylo_cgi/index.cgi).

2.5. Gene Expression Analysis

To investigate the expression pattern of PdDAM6 during bud dormancy progression, real time qPCR experiments were executed with a One Step Plus real-time PCR system (Applied Biosystems, Foster City, CA, USA). Specific primers were designed based on an almond PdDAM6 sequence using Primer3 software (Forward Primer: 5′ AGGAAATACTGGACCTGCGT-3′; Reverse Primer: 5′-GGTGGAGGTGGCAATTATGG-3′). Reaction efficiency was checked by the standard curve method. For all real-time qPCR reactions, a 10 µL mix was made including: 5 µL Power SYBR®Green PCR Master Mix (Applied Biosystems, Foster City, CA, USA), 20 ng of cDNA, and 0.5 µL of each primer (5 µM). High-fidelity PCR was performed using KOD Hot Start DNA polymerase (Novagen, Berlin, Germany), and the product was cloned into *E. coli* using a Zero Blunt Topo PCR Cloning Kit (Life Technologies, Carlsbad, CA, USA) for sequencing. PCR was performed in a 30 µL mix according to the manufacturer's instructions with 150 ng of cDNA from each almond cultivar and 10 µM primers. The PCR reaction was incubated at 94 °C for 2′ for the initial denaturalisation step, followed by 35 cycles of 94 °C for 30″, 62 °C for 1′ and 68 °C for 1′. A final extension step at 68 °C was set for 10′. Each biological sample was implemented in duplicate. RPII was used as reference gene for data normalization using primers designed by Tong et al. [25] (Forward Primer: 5′-TGAAGCATACACCTATGATGATGAAG; Reverse primer: 5′ CTTTGACAGCACCAGTAGATTCC-3′) and the levels of relative expression were calculated by the 2–ΔΔCt method [26] taking Ct value from November the 10th samples as the reference expression level.

3. Results

3.1. Chilling Requirements Evaluation

Chilling accumulation in field conditions during the two seasons of study (2015–2016 and 2016–2017) calculated as chill units (CUs) according to the Richardson model and in chill portions (CPs) according to the dynamic model is shown in Figure 2.

During the first year of the study (2015–2016), an important reduction of chill accumulation was observed, mainly in terms of chill units (CUs). As shown in Table 1, the dormancy release date was observed earlier during the 2016–2017 season, when a higher amount of chill units accumulated. Regarding flowering time, an important advance was observed in the late cultivar 'Penta'.

Table 1. Chill accumulated percentage (chill portions) in field conditions during the seasons 2015–2016 and 2016–2017.

Season		Stage A		Stage B		Flowering Date
		Date	Chill Accumulation	Date	Chill Accumulation	
'Desmayo Largueta'	**2015/2016**	November 10	0	December 21	16	January 28
	2016/2017	November 10	0	December 15	24	January 27
'Penta'	**2015/2016**	November 10	0	February 10	41	March 25
	2016/2017	November 10	0	February 2	54	March 12

3.2. cDNA Isolation, Cloning and Phylogenetic Analysis

Phylogenetic analysis of PdDAM6 protein sequence from 'Desmayo Largueta' and 'Penta' sequences confirmed that *PdDAM6* is indeed a member of the DAM family transcription factors. In addition, the phylogenetic tree showed that PdDAM6 branch (including sequences from 'Desmayo Largueta' and 'Penta' cultivars) is closer to *Prunus persica* DAM6 (*PpDAM6*) and *Prunus pseudocerasus* (*PpsDAM6*) rather than to DAM5 protein group (Figure 3).

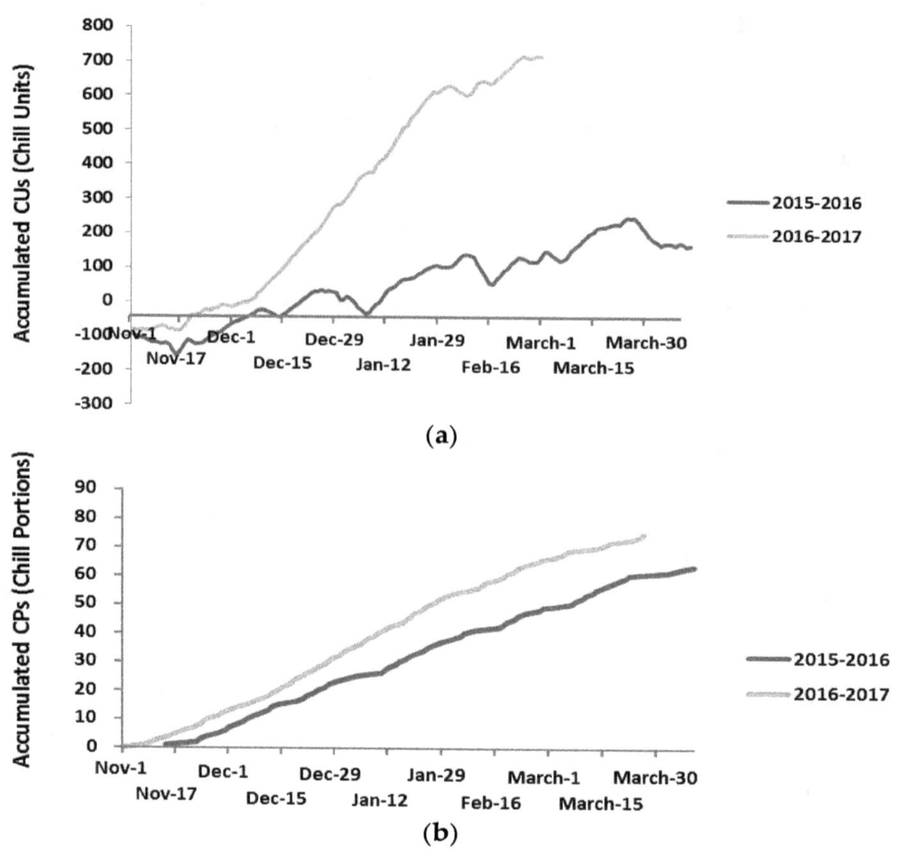

Figure 2. (**a**) Chilling accumulation in field conditions during the two seasons of study (2015–2016 and 2016–2017) calculated as chill units (CUs) according to the Richardson model; (**b**) calculated in chill portions (CPs) according to the dynamic model.

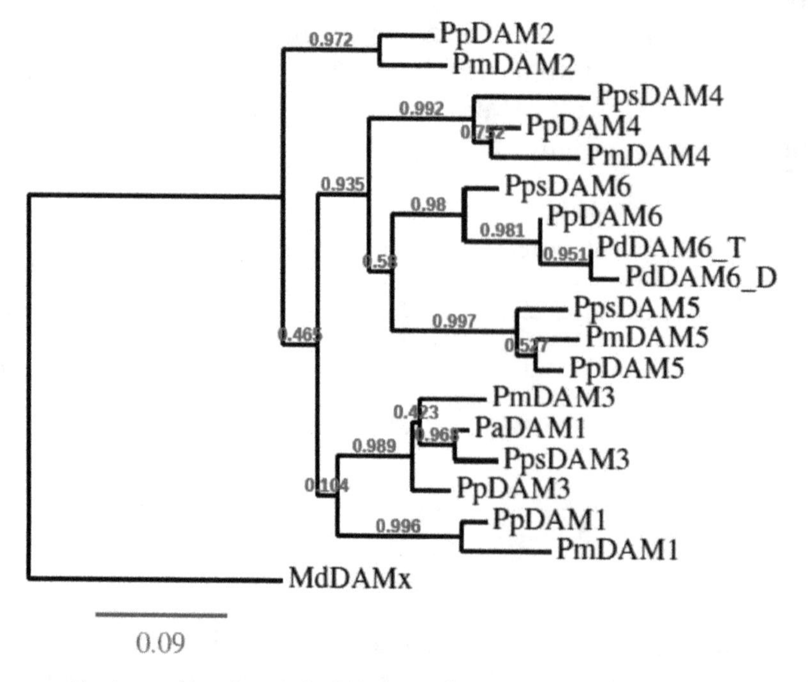

Figure 3. Phylogenetic tree showing relationships between homologous proteins to 'Desmayo Largueta' (PdDAM6_D) and 'Penta' (PdDAM6_P) sequence. *Prunus persica* (Pp), *Prunus mume* (Pm), *Prunus pseudocerasus* (Pps) and *Prunus avium* (Pa). aLRT (approximate likelihood ratio) values are indicated in red. *Malus domestica* DAM protein sequence (MdDAMx) was used as outgroup to root the phylogenetic tree.

3.3. Expression Analysis

Expression analysis of *PdDAM6* showed a progressive decrease in mRNA levels until the dormancy period was completely overcome, in both 'Desmayo Largueta' and 'Penta' cultivar samples (Figure 4).

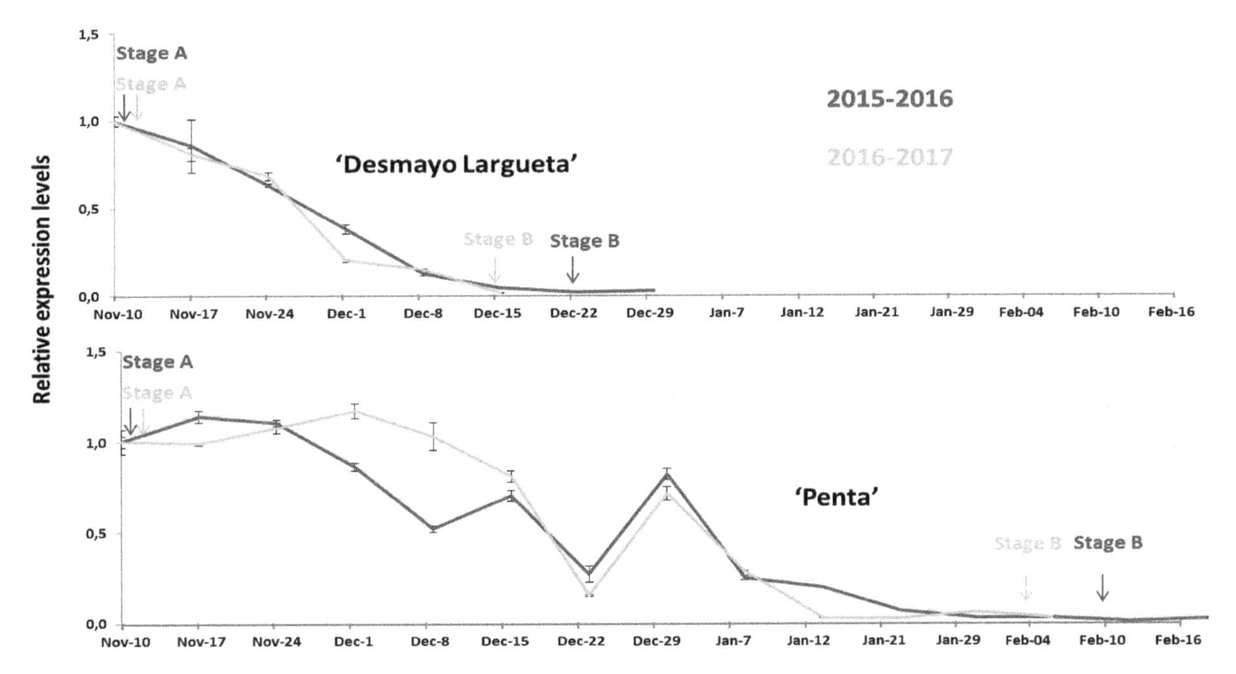

Figure 4. Relative gene expression of *PdDAM6* gene evaluated by qPCR (quantitative Polymerase Chain Reaction) 'Desmayo Largueta' and 'Penta' almond cultivars during the seasons 2015–2016 and 2016–2017. Standard deviations are indicated with vertical bars.

4. Discussion

As shown by Prudencio et al. [6], the estimation of chill accumulation under different climatic conditions showed that the dynamic model presents less variation than the Richardson model. As expected, the chilling requirements of the almond cultivars were related to their flowering time. However, in general, these values were lower than in previous evaluations performed by our group [3,8] mainly in the case of the warmer year.

The first full-length cDNA from the DAM gene family was obtained for almond. The clone designated *PdDAM6* (*Prunus dulcis DAM6*) was obtained from 'Desmayo Largueta' and 'Penta' almond cultivars, which display different phenotypes regarding chilling requirement and flowering time. Phylogenetic and expression analysis was performed to further characterize the sequences and to study the biological role of DAM proteins during flower bud dormancy progression in almond.

Our results clearly indicated that the level of expression of *DAM6* in both almond cultivars with different chilling requirements and flowering time decreased concomitantly with chill accumulation and dormancy progression, although for the late cultivar 'Penta', a relative increase was observed prior to dormancy release. These results supported that obtained by Leida et al. [14] and Jiménez et al. [15], highlighting the role of this gene in flower bud dormancy maintenance. In addition, a down-regulation of *DORMANCY-ASSOCIATED MADS-box6* has been observed in Japanese apricot [17,18] during dormancy release.

Monitoring bud transition from endodormany to ecodormancy should be of great interest in terms of the use and optimization of biostimulants to promote flowering in fruit tree species [27,28] in the present climate change and warming context. The moment of application of these biostimulants is critical for success and depends on the endodormancy stage of the bud and its transition to ecodormancy [29] or the forcing strategies [30]. Treatments with these biostimulants should be applied at the optimum time for breaking bud dormancy, as they can be null or even toxic depending on the stage of the bud [29]. Monitoring almond flower bud dormancy through *DAM* expression could be used to determione the suitable moment to apply these biostimulants.

5. Conclusions

The estimation of chill accumulation using different models showed that the 2015–2016 season was warmer than the 2016–2017 season, and this was reflected in the dormancy release date of the cultivars. This illustrates the risk of growing extra-late cultivars in warm-winter areas, as production could be negatively affected if chilling requirement is not satisfied. The endodormancy to ecodormancy transition involves a transcriptional reprogramming in which genes acting on dormancy maintenance would be downregulated. This seems to be the case of *PdDAM6* for almond.

Author Contributions: A.S.P. and P.M.-G. participated in the design and coordination of the study. F.D. and P.M.-G. collaborated in the fieldwork. A.S.P. carried out the qPCR and cloning protocols. A.S.P., F.D. and P.M.-G. carried out data analysis. A.S.P., F.D. and P.M.-G. participated in the manuscript elaboration and discussion.

References

1. Lloret, A.; Badenes, M.L.; Ríos, G. Modulation of Dormancy and Growth Responses in Reproductive Buds of Temperate Trees. *Front. Plant Sci.* **2018**, *9*, 1368. [CrossRef] [PubMed]
2. Lang, B.A.; Early, J.D.; Martin, G.C.; Darnell, R.L. Endo-, para- and ecodormancy: Physiological terminology and classification for dormancy research. *Hort. Sci* **1987**, *22*, 371–377.
3. Egea, J.; Ortega, E.; Martínez-Gómez, P.; Dicenta, F. Chilling and heat requirements of almond cultivars for flowering. *Environ. Exp. Bot.* **2003**, *50*, 79–85. [CrossRef]
4. Campoy, J.A.; Ruiz, D.; Egea, J. Dormancy in temperate fruit trees in a global warming context: A review. *Sci. Hort.* **2011**, *130*, 357–372. [CrossRef]
5. Martínez-Gómez, P.; Prudencio, A.S.; Gradziel, T.M.; Dicenta, F. The delay of flowering time in almond: A review of the combined effect of adaptation, 406 mutation and breeding. *Euphytica* **2017**, *213*, 197. [CrossRef]
6. Prudencio, A.S.; Martínez-Gómez, P.; Dicenta, F. Evaluation of breaking dormancy, flowering and productivity of extra-late and ultra-late flowering almond cultivars during cold and warm seasons in South-East of Spain. *Sci. Hort.* **2018**, *235*, 39–46. [CrossRef]
7. Dicenta, F.; Sánchez-Pérez, P.; Batlle, I.; Martínez-Gómez, P. Late-blooming almond cultivar development. In *Almond: Botany, Production and Uses*; Rafael Socias i Company, Gradizel, T.M., Ed.; CABI: Boston, MA, USA, 2017.
8. Sánchez-Pérez, R.; Dicenta, F.; Martínez-Gómez, P. Inheritance of chilling and heat requirements for flowering in almond and QTL analysis. *Tree Genet. Gen.* **2012**, *8*, 379–389. [CrossRef]
9. Sánchez-Pérez, R.; Del Cueto, J.; Dicenta, F.; Martínez-Gómez, P. Recent advancements to study flowering time in almond and other *Prunus* species. *Front. Plant Sci.* **2014**, *5*, 334. [PubMed]
10. Bianchi, V.; Rubio, M.; Trainotti, L.; Verde, I.; Bonghi, C. ; Martínez-Gómez, P. *Prunus* transcription factors: Breeding perspectives. *Front. Plant Sci.* **2015**, *6*, 443. [PubMed]
11. Bielenberg, D.G.; Wang, Y.; Fan, S.; Reighard, R.; Abbott, A.G. A Deletion Affect ing Several Gene Candidates is Present in the Evergrowing Peach Mutant. *J. Hered.* **2004**, *95*, 436–444. [CrossRef] [PubMed]
12. Bielenberg, D.G.; Wang, Y.; Li, Z.; Zhebentyayeva, T.; Fan, S.; Reighard, G.L. Sequencing and annotation of the evergrowing locus in peach [*Prunus persica* (L.) Batsch] reveals a cluster of six MADS-box transcription factors as candidate genes for regulation of terminal bud formation. *Tree Genet. Gen.* **2008**, *4*, 495–507. [CrossRef]
13. Leida, C.; Terol, J.; Martí, G.; Agustí, M.; Llácer, G.; Badenes, M.L. Identification of genes associated with bud dormancy release in Prunus persica by suppression subtractive hybridization. *Tree Physiol.* **2010**, *30*, 655–666. [CrossRef] [PubMed]

14. Leida, C.; Conesa, A.; Llácer, G.; Badenes, M.L.; Ríos, G. Histone modifications and expression of DAM6 gene in peach are modulated during bud dormancy release in a cultivar-dependent manner. *New Phytol.* **2012**, *193*, 67–80. [CrossRef] [PubMed]
15. Jiménez, S.; Reighard, G.L.; Bielenberg, D.G. Gene expression of *DAM5* and *DAM6* is suppressed by chilling temperatures and inversely correlated with bud break rate. *Plant Mol. Biol.* **2010**, *73*, 157–167. [CrossRef] [PubMed]
16. Yamane, H.; Ooka, T.; Jotatsu, H.; Sasaki, R.; Tao, R. Expression analysis of *PpDAM5* and *PpDAM6* during flower bud development in peach (*Prunus persica*). *Sci. Hort.* **2011**, *129*, 844–848. [CrossRef]
17. Sasaki, R.; Yamane, H.; Ooka, T.; Jotatsu, H.; Kitamura, Y.; Akagi, T.; Tao, R. Functional and Expressional Analyses of *PmDAM* Genes Associated with Endodormancy in Japanese Apricot. *Plant Physiol.* **2011**, *157*, 485–497. [CrossRef] [PubMed]
18. Kitamura, Y.; Takeuchi, T.; Yamane, H.; Tao, R. Simultaneous down-regulation of *DORMANCY-ASSOCIATED MADS-box6* and *SOC1* during dormancy release in Japanese apricot (*Prunus mume*) flower buds. *J. Hort. Sci. Biotechnol.* **2016**, *5*, 476–482. [CrossRef]
19. Felipe, A.J. Phenological states of almond (In Italian). In Proceedings of the Third GREMPA Colloquium, Bari, Italy, 3–7 October 1977; pp. 101–103.
20. Richardson, E.A. ; S.D. Seeley, D.R.; Walker, J.L.; Anderson, M.; Ashcroft, G.L. Pheno-climatology of spring peach bud development. *Hort. Sci.* **1975**, *10*, 236–237.
21. Erez, A.; Couvillon, G.A.; Hendershott, C.H. The effect of cycle length on chilling negation by high temperatures in dormant peach leaf buds. *J. Am. Soc. Hort. Sci.* **1979**, *104*, 573–576.
22. Fishman, S.; Erez, A.; Couvillon, G.A. The temperature dependence of dormancy breaking in plants: Computer simulation of processes studied under controlled temperatures. *J. Theor. Biol.* **1987**, *126*, 309–321. [CrossRef]
23. Fishman, S.; Erez, A.; Couvillon, G.A. The temperature dependence of dormancy breaking in plants: Mathematical analysis of a two-step model involving a cooperative transition. *J. Theorl Biol.* **1987**, *124*, 473–483. [CrossRef]
24. Le Provost, G.; Herrera, R.; Paiva, J.A.; Chaumeil, P.; Salin, F.; Plomion, C. A micromethod for high throughput RNA extraction in forest trees. *Biol. Res.* **2007**, *40*, 291–297. [CrossRef] [PubMed]
25. Tong, Z.; Gao, Z.; Wang, F.; Zhou, J.; Zang, Z. Selection of reliable reference genes for gene expression studies in peach using real-time PCR. *BMC Mol. Biol.* **2009**, *10*, 71. [CrossRef] [PubMed]
26. Livak, K.J.; Schmittgen, T.D. Analysis of relative gene expression data using real-time quantitative PCR and the 2(-Delta Delta C(T)) Method. *Methods* **2001**, *25*, 402–408. [CrossRef] [PubMed]
27. Ionescu, I.A.; Moller, B.L.; Sánchez-Pérez, R. Chemical control of Flowering time. *J. Exp. Bot.* **2017**, *68*, 369–382. [CrossRef] [PubMed]
28. Ionescu, I.A.; López-Ortega, G.; Burow, M.; Bayo-Canha, A.; Junge, A.; Gericke, O.; Moller, B.L.; Sánchez-Pérez, R. Transcriptome and Metabolite Changes during Hydrogen Cyanimide-Induced Floral Bud break in Sweet Cherry. *Front. Plant Sci.* **2017**, *8*, 1233. [CrossRef] [PubMed]
29. Erez, A. Means to compensate for insufficient chilling to improve bloom and leafing. *Acta Hort.* **1995**, *395*, 81–95. [CrossRef]
30. Kauffman, H.; Blanke, M. Substitution of winter chilling by spring forcing for flowering using sweet cherry as model crop. *Sci. Hort.* **2018**, *244*, 75–81. [CrossRef]

15

Effects of Biochar on Container Substrate Properties and Growth of Plants

Lan Huang [1] and Mengmeng Gu [2,*]

[1] Department of Horticultural Sciences, Texas A&M University, College Station, TX 77843, USA; huanglan92@tamu.edu
[2] Department of Horticultural Sciences, Texas A&M AgriLife Extension Service, College Station, TX 77843, USA
* Correspondence: mgu@tamu.edu.

Abstract: Biochar refers to a processed, carbon-rich material made from biomass. This article provides a brief summary on the effects of biochar on container substrate properties and plant growth. Biochar could be produced through pyrolysis, gasification, and hydrothermal carbonization of various feedstocks. Biochar produced through different production conditions and feedstocks affect its properties and how it performs when incorporated in container substrates. Biochar incorporation affects the physical and chemical properties of container substrates, including bulk density, total porosity, container capacity, nutrient availability, pH, electrical conductivity and cation exchange capacity. Biochar could also affect microbial activities. The effects of biochar incorporation on plant growth in container substrates depend on biochar properties, plant type, percentage of biochar applied and other container substrates components mixed with biochar. A review of the literature on the impact of biochar on container-grown plants without other factors (such as irrigation or fertilization rates) indicated that 77.3% of the studies found that certain percentages of biochar addition in container substrates promoted plant growth, and 50% of the studies revealed that plant growth decreased due to certain percentages of biochar incorporation. Most of the plants tested in these studies were herbaceous plants. More plant species should be tested for a broader assessment of the use of biochar. Toxic substances (heavy metals, polycyclic aromatic hydrocarbons and dioxin) in biochars used in container substrates has rarely been studied. Caution is needed when selecting feedstocks and setting up biochar production conditions, which might cause toxic contaminants in the biochar products that could have negative effects on plant growth.

Keywords: container substrates; physical properties; chemical properties; biomass

---

1. Introduction

Biochar refers to processed, carbon-rich material derived from biomass [1–3]. Recent research has shown that biochar can be used as a replacement for commonly-used container substrates [4–8]. Container substrates are often soilless, making it easy to achieve consistency. Primary substrate components include peat moss, vermiculite, perlite, bark, and compost [9]. Peat moss is an excellent substrate component; it has essential characteristics such as low pH, high cation exchange capacity (CEC), and appropriate aeration and good container capacity [10–12], which are ideal for horticultural container application. However, intensive extraction of peat from peatlands can damage natural habitats and release CO_2 into the atmosphere if the disturbed peatland is left unrestored [13]. The United Kingdom government has thus proposed reducing the use of peat [14]. The cost of this commonly-used substrate is also high due to the extreme cost of transportation, fuel for extraction, and processing [9,15]. Therefore, it is beneficial and necessary to search for alternative environmentally-friendly and local substrate components [9,16]. Research has shown that biochar could be a potential alternative to commonly-used substrates. Using biochars (a byproduct of bioenergy production) in agriculture adds value to bioenergy production [17]. Biochar could offer economic advantages over other commonly-used substrates, if produced on site. Extensive research has shown that replacing a certain percentage of commonly-used container substrates with biochar could increase plant growth in certain conditions [18–22].

However, biochars are variable, and their impact on container substrates could vary. It would be of interest to examine the characteristics of biochars, their incorporation in container substrates, and their effects on diverse types of container-grown plants. In this review, we provide a brief summary of the effects of biochar on container substrate properties and plant growth, and discuss the potential mechanism behind their effects. This review examines factors related to the impact of biochar, which include feedstock sources, production conditions, percentage of biochar applied, other substrate components mixed with biochar, and plant species. These factors can help address the general hypothesis that incorporation of biochar may not always have beneficial effects on container substrate properties or plant growth.

2. Biochar Production

There are many variables prior to, during, and after production of biochar. These factors will eventually affect biochar properties and its effect on plant growth and container properties when incorporated in container substrates.

2.1. Biochar Production Methods

There are three main processes to produce biochar: pyrolysis, gasification and hydrothermal carbonization. Pyrolysis is the thermal decomposition of biomass by heating (around 400 °C to 600 °C) without oxygen [23–25]. Compared to pyrolysis, gasification is conducted under small amounts of oxygen at relatively higher temperatures (around 700 °C to 1200 °C) [2]. Gasification produces smaller quantities of biochar with lower carbon (C) content than pyrolysis [2,25,26]. Hydrothermal carbonization uses water and catalysts at lower temperatures (180 to 300 °C) under high pressure to convert biomass to a different type of biochar product, hydrochar [27,28]. Hydrochars are acidic, and have low surface areas, less aromatic compounds, and higher CEC than those produced by pyrolysis and gasification [28,29]. Production temperature significantly influences the characteristics of biochars (Table 1). Biochar made from pruning waste at 500 °C had higher pH and different container capacity, total porosity, electrical conductivity (EC) and CEC, when compared to biochar produced at 300 °C [20]. Biochars made from different production processes can have different physical and chemical properties.

Utilizing biochar in agriculture adds values to biomass pyrolysis and gasification. The main purpose of fast pyrolysis is to produce syngas and bio-oil [17,23] and gasification syngas [30], with biochar being the byproduct. Syngas mainly includes carbon monoxide and hydrogen [31]. It could be used to provide energy for other pyrolysis processes. Bio-oil could be burned to produce heat or further processed to be used as fuel [32]. A specific process and its heating rate could be modified to produce desirable products. For example, gasification has higher yields of syngas and energy than pyrolysis [33]. Liquid bio-oil produced by pyrolysis has higher energy density and is cheaper and easier to transport; however, it is corrosive, which makes it difficult to store for a long time [31]. Slow pyrolysis produces more biochar and syngas, and fast pyrolysis more bio-oil [34]. The residence time (the amount of time taken in the pyrolysis procedure) of slow pyrolysis is from 5 min to 30 min, while that of fast pyrolysis is from seconds to less than a second, and the temperatures are higher [23,35]. Raising pyrolysis temperature can decrease biochar yield [36,37]. Production conditions can be adjusted on the basis of whether the desirable products are biochar or bioenergy products (bio-oil or syngas). Low temperatures and slow pyrolysis could be used to produce more biochar than the other products.

Pre-treatment of feedstocks has been reported to have a significant influence on biochar ash content, yield, and properties. Pre-treatment of biomass, such as washing with water or acid, could help remove some ash culprits in feedstocks to reduce fouling, and improve the quality of biomass feedstock and final biochar products [37,38]. Rahman et al. [37] tested the effectiveness of different pre-treatments by comparing the EC of the initial washing medium and leachate collected after treatments. The result showed that the leachate EC of palm kernel shell increased, when pre-treated with dilute acid, dilute alkali, and distilled water. The highest increase in EC was found using dilute acid pre-treatment as a result of the removal of soil and alkaline metal by the acid solution and degradation of the biomass chemical composition. The ash content of palm kernel shell was reduced when pretreated with distilled water or diluted acid. The ash content increased with alkaline pre-treatment since abundant sodium ions in alkaline medium prevented ions from leaching into the medium and ions were bound and tied up by the biomass particles, which resulted in a high amount of ash content [37]. Torrefaction pre-treatment, which is a low temperature thermal conversion conducted without oxygen aiming to reduce moisture content of the biomass, could increase biochar yield during pyrolysis because the pretreatment predisposed carbon and oxygen content to remain as solids [39]. Another study showed that paper mill sludge as biochar feedstock, pre-treated with phosphoric acid and torrefaction, followed by pyrolysis, resulted in reduced volatile matter content, increased inorganic matter, and increased biochar yield [40]. It was shown that biochar made from feedstock with pretreatments such as light bio-oil or phosphoric acid may have larger surface areas and more porous structure [41,42], which could influence the effects of biochar on air space, nutrient and water-holding ability, and microbial activity. Biochar made from bark pre-treated with tannery slurry as an alkaline treatment could have a higher $NH4^+$ absorption capacity, as well as more surface functional groups (carboxyl and carbonyl groups) formed than untreated ones [43], causing increased CEC of biochar. Silica enrichment was also found in biochar made from rick husk pretreated with bio-oil or HCl [42].

In addition to pre-treatments of feedstocks, post-treatments could also change biochar properties. Some biochars could contain toxic compounds such as polycyclic aromatic hydrocarbons (PAHs) during production. Drying biochars at temperature of 100 °C, 200 °C, and 300 °C significantly decreased the amount of PAHs in biochars, which indicated that the release of PAHs from biochars was due to the increased opening of the pores and diffusion of PAHs from the pores after the thermal treatment [44]. Biochar could be treated and mixed with other substances. Dumroese et al. [7] dry-blended biochar with wood flour, polylactic acid, and starch to form pelleted biochar, which is preferred over the original fine-textured and dusty form for its handling convenience and even incorporation. McCabe et al. [45] evenly blended soybean-based bioplastics with biochar in a pelletized form as a source of nutrients in container substrates.

2.2. Biochar Feedstocks

In addition to production conditions, biochars could be made from varying feedstocks, which would contribute to differences in physical and chemical properties (Table 1). The feedstocks could be waste materials such as green waste [18], forest waste [46,47], wheat straw [5], sugarcane bagasse [48], rice hull [49], crab shell [50] and *Eucalyptus saligna* wood chips (byproduct of construction, fuel-wood and pulp wood) [51]. Biochars could also be made from non-waste materials such as holm oak [52], conifer wood [53], citrus wood [54] and pine wood [6,55–57]. The crab shell biochar and oak chip biochar have different pH ECs, and C, nitrogen (N), phosphorus (P) and potassium (K) content, although they were made by the same production method and temperature [50]. Straw biochar had a higher pH, exchangeable cations, and K content compared to a wood biochar [5]. The biochars made from sewage sludges of two different municipal plants also had slightly different pHs and N content [19]. Biochar could have high P and K content and could be used as P and K fertilizers, when made from rice hulls with high content of the minerals [49]. It was shown that biochar properties were related to the properties of the original feedstock [49]. The biochars made from different feedstocks could have different physical and chemical properties, which should be taken into account when they are incorporated in containers.

Table 1. Summary of the feedstock, production condition and properties of the biochars used in container substrates.

Biochar Feedstock	Production Temp (°C)	CC (%)	AS (%)	TP (%)	BD (g cm^{-3})	pH	EC (dS m^{-1})	CEC (cmol kg^{-1})	N (%)	C (%)	P (%)	K (%)	Na (%)	Ca (%)	Mg (%)	S (%)	Reference
Citrus wood [z]	n	n	n	n	n	7.6	1.6	n	0.6	70.6	0.0008	0.37	0.32	0.02	0.01	0.07	[54]
Coir (coconut husk fiber)	450	64	33	97	0.14	8.2	1.0	153	1.3	n	0.17	1.89	4.83	0.33	0.73	n	[21]
Conifer wood	450	n	n	92	0.64	8.5	0.4	n	n	n	n	n	n	n	n	n	[53]
Crab shell [z]	200–250	n	n	n	n	8.8	0.005	n	3.6	28.7	0.03	0.61	0.04	0.18	0.08	n	[50]
Eucalyptus saligna wood chip	550	n	n	n	n	8.8	0.2	n	0.3	83.6	0.02	0.24	n	2.13	0.11	0.05	[51]
Forest waste	n	n	n	n	n	9.6	0.7	n	0.7	59.5	0.08	0.87	0.04	2.90	0.24	0.07	[46,47]
Green waste	550	n	n	n	n	7.7	n	250	0.3	77.5	n	n	n	n	n	0.00	[58]
Green waste (willow, pagoda tree and poplar)	n	27	22	49	0.44	8.0	0.9	n	1.2	50.4	0.01	0.47	n	n	n	n	[18]
Green waste (tomato crop)	550	n	28	n	0.13	10.4	3.3	524	n	55.0	n	n	n	n	n	n	[59]
Hardwood pellets	n	n	n	n	0.38	8.0	1.1	n	n	n	0.0005	0.04	0.001	0.02	0.002	n	[5]
Holm oak	650	51	29	80	0.32	9.3	0.5	n	0.9	n	0.18	0.77	n	3.76	0.40	n	[52]
Mixed hardwood (oak, elm, and hickory)	450	n	n	n	0.28	n	n	n	n	n	0.29	3.59	0.02	38.28	0.97	n	[60]
Mixed hardwood	n	60	24	85	0.15	11.2	2.0	n	0.2	n	0.05	0.64	0.01	2.75	0.13	0.02	[61]
Mixed softwood [y]	800	n	n	n	n	10.9	0.5	19	n	n	0.02	n	n	n	n	n	[62]
Oak chip [z]	200–250	n	n	n	n	5.1	0.3	n	0.1	52.2	0.09	0.10	0.06	1.03	0.08	n	[50]
Olive mill waste	500	n	n	n	n	9.7	9.2	n	0.6	59.5	0.90	6.42	0.05	3.40	0.61	0.17	[47]
Pine chip [z]	200–250	n	n	n	n	6.4	0.03	n	0.3	53.7	0.05	0.65	0.05	0.23	0.08	n	[50]

Table 1. *Cont.*

Biochar Feedstock	Production Temp (°C)	CC (%)	AS (%)	TP (%)	BD (g cm^{-3})	pH	EC (dS m^{-1})	CEC (cmol kg^{-1})	N (%)	C (%)	P (%)	K (%)	Na (%)	Ca (%)	Mg (%)	S (%)	Reference
Pine cone ^{z}	200–250	n	n	n	n	5.1	1.2	n	0.6	53.2	0.01	0.16	0.04	0.36	0.05	n	[50]
Pine wood	450	49	34	83	0.17	n	n	n	0.4	48.1	n	0.10	n	0.50	0.30	n	[6,57]
Pine wood	450	47	36	83	0.18	5.4	0.2	n	n	n	n	n	n	n	n	n	[55,56]
Poplar ^{y}	1100–1200	57	34	91	n	9.7	0.2	n	0.7	n	0.51	0.98	n	4.31	7.64	n	[63]
Pruning wastes	300	17	4	21	0.18	7.5	0.3	26	1.2	66.2	0.004	n	n	n	n	n	[20]
Pruning wastes	500	35	4	39	0.18	10.3	1.0	16	1.2	77.7	0.01	n	n	n	n	n	[20]
Rice husk	500	n	n	n	n	10.2	0.8	50	0.3	20.5	n	n	n	n	n	n	[64]
Rice husk ^{z}	n	n	n	n	0.30	7.3	n	n	1.1	n	0.10	0.50	n	n	n	n	[65]
Rice husk ^{z}	200–250	n	n	n	n	6.3	0.4	n	0.6	45.4	1.21	0.27	0.73	15.80	1.04	n	[50]
Rice hull ^{y}	815–871	n	n	n	0.20	10.5	n	n	0.2	17.7	0.30	0.98	n	0.35	0.15	0.03	[49,66,67]
Sewage sludge ^{x}	450	n	n	n	n	7.9	1.1	n	1.1	n	n	n	n	n	n	n	[19]
Sewage sludge ^{x}	450	n	n	n	n	7.5	1.1	n	3.1	n	n	n	n	n	n	n	[19]
Southern yellow pine	400	n	n	n	n	6.0	n	n	n	n	0.03	0.29	n	0.06	0.12	0.08	[68]
Spruce wood ^{y}	1100–1200	29	63	92	n	11.1	0.3	n	0.2	n	0.05	0.74	n	1.34	0.17	n	[63]
Sugarcane bagasse ^{z}	343	n	n	n	0.11	5.8	n	n	n	n	n	n	n	n	n	n	[48]
Sugarcane bagasse ^{z}	343	n	n	n	0.11	6.1	n	n	n	n	n	n	n	n	n	n	[48]
Switchgrass ^{z}	1000	n	n	n	0.10	10.8	3.5	n	1.3	79.0	1.20	6.60	n	n	n	n	[69]
Wheat straw	600	n	n	n	0.31	10.0	1.0	n	1.0	79.3	n	n	n	n	n	n	[70]
Wheat straw	n	n	n	n	0.24	9.5	2.5	n	n	n	0.003	0.10	0.002	0.004	0.0009	n	[5]

Note: Production temp: production temperature; CC: container capacity; AS: air space; TP: total porosity; BD: bulk density; EC: electrical conductivity; CEC: cation exchange capacity. Pyrolysis was the biochar production method, unless indicated otherwise. "n" means not available. ^{z}: Biochar production method was not available. ^{y}: Biochar was produced from gasification. ^{x}: Two different sewage sludges were selected from two municipal plants.

3. Effects of Biochar on Container Substrates

3.1. Physical Properties

3.1.1. Bulk Density

The addition of biochar affects the physical properties of container substrates. Biochars have a higher bulk density than commonly-used substrate components, such as peat moss, perlite, and vermiculite. Using biochar to replace certain percentages of peat could thus increase the bulk density of the substrates [5,7,18,71,72].

3.1.2. Container Capacity, Air Space and Total Porosity

Biochar incorporation in container substrates may affect container capacity, air space, and total porosity. Particle size distribution of the substrate components is important for determining their physical properties [73]. Due to the differing particle sizes of biochars and substrate components, the effects of biochar incorporation on the physical properties of a container substrate will vary. Container capacity is the maximum percent volume of water a substrate can hold after gravity drainage [74]. Container substrates absorb water in small pores (micropores) between, or inside component particles [10]. Méndez et al. [75] showed that the incorporation of 50% (by vol.) biochar with peat increased container capacity, compared to those with 100% peat substrate due to increased micropores after biochar incorporation. Similar to these results, Zhang et al. [21] also reported that mixing 20% or 35% (w/w) biochar with compost made from green waste increased container capacity. Yet, some research has shown that the incorporation of biochar in container substrates had no effect on container capacity [5,18]. The differing results after biochar incorporation could be due to the different particle sizes of the biochars and the substrate components used. Besides container capacity, biochar incorporation could also affect air space. Air space is the proportion of air-filled large pores (macropores) after drainage [10]. Méndez et al. [75] showed that the incorporation of 50% (by vol.) biochar with peat increased the air space compared to 100% peat substrate. In this study, the percentage of particle size larger than 2 mm was 29% (w/w) for biochar but 8.8% for peat. Thus, the increased air space was caused by an increased number of macropores due to the incorporation of biochar with larger particle size. Zhang et al. [21] confirmed this by showing that mixing biochar with compost increased the percentage of particles larger than 2 mm and thus increased the air space. Total porosity is the sum of air space and container capacity. The effect of biochar on total porosity is related to its effect on air space and container capacity. Substituting peat with 50% biochar (by vol.) made from green waste had no effect on total porosity [18]. Méndez et al. [75] concluded that the addition of biochar produced from deinking sludge increased the total porosity. Zhang et al. [21] also showed that mixing biochar with compost increased the total porosity. Vaughn et al. [5] showed that the effects of biochar on total porosity were mixed and there was no specific trend, when mixing biochar with peat. In summary, biochar incorporation could impact total porosity, air space, and container capacity.

3.2. Chemical Properties

3.2.1. pH

In general, biochar is effective at increasing the pH of container substrates since the pH of biochars used in most research is neutral to basic [21,53,58,59]. Biochar could buffer acidity due to the negative charge on the surface of biochar [76]. However, the pH of biochars could be acidic. The pH of the biochar depends on the nature of the feedstock and the temperatures during biochar production. The lower the temperature of production, the lower the pH of the biochar. The pH of oak wood biochar was 4.8 when produced at 350 °C [24]. Khodadad et al. [77] also showed pH of biochar made from pyrolysis of oak and grass at 250 °C was 3.5. Lima et al. [78] showed that the pH was around 5.9 for biochars made from pecan shell at 350 °C and switchgrass at 250 °C.

3.2.2. Electrical Conductivity

Biochar incorporation could increase container substrate EC due to high EC of the biochar used. The EC of biochar was affected by the biochar functional groups (such as fused-ring aromatic structures and anomeric O-C-O carbons), metal oxide precipitates and binding of metals [24,79]. Hossain et al. [80] also found that as pyrolysis temperature increased, EC of the sludge biochars decreased. When incorporating biochar in container substrate, Vaughn et al. [5] showed that mixing 5%, 10%, and 15% (by vol.) pelletized wheat straw and hardwood biochars with container substrates containing peat moss and vermiculite increased the EC. Tian et al. [18] also found that adding 50% (by vol.) biochar made from green waste to peat moss media significantly increased EC. The increased substrate EC after biochar incorporation could be due to the high pH, large surface area, and charge density of the biochar [70].

3.2.3. Cation Exchange Capacity

Biochar incorporation could affect CEC and nutrient availability, which is related to the original properties of biochar itself. Surface functional groups, such as carboxylate, carbonyl and ether are responsible for the CEC of biochar [81]. Different biochars have different chemical functional groups. Vaughn et al. [5] found that some volatile materials were removed and wood cellulosic polymers were carbonized in wood biochar after pyrolysis, while wheat straw biochar was less carbonized and had more chemical functionality, which serves as exchange sites for nutrient absorption. It was shown that CEC was higher in a 25% biochar and 75% peat moss mix (by vol.) than that in 100% peat moss [22]. Some biochars can even provide nutrients to the plants due to the high concentration of certain nutrients in the original feedstocks. Some forms of biochars can serve as a source of P and K, which leads to increased availability of these minerals in container substrates and improved fertility [49,66,82].

3.3. Effects on the Microbial Activities

Biochar incorporation may affect microbial activity and biomass in containers. Adding biochar can increase pH, available water content, and influx of nutrients as discussed above, thus stimulating microbial communities and increasing microbial biomass. Warnock et al. [83] also indicated that porous biochar with a high surface area could provide shelter for microorganisms. Saito [84] showed that biochar could serve as a microhabitat for arbuscular mycorrhizal fungi. Higher mycorrhizal colonization and plant growth were shown in mixes of biochar and soil in container experiments [85]. However, only a limited amount of research investigated the effects of biochar on microbial activity or inoculation with mycorrhizae in soilless substrates. Increased mycorrhizal colonization was found in containers containing sand and clay in a ratio of 3:1 (by vol.) with activated biochar (2 g per container) [86]. Inoculation with arbuscular mycorrhizas fungus significantly increased *Pelargonium zonale* plant growth in containers with 0%, 30% or 70% (by vol.) biochar with the rest being peat [87]. Biochars produced at different temperatures may have different surface areas and adsorption abilities [88], which could lead to different levels of nutrient retention and effects on microbial activities.

4. Effects of Biochar on Plant Growth in Container Substrates

There is an increasing amount of research on the effects of biochar on container-grown plant growth that shows the potential for biochar to be a replacement for commonly-used soilless container substrate components including peat moss, bark, vermiculite, perlite, coir, etc. Mixing biochar in container substrates may have a positive impact on plant growth due to beneficial effects like improved container physical and chemical properties and enhanced nutrient and water retention, as mentioned above. Tian et al. [18] found that mixing biochar made from green waste with peat (50% each, by vol.) increased total biomass and leaf surface area of *Calathea rotundifolia* cv. Fasciata when compared to that of peat substrates alone, because of improved substrate properties and increased nutrient retention after biochar incorporation. Replacing 10% (by vol.) of peat with sewage sludge biochar enhanced lettuce (*Lactuca sativa*) biomass production by 184%–270% when compared to 100% peat-based substrate, due to increased N, P and K concentrations and microbial activities [19]. Incorporation of biochar produced from pruning waste at 300 °C (pH = 7.53) and 500 °C (pH = 10.3) into peat substrates at the ratio of 50% and 75% (by vol.) increased lettuce biomass when compared to those in peat alone (pH = 6.14), probably because the increased pH after biochar incorporation was more ideal for many crops [20]. Graber et al. [54] tested the effects of mixing three ratios of citrus wood biochar (1%, 3% or 5%, w/w) with commercial container substrates (a mixture of coconut fiber and tuff at a 7:3 ratio by vol.) on the growth of peppers (*Capsicum annuum*) and tomatoes (*Solanum lycopersicum*). The effects included increased leaf area, shoot dry weight (after detaching the fruits), numbers of flowers and fruit of pepper and increased plant height and leaf size of tomato plants compared to those in commercial container substrates. Graber et al. [54] indicated two possible reasons for the responses, increased beneficial microbial populations or low doses of biochar chemicals stimulating plant growth (hormesis). Mixing 20% or 35% (w/w) biochar made from coir in composted green waste medium increased plant height, root and shoot length, and root fresh and dry weight of *Calathea insignis* when compared to one without any biochar incorporation, effects due to increased water retention, optimized total porosity, aeration porosity, water-holding porosity, nutrients, and microbial activities [21]. Overall, increased plant growth after biochar incorporation could be attributed to increased availability of nutrients and improved water retention, both desirable substrate properties.

However, biochar incorporation may not always improve plant growth. Not all biochars are the same (Table 1). The effects of biochars on container-grown plants are variable (Tables 2–5) depending on multiple factors. There are distinct interactions between biochar and different substrate components. Different biochars, biochar incorporation rate, and other components mixed with biochar can contribute to differing results. Furthermore, individual plant responses to biochar also vary. Across studies of the effects of biochar alone on plant growth, without other factors such as irrigation or fertilization rates, (Tables 2–4), 77.3% reported that some biochar addition to container substrates could promote plant growth, and 50% revealed that plant growth or dry weight was suppressed by some biochar in container substrates. Most studies (69.4%) in Tables 2–5 investigated plant growth in container substrates with biochar for 12 weeks or less than 12 weeks. The length of the experiments in these studies varied from 3 weeks to 7 months. Many mechanisms of biochar-plant interactions are not fully understood.

4.1. *Different Plant Species*

The impact of biochar on plant growth differs by species since different plants have different suitable growth conditions or different tolerance to certain stresses. Mixing potato anaerobic digestate with acidified wood pellet biochar (1:1, by vol.) led to higher fresh and dry weight of tomatoes than a peat: vermiculite control, but led to lower fresh and dry weight of marigold (*Calendula officinalis*) plant [71]. The EC of potato anaerobic digestate is high (7.1 dS m^{-1}). The different fresh and dry weight responses of tomato and marigold could be due to the salt tolerances of these two plants [71]. Choi et al. [57] also showed that mixes with 20% pine bark and 80% biochar (by vol.) led to higher chrysanthemum (*Chrysanthemum nankingense*) fresh and dry weight, but lower tomato plant fresh and dry weight when compared to the control. The reduced tomato plant fresh weight and dry weight was because tomato usually requires more nutrients than other plants and biochar can hold or capture nutrients. Furthermore, 80% biochar mixes had no effect on lettuce (*Lactuca sativa*) and basil (*Ocimum basilicum*) fresh and dry weights. Altland and Locke [67] also showed that mixes of 20% (by vol.) gasified rice hull biochar with Sunshine Mix #2 fertilized with 100 mg L^{-1} N using ammonium nitrate and 0.9 kg m^{-3} Micromax caused a smaller *Pelargonium x hortorum* shoot dry weight but increased shoot dry weight of tomato plants when compared to the control (Sunshine Mix #2) fertilized at the rate of 100 mg L^{-1} N with a commercial complete fertilizer with micronutrients.

Table 2. Summary of the effects of biochar made from different feedstocks mixed with other substrate components on container-grown plants, with percentage of biochar in container substrates less than 50% (by vol.).

Plant Species	Non-Biochar Components	Biochar Feedstock	Percentage (%, by vol.) of Biochar and the Effects on Plants' Dry Weight/Growth Index (DW/GI) [z]								Reference
			1	5	10	15	20	25	30	40	
Buxus sempervirens × *Buxus microphylla*	Pine bark and 24 g osmocote 18N–6P–12K	Switchgrass			=/n			=/n			[69]
Calendula offcinalis	Coir	Forest waste			=/n			=/n			[46]
Chrysanthemum nankingense	Pine bark	Pine wood					=/=			=/=	[57]
Cucumis melo	Sunshine commercial growing medium	Standard sugarcane bagasse						=/= [y]			[48]
		Sugarcane bagasse using a pneumatic transport system						=/= [y]			
Cucurbita pepo	Sunshine commercial growing medium	Standard sugarcane bagasse						=/= [y]			
		Sugarcane bagasse using a pneumatic transport system						=/= [y]			
Euphorbia × *lomi*	Peat	Conifer wood					n/= [y]			n/+ [y]	[53]
Euphorbia pulcherrima	Sunshine Mix #1	Pine wood					+/=			=/=	[55]
Hydrangea paniculata	Pine bark and 24 g osmocote 18N–6P–12K	Switchgrass			=/n			-/n			[69]
Lactuca sativa	Peat	Sewage sludge			+/+ [y]						[19]
Lactuca sativa 'Black Seeded Simpson'	Pine bark	Pine wood					n/=			n/=	[57]
Lilium longiflorum	Sunshine Mix #1	Pine wood					= [x]/= [y]			= [x]/= [y]	[56]
Ocimum basilicum 'Genovese'	Pine bark	Pine wood					=/n			=/n	[57]
Ocimum basilicum	Peat	Softwood from spruce wood							=/n		[63]
		Harwood from poplar							=/n		
Ocimum basilicum	5% vermicompost (VC) with the rest being Berger BM7	Mixed hardwood					=/=			=/=	[61]
	10% VC with the rest being Berger BM7						=/=			=/=	
	15% VC with the rest being Berger BM7						=/=			+/=	
	20% VC with the rest being Berger BM7						=/=			+/=	

Table 2. *Cont.*

Plant Species	Non-Biochar Components	Biochar Feedstock	Percentage (%, by vol.) of Biochar and the Effects on Plants' Dry Weight/Growth Index (DW/GI) [z]								Reference
			1	5	10	15	20	25	30	40	
Petunia hybrida	Coir	Forest waste			=/n			=/n			[46]
Solanum lycopersicum. 'Red Robin'	50% vermiculite with the rest being peat and biochar	Pelletized wheat straw		=/+ [y]	=/+ [y]	=/+ [y]					[5]
		Hardwood pellets		=/+ [y]	=/+ [y]	=/+ [y]					
Solanum lycopersicum 'Cuarenteno'	Coir	Forest waste			=/n			=/n			[47]
		Olive mill waste			+/n			=/n			
Solanum lycopersicum 'Gransol Rijk Zwaan'		Forest waste			=/n			-/n			
		Olive mill waste			=/n			=/n			
Solanum lycopersicum 'Hope'	Pine bark	Pine wood					=/=			+/=	[57]
Solanum lycopersicum 'Roma'	5% VC with the rest being Berger BM7	Mixed hardwood					=/=			=/=	[61]
	10% VC with the rest being Berger BM7						+/=			=/=	
	15% VC with the rest being Berger BM7						=/+			+/=	
	20% VC with the rest being Berger BM7						=/=			=/=	
Tagetes erecta 'Inca II Yellow Hybrid'	50% vermiculite with the rest being peat and biochar	Pelletized wheat straw		=/+ [y]	=/+ [y]	=/+ [y]					[5]
		Hardwood pellets		=/= [y]	=/+ [y]	=/+ [y]					

[z]: "+" means increased; "=" means there was no significant difference; "-" means decreased; "n" means not available. [y]: Result for this was for plant height not growth index. [x]: Result for this was for leaf dry height not total dry weight.

Table 3. Summary of the effects of biochar made from different feedstocks mixed with other substrate components on container-grown plants, with percentage of biochar in container substrates ranging from 50% to 100% (by vol.).

Plant Species	Non-Biochar Components	Biochar Feedstock	Percentage (%, by vol.) of Biochar and Its Effect on Plants' Dry Weight/Growth Index (DW/GI) [z]								Reference
			50	60	66	70	75	80	90	100	
Anethum graveolens	Perlite	Rice husk	+/+ [y]								[64] [w]
Brassica rapa ssp. *pekinensis*			+/+ [y]								
Brassica rapa var. *rosularis*			+/+ [y]								

Table 3. *Cont.*

Plant Species	Non-Biochar Components	Biochar Feedstock	Percentage (%, by vol.) of Biochar and Its Effect on Plants' Dry Weight/Growth Index (DW/GI) ^{z}								Reference
			50	60	66	70	75	80	90	100	
Calathea rotundifolia cv. Fasciata	Peat	Green waste	+/n								[18]
Chrysanthemum nankingense	Pine bark	Pine wood		=/n				+/n		=/n	[57]
Cucumis melo	Sunshine commercial growing medium	Standard sugarcane bagasse	=/=^{y}				-/=^{y}			-/=^{y}	[48]
		Sugarcane bagasse using a pneumatic transport system	=/+^{y}				=/=^{y}			-/=^{y}	
Cucurbita pepo	Sunshine commercial growing medium	Standard sugarcane bagasse	=/=^{y}				=/=^{y}			=/=^{y}	
		Sugarcane bagasse using a pneumatic transport system	+/+^{y}				=/=^{y}			=/=^{y}	
Euphorbia × *lomi*	Peat	Conifer wood		+/+^{y}				+/+^{y}		n/=^{y}	[53]
Euphorbia pulcherrima	Sunshine Mix #1	Pine wood		-/=				-/=		-/-	[55]
Lactuca sativa	Perlite	Rice husk	+/+^{y}								[64] ^{w}
Lactuca sativa	Peat	Deinking sludge	+/n								[75]
	Coir		-/n								
Lactuca sativa	Peat	Pruning wastes	+/n				+/n				[20]
Lactuca sativa 'Black Seeded Simpson'	Pine bark	Pine wood		n/=				n/=		n/=	[57]
Lilium longiflorum	Sunshine Mix #1	Pinewood		=^{x}/=^{y}				=^{x}/=^{y}			[56]
Malva verticillata	Perlite	Rice husk	+/+^{y}								[64]^{w}
Ocimum basilicum	5% VC with the rest being Berger BM7	Mixed hardwood		=/=							[61]
	10% VC with the rest being Berger BM7			=/=							
Ocimum basilicum	15% VC with the rest being Berger BM7	Mixed hardwood		+/=							[61]
	20% VC with the rest being Berger BM7			=/=							
Ocimum basilicum	5% chicken manure compost (CM) with the rest being Berger BM7	Mixed hardwood		=/=		=/=		-/-			[61]
Ocimum basilicum	5% VC with the rest being Berger BM7			=/=		=/=		-/-			

Table 3. *Cont.*

Plant Species	Non-Biochar Components	Biochar Feedstock	Percentage (%, by vol.) of Biochar and Its Effect on Plants' Dry Weight/Growth Index (DW/GI) [z]								Reference
			50	60	66	70	75	80	90	100	
Ocimum basilicum 'Genovese'	Pine bark	Pine wood		=/n				=/n		-/n	[57]
Solanum lycopersicum 'Red Robin'	Potato digestate	Wood pellet	+/= [y]								[71]
		Pelletized wheat straw	+/= [y]								
		Pennycress presscake	=/= [y]								
Solanum lycopersicum 'Gransol Rijk Zwaan'	Coir	Forest waste	-/n				-/n			-/n	[47]
		Olive mill waste	-/n				-/n			-/n	
Solanum lycopersicum 'Cuarenteno'	Coir	Forest waste	-/n				-/n			-/n	
		Olive mill waste	=/n				=/n			=/n	
Solanum lycopersicum	Faecal sludge based compost	Rice husk	=/= [y]							-/- [y]	[65]
Solanum lycopersicum 'Roma'	5% VC with the rest being Berger BM7	Mixed hardwood		=/=							[61]
	10% VC with the rest being Berger BM7			=/=							
	15% VC with the rest being Berger BM7			=/=							
	20% VC with the rest being Berger BM7			=/=							
Solanum lycopersicum 'Tumbling Tom Red"	5% CM with the rest being Berger BM7	Mixed hardwood		+/=		=/=		-/-			[61]
	5% VC with the rest being Berger BM7			=/=		=/=		=/=			
Solanum lycopersicum 'Hope'	Pine bark	Pine wood		=/=				-/=		-/=	[57]
Tagetes erecta	Potato digestate	Wood pellet	-/= [y]								[71]
		Pelletized wheat straw	=/= [y]								
		Pennycress presscake	-/= [y]								

[z]: "+" means increased; "=" means there was no significant difference; "-" means decreased; "n" means not available. [y]: Result for this was for plant height not growth index. [x]: Result for this was for leaf dry height not total dry weight. [w]: Hydroponic experiment.

Table 4. Summary of the effects of biochar made from different feedstocks mixed with other substrate components on container-grown plants, with percentage of biochar in container substrates measured by weight.

Plant Species	Non-Biochar Components	Biochar Feedstock	Percentage (%, by weight) of Biochar and Its Effect on Plants' Dry Weight/Growth Index (DW/GI) [z]										Reference
			1	2.5	3	5	10	20	35	40	60	80	
Acmena smithii	Growing medium (pine bark, coir, clinker ash and coarse sand) with 3 kg m^{-3} controlled-release fertilizer (CRF)	*Eucalyptus saligna* wood chip		=/n		=/n	=/n						[51]
	Growing medium (pine bark, coir, clinker ash and coarse sand) with 6 kg m^{-3} CRF			+/n		=/n	=/n						
	Growing medium (pine bark, coir, clinker ash and coarse sand) with no CRF			+/n		=/n	=/n						
Calathea insignis	Composted green waste medium	Coir (coconut husk fiber)						+/+ [y]	+/+ [y]				[21]
	0.5% humic acid (w/w) with the rest being green waste compost							+/+ [y]	+/+ [y]				
	0.7% humic acid (w/w) with the rest being green waste compost							+/+ [y]	+/= [y]				
Capsicum annuum	A mixture of coconut fiber and tuff at a ratio of 7:3 (by vol.)	Citrus wood	n/= [y]		n/= [y]								[54]
Capsicum annuum	Sphagnum peatmoss-based medium in 50-cell transplant trays	Hardwood including oak, elm, and hickory						=/= [y]		=/= [y]	-/- [y]	-/- [y]	[60]
	Sphagnum peatmoss-based medium in 72-cell transplant trays							=/+ [y]		=/= [y]	=/= [y]	-/- [y]	
	Sphagnum peatmoss-based medium in 98-cell transplant trays							=/= [y]		=/= [y]	=/= [y]	-/- [y]	
Solanum lycopersicum	A mixture of coconut fiber and tuff at a ratio of 7:3 (by vol.)	Citrus wood	n/+ [y]		n/+ [y]								[54]
Viola × *hybrida*	Growing medium (pine bark, coir, clinker ash and coarse sand) blended with 3 kg m^{-3} CRF	*Eucalyptus saligna* wood chip		=/n		=/n	=/n						[51]
	Growing medium (pine bark, coir, clinker ash and coarse sand) blended with 6 kg m^{-3} CRF			+/n		=/n	+/n						
	Growing medium (pine bark, coir, clinker ash and coarse sand) with no CRF			-/n		=/n	-/n						
Viola × *wittrockiana*	Growing medium (pine bark, coir, clinker ash and coarse sand) blended with 3 kg m^{-3} CRF	*Eucalyptus saligna* wood chip		-/n		=/n	=/n						[51]
	Growing medium (pine bark, coir, clinker ash and coarse sand) blended with 6 kg m^{-3} CRF			+/n		=/n	=/n						
	Growing medium (pine bark, coir, clinker ash and coarse sand) with no CRF			+/n		=/n	=/n						

[z]: "+" means increased; "=" means there was no significant difference; "-" means decreased; "n" means not available. [y]: Result for this was for plant height not growth index.

Table 5. Other studies testing the effects of biochar mixed with other substrate components on container-grown plants.

Plant Species	Non-Biochar Components	Biochar Feedstock	Biochar Percentage (by vol.)	Effects on Plant Growth z	Other Information	Reference
Agrostis stolonifera	Sand	Southern yellow pine	15%	Plant height (=)/DW (=)/FW (=)	Control was mixes with 85% sand and 15% peat	[68]
	85% sand and 10% peat, vermicompost, yard-waste compost, Organimix compost, humus or worm castings		5%	Plant height (=)/DW (=)/FW (=)		
	85% sand and 10% anaerobic biosolids		5%	Plant height (+)/DW (+)/FW (+)		
Helianthus annuus	Pig slurry compost	Holm oak	40% or 80%	Shoot DW (+)	Compared to mixes with 40% or 80% coir with the rest being pig slurry compost, respectively	[52]
	Pig slurry compost		60%	Shoot DW (=)	Control was mixes with 60% coir with the rest being pig slurry compost	
	No		100%	Shoot DW (=)	Control was 100% coir	
Ipomoea aquatica	Spent pig litter compost, vermiculite, perlite and peat	Wheat straw	2%, 4% or 8%	Germination rate (=)		[70]
			10%, 12%, 14% or 16%	Germination rate (-)		
Lactuca sativa	Two parts of single wood species sawdust to one-part poultry manure	Rice husk	50% or 66%	Plant height (-)	Half irrigation (0.1125mm)	[92]
			50%	Plant height (=)	Full irrigation (0.225mm)	
			66%	Plant height (+)	Full irrigation (0.225mm)	
Pelargonium × hortorum 'Maverick Red'	Sunshine Mix #2	Rice hull	1% or 10%	Shoot DW (-)	Plants in biochar-added substrates were fertilized with 100 mg L^{-1} N. Control was Sunshine Mix #2 with a fertilizer (20N-4.4P-16.6K-0.15Mg-0.02B-0.01Cu-0.1Fe-0.05Mn-0.01Mo-0.05Zn) at the rate of 100 mg L^{-1} N	[66]
	Sunshine Mix #2 with a micronutrient package (Micromax, The Scotts Co., Marysville, OH) at 0.9 kg m^{-3}	Rice hull	5%,10% or 15%	Shoot DW (=)		[67]
			20%	Shoot DW (-)		
Pelargonium zonale	Peat	N/A	30% or 70%	Plant height (=)/DW (=)	140 mg L^{-1} slow released fertilizer applied	[87]
			30%	Plant height (=)/DW (=)	210 mg L^{-1} slow released fertilizer applied	
			70%	Plant height (-)/DW (-)		
Solanum lycopersicum 'Megabite'	Sunshine Mix #2 with a micronutrient package (Micromax, The Scotts Co., Marysville, OH) at 0.9 kg m^{-3}	Rice hull	5%	Shoot DW (=)	Plants in biochar-added substrates were fertilized with 100 mg L^{-1} N. Control was Sunshine Mix #2 with a fertilizer (20N-4.4P- 16.6K-0.15Mg-0.02B-0.01Cu-0.1Fe-0.05Mn-0.01Mo-0.05Zn) at the rate of 100 mg L^{-1} N	[67]
			10%, 15% or 20%	Shoot DW (+)		
Solanum lycopersicum	Pine (*Pinus radiata* D. Don) sawdust	Tomato crop green waste	25%, 50%, 75% or 100%	Shoot fresh weight (FW) (=)/Fruit number (=)/ Yield (=)	Control was 100% pine sawdust.	[59]

Table 5. *Cont.*

Plant Species	Non-Biochar Components	Biochar Feedstock	Biochar Percentage (by vol.)	Effects on Plant Growth [z]	Other Information	Reference
Sylibum marianum	Pig slurry compost	Holm oak	40%, 60% or 80%	Shoot DW (=)	Compared to mixes with 40%, 60%, or 80% coir with the rest being pig slurry compost, respectively	[52]
	No		100%	Shoot DW (-)	Control was 100% coir	
Tagetes erecta	30% perlite with the rest being peat and biochar	Mixed softwood	10%, 20%, 30%, 40%, 50%, 60% or 70%	Plant height (=)/DW (=)	No pH adjustment; control was 70:30 peat: perlite mixture.	[62]
			10% or 70%	Plant height (-)/DW (=)	pH adjusted to 5.8; control was 70:30 peat: perlite mixture.	
			20%, 30%, 40%, 50% or 60%	Plant height (=)/DW (=)		
Zelkova serrata	Growing medium mixture of peat moss, perlite, and vermiculite at a ratio of 1:1:1 by vol.	Pine chip	20%	Plant height (=)/Stem DW (=)	0.5 or 1 g/L fertilization	[50]
		Oak chip	20%	Plant height (=)/Stem DW (=)	0.5 or 1 g/L fertilization	
		Pine cone	20%	Plant height (-)/Stem DW (-)	0.5 g/L fertilization	
				Plant height (=)/Stem DW (=)	1 g/L fertilization	
		Rice husk	20%	Plant height (+)/Stem DW (+)	0.5 g/L fertilization	
				Plant height (=)/Stem DW (=)	1 g/L fertilization	
		Crab shell	20%	Plant height (-)/Stem DW (-)	0.5 or 1 g/L fertilization	

[z]: "+" means increased; "=" means there was no significant difference; "-" means decreased; "n" means not available.

(20N-4.4P-16.6K-0.15Mg-0.02B-0.01Cu-0.1Fe-0.05Mn-0.01Mo-0.05Zn). The effects of biochar on container-grown plants could be different due to different plant materials.

Most of the plant species used in testing biochars in container substrates have been herbaceous. Only six woody plants have been tested, including Japanese zelkova (*Zelkova serrata*), lilly pilly (*Acmena smithii*), 'Green Velvet' boxwood (*Buxus sempervirens* × *Buxus microphylla*), Pinky Winky hardy hydrangea (*Hydrangea paniculata*), myrtle (*Myrtus communis*) and mastic tree (*Pistacia lentiscus*). Across all studies, the most frequently tested species have been tomato and lettuce. About 30.5% of the studies used tomato plants to test biochars in container substrates and 19.4% used lettuce. Research is needed to test more plant species.

4.2. Different Biochar and Biochar Percentage in Container Substrates

The impact of biochar on plant growth depends on the properties of the biochar used and the percentage of biochar in the substrates. Those factors impact the overall physical and chemical properties of the container substrates, such as pH, container capacity and CEC. Belda et al. [89] showed that mixing 10%, 25% or 50% (by vol.) forest waste biochar with coir led to higher *Myrtus communis* and *Pistacia lentiscus* stem length and dry weight than using olive mill waste biochar. It was shown that *Zelkova serrata* plants in mixes that contained 20% rice husk biochar with the rest of the mixture composed of peat moss, perlite, and vermiculite at a ratio of 1:1:1 (by vol.) were 6 times larger than those in mixes with crab shell biochar, which could be due to the high concentration of nutrients, nutrient absorption ability and water retention ability of rice husk biochar [50]. Webber et al. [48] showed that pneumatic sugarcane bagasse biochar and standard sugarcane bagasse biochar led to different effects on plant growth, due to different physical and chemical compositions of the two biochars, produced by different conditions. Pumpkin (*Cucurbita pepo*) and muskmelon (*Cucumis melo*) both had increased plant height in mixes with 50% pneumatic sugarcane bagasse biochar with the rest being Sunshine commercial growing media (by vol.) compared to the control, while both in mixes with 50% standard sugarcane bagasse biochar showed similar plant height to the control. Webber et al. [48] also indicated that different biochar percentages could affect the results and showed that mixes with 75% or 100% biochar decreased muskmelon plant dry weight, but mixes with 25% or 50% biochar had no effect. Similarly, the aboveground dry weight of *Viola* × *hybrida* showed no significant effects after the incorporation of 5% (w/w) *Eucalyptus saligna* wood chip biochar to growing medium containing pine bark, coir, clinker ash and coarse sand, but aboveground dry weight decreased when mixing 10% (w/w) biochar with the growing medium, when compared to the control [51]. The decreased plant dry weight was due to reduced concentrations of S, P, and Ca caused by the binding ability of the biochar [51]. Fan et al. [70] found that the germination rate of water spinach (*Ipomoea aquatica*) decreased when the biochar incorporation rate in mixes containing spent pig litter compost, vermiculite, perlite and peat increased to 10%, 12%, 14% or 16% (by vol.) due to the high and unsuitable pH and EC after biochar incorporation, while there was no effect on the germination rate if the biochar incorporation rate was 2%, 4% or 8% (by vol.). Conversa et al. [87] showed that mixing peat with biochar at the ratio of 70:30 (by vol.) with slow released fertilizer at a rate of 140 and 210 mg L^{-1} led to increased *Pelargonium* leaf number and similar shoot dry weight compared to the control. However, mixing peat with biochar at the ratio of 30:70 (by vol.) with a high rate of slow release fertilizer (210 mg L^{-1}) showed decreased *Pelargonium* plant growth and flowering traits due to osmotic stress caused by high EC and decreased mycorrhizal activity with this high biochar rate [87]. Awad et al. [64] also showed that mixes with 50% (by vol.) biochar with the rest being perlite led to increased dry weight and growth of Chinese cabbage (*Brassica rapa* ssp. *pekinensis*), dill (*Anethum graveolens*), curled mallow (*Malva verticillata*), red lettuce, and tatsoi (*Brassica rapa* var. *rosularis*) while 100% rice husk biochar decreased plant growth due to high pH of the substrate, low air space, and decreased N availability due to biochar's N absorption ability.

Across studies that mixed biochar in container substrates by volume and tested the effects of biochar on plant growth without other factors (Tables 2 and 3), 72.2% incorporated biochar at 50% or more (by vol.) in container substrates. This suggested that the substitution of the commonly-used substrates or substrate components with a large proportion of biochar is highly desired and, based on the results, achievable. About 36.4% of the studies (Tables 2 and 3) showed that mixing high percentages of biochar (at least 50% by vol.) in container media could improve the growth of some species when compared to the control. All container substrates with biochar percentages lower than 25% (by vol.) led to similar or higher plant growth or dry weight when compared to the control. A biochar incorporation rate as high as 100% (by vol.) in container substrate often led to similar plant growth to the control [48,53,57].

The physical and chemical properties of biochar could determine whether a large proportion of biochar could be used in container substrates to grow plants. When the physical and chemical properties of biochar or substrates with high percentages of biochar are similar to the commercial substrates or are in the ideal range for container-grown plant growth, a high percentage of biochar could be incorporated into the container substrate. The recommended ranges for the physical properties of most substrates used in commercial container plant production are 50%–85% for total porosity, 10%–30% for air space, 45%–65% for container capacity and 0.19 to 0.7 g cm^{-3} for bulk density [72]. Choi et al. [57] has achieved using 100% biochar substrates to replace the 100% pine bark substrates to grow chrysanthemum and lettuce. The container capacity and air space of the biochar were similar to the bark [57]. Although the total porosity of the biochar used was different from that of the bark, it was in the recommended range for container plant production [57]. Guo et al. [56] also succeeded using up to 80% biochar in peat-based commercial substrates, and the physical properties of the biochar substrates were in, or close to, the recommended range for container plant production. Among all properties, pH could be a limiting factor determining the potential use of biochar in containers. Webber et al. [48] made two kinds of biochars, pneumatic sugarcane bagasse biochar and standard sugarcane bagasse biochar, and indicated that these two biochars could be used in containers as high as 100% to grow pumpkin seedlings for 20 days. The pH of these two bicohars were 5.8 and 6.05, respectively. If the pH of the biochar is high, other acidic components should be added to reduce the pH or a high percentage of biochar in a container may not be achievable. It was shown that the addition of 80% (by vol.) biochar (pH = 8.5) to peat (pH = 5.7) increased plant growth due to neutral pH and improved water holding and air structure after biochar addition [53].

4.3. Other Substrate Components Mixed with Biochar in Container Substrates

The other substrate components used with biochar could affect plant growth due to their different physical and chemical properties and their effects on the overall container substrate properties. Substrate components mixed with biochar have included peat, vermiculite, perlite, coir, pine bark, pine sawdust, commercial growing media, compost, composted green waste and potato digestate (Tables 2–5). Gu et al. [90] showed that gomphrena (*Gomphrena globosa*) grown in 5%, 10%, 15%, 20%, 25% and 30% (by vol.) pinewood biochar mixed with the peat-based Sunshine Mix #1 had greater width and height, higher fresh weight and dry weight than those grown in biochar mixed with bark substrates at 43 days after transplanting. The reason for this result could be that peat-based substrates have more organic matter and higher water and nutrient holding capacity than bark-based substrates. Ain Najwa et al. [91] also indicated that the fruit number and fresh weight of tomato in mixes with coco peat and 150 g biochar were higher than in mixes with oil palm fruit bunch (a newly developed organic medium) and 150 g biochar due to different physical and chemical properties of these two substrates. Vaughn et al. [68] showed that creeping bentgrass (*Agrostis stolonifera*) had higher fresh and dry weight and shoot height in mixes with 85% sand, 10% anaerobic biosolids and 5% biochar (by vol.) than the one in mixes with 85% sand, 10% peat and 5% biochar (vol.), due to higher nitrate concentration caused by biosolid incorporation. Méndez et al. [75] also demonstrated that the total biomass and shoot and root weight of lettuce were higher in deinking sludge biochar with peat (50:50 by vol.) than those in biochar mixed with coir (50:50 by vol.). The lower plant biomass in coir with biochar incorporation may be due to the lower CEC, N and P in coir when compared to peat. Fan et al. [70]

investigated the effects of mixed wheat straw biochar with or without superabsorbent polymer on the substrates containing spent pig litter compost, vermiculite, perlite and peat. The germination rate of water spinach decreased when the biochar incorporation rate in the medium without superabsorbent polymer was 10%, 12%, 14% or 16% (by vol.) due to the high and unsuitable pH and EC after biochar incorporation. However, there was no difference on germination rate between the mixes with different percentages of biochar (from 0% to 16% by vol.) when biochar was applied together with superabsorbent polymer. The reason was that the incorporation of superabsorbent polymer increased the porosity and water-holding capacity and also effectively prevented an excessive increase of pH and EC at the high biochar rates [70]. Margenot et al. [62] also showed that mixes with 10%, 20%, 30%, 40%, 50%, 60% or 70% softwood biochar and 30% perlite with the rest being peat (by vol.) led to similar seed germination and plant height compared to control (mixes with 30% perlite and 70% peat by vol.). However, if other components such as calcium hydroxide were added to increase the pH of 10% biochar mixes to 5.8 or pyroligneous acid to decrease substrate (mixes with more than 10% biochar) pH, lower seed germination resulted in mixes with 50%, 60% or 70% biochar and lower plant height in mixes with 10% or 70% biochar.

5. Effect of Potentially Toxic Contaminants in Biochar on Plant Growth

Biochar may contain potentially toxic substances, such as heavy metals and organic contaminants (PAH and dioxin), which are affected by the production conditions and feedstocks used. The incorporation of biochar with a high content of these contaminants is a concern. Various studies have shown reduced plant growth caused by the toxicity of PAHs [93,94], dioxins [95] and heavy metals [96,97]. The utilization of biochar that contains toxic substances could be detrimental, and could influence plant growth and development, leach into groundwater, and have noxious effects on soil function and microorganisms. However, toxic substances (heavy metals, PAHs and dioxin) in biochars used in container substrates have rarely been tested. Attention is needed when choosing biochar feedstocks and biochar production conditions to avoid or minimize the production of toxic substances.

Biochar could contain heavy metals from contaminated feedstocks; however, heavy metals could be transformed to more stable forms after pyrolysis, thus having less effect on plant growth. Heavy metals may remain in biochar made from contaminated feedstock such as cadmium (Cd), copper (Cu), lead (Pb), and zinc (Zn) as observed with contaminated willow leaves and branches [98] or sewage [99]. However, the heavy metals in biochar might have low bioavailability after pyrolysis and a lower risk to plant growth. Jin et al. [100] found most of the heavy metals in sludge biochar after pyrolysis at 400 to 600 °C, including Cu, Zn, Pb, chromium (Cr), manganese (Mn) and nickel (Ni), were in their oxidized and residual forms, which had low bioavailability and thus risks. Similarly, Devi and Saroha [101] found that the bioavailability of heavy metals (Cr, Cu, Ni, Zn and Pb) in paper mill sludge biochar derived from pyrolysis at 200 °C to 700 °C was reduced due to transformation into more stable forms. Buss et al. [102] investigated the effects of 19 types of biochar produced from marginal biomass containing contaminants (such as Cu, Cr, Ni and Zn) on plant growth and found that only five types of biochar in the study showed suppressive effects on plant growth after adding 5% (by weight) of biochar in sand due to high K and pH, not heavy metals.

Although PAHs could be formed in biochars due to production conditions, the amount of PAH in biochars used in many studies has been low and may have had low toxicity for plant growth. Large quantities of PAHs are formed in reactions at high temperatures, especially over 750 °C [103], although no research was found using biochar produced over 750 °C in container substrates. There is also evidence that small amounts of PAHs can be formed in pyrolysis reactors operating between 400 °C and 600 °C [103,104], which is the temperature range that most biochars suitable as container substrate component were produced [6,19–21,51,70,75,90]. Research has shown that PAHs in biochar produced from slow pyrolysis between temperature 250 °C and 900 °C had very low bioavailability [105]. Wiedner et al. [29] also found that all biochars made from gasification of poplar, wheat straw, sorghum and olive, and from pyrolysis of draff (the waste product from the production of beer after separating liquid malt) and miscanthus contained very low content of PAH (below 1.7 mg kg^{-1}) and biochar made from woodchip gasification (15 mg kg^{-1} PAH). Although biochars produced at certain conditions, especially over 750 °C, could contain PAHs, no research was found using these biochars in container substrates to test their effects on substrate properties and plant growth.

Dioxins could be formed in biochar if the feedstock contains chlorine in certain conditions, but dioxin concentration in biochars could be very low and have a negligible effect on plant growth. Dioxins refer to compounds such as polychlorinated dibenzo dioxins (PCDDs) and polychlorinated dibenzo furans (PCDFs), which are persistent organic pollutants [106]. Dioxins could be formed only in biochars made from feedstock containing chlorine, such as straws, grasses, halogenated plastics and food waste containing sodium chloride under specific conditions [103,106]. Dioxins could be produced during two pathways: "precursor" pathway, which begins with the synthesis of dioxin precursors from feedstock containing chlorine at temperatures between 300 °C and 600 °C; and the "de novo" pathway, which occurs between 200 °C and 400 °C in a catalytic reaction with oxygen and carbon [106–108]. However, the dioxin in biochar made from feedstock with chlorine could be very low. Hale et al. [105] investigated the biochars produced at 250 °C to 900 °C via slow pyrolysis, fast pyrolysis and gasification and found that total dioxin concentrations in biochars tested were very low (92 pg g^{-1}) and bioavailable concentrations were below detection limit [105]. Wiedner et al. [29] found that the dioxins in four biochars produced from gasification of poplar and olive residues and pyrolysis of draff and wood chips and two other hydrochars made from leftover food and sewage sludge were all under the limit of detection, except the one made from sewage sludge (14.2 ng kg^{-1}). No evidence was found testing the effect of biochars with dioxin in container substrates on plant growth.

6. Discussion

The incorporation of biochar into container substrates could affect physical and chemical properties of the container substrates and thus contribute to the growth of container-grown plants. Most biochars have a higher bulk density than commonly-used substrates, and thus the incorporation of biochar could increase the bulk density of the container substrate. The effect of biochar on container capacity, air space, and total porosity of the container substrates depends on the particle size distribution of the biochar and the other components in the container. The liming effect of alkaline biochars could adjust the container substrate with low pH to an optimal pH. In addition, biochar incorporation could increase EC, nutrient availability, and CEC.

The effects of biochar on plant growth in container substrates varies as not all biochars are the same. The characteristics of biochars differ according to the feedstock used and the pyrolysis process. Many factors, such as plant species and the ratio of biochar to other container substrate components, can contribute to different results on container substrate properties and plant growth. Across studies testing the effects of biochar on plant growth but not other factors (such as irrigation or fertilization rates) (Tables 2–4), 77.3% of the studies found that plant growth could be increased by the incorporation of certain percentages of biochar in container substrates, and 50% revealed that certain percentages of biochar addition could decrease plant growth. Among studies mixing biochar with container substrates by volume and testing the effects of biochar on plant growth without other factors (Tables 2 and 3), 36.4% showed that container substrates with high percentages of biochar (at least 50% by vol.) could improve plant growth under certain conditions compared to the control. All the container substrates with biochar percentages lower than 25% (by vol.) led to similar or higher plant growth or dry weight when compared to the control. A biochar incorporation rate as high as 100% (by vol.) in container substrates could lead to similar plant growth to the control. The physical and chemical properties of the biochar could determine whether a large proportion of biochar could be used in container substrates to grow plants.

There is no universal standard for using biochar in container substrates for all plants. Many mechanisms of biochar are not fully understood. Research on biochar in container substrates is still in an exploratory state. Most research has focused on testing whether biochar could be used to substitute for commonly-used substrates such as peat, perlite and bark in containers to grow plants, and compared plant growth with a control that had no biochar addition. There is very limited research that tests other properties such as the effect of biochar on disease suppression in container substrates. Research has shown that biochar could impact greenhouse gas emissions in soil, but limited research has been conducted on soilless container substrates. A limited number of published studies have investigated the effect of biochar on microbial activity or inoculation with mycorrhizae in containers. Most of the species used in reported studies testing biochar in container substrates have been herbaceous plants. More plant species should be used to test the effects of biochar to broaden its use. Future studies could be focused on biochars with promising results, to fine-tune the pyrolysis process and incorporate formulae for diverse container substrates.

Author Contributions: This review is a product of the combined effort of both authors. L.H. wrote the original draft and improved it based on M.G.'s advice and assistance. M.G. reviewed, edited and revised the manuscript.

References

1. Lehmann, J. A handful of carbon. *Nature* **2007**, *447*, 143–144. [CrossRef] [PubMed]
2. Hansen, V.; Hauggaard-Nielsen, H.; Petersen, C.T.; Mikkelsen, T.N.; Müller-Stöver, D. Effects of gasification biochar on plant-available water capacity and plant growth in two contrasting soil types. *Soil Tillage Res.* **2016**, *161*, 1–9. [CrossRef]
3. Johannes, L.; Stephen, J. Biochar for environmental management: An introduction. In *Biochar for Environmental Management-Science and Technology*; Earthscan: London, UK, 2009; pp. 1–10.
4. Vaughn, S.F.; Kenar, J.A.; Eller, F.J.; Moser, B.R.; Jackson, M.A.; Peterson, S.C. Physical and chemical characterization of biochars produced from coppiced wood of thirteen tree species for use in horticultural substrates. *Ind. Crop. Prod.* **2015**, *66*, 44–51. [CrossRef]
5. Vaughn, S.F.; Kenar, J.A.; Thompson, A.R.; Peterson, S.C. Comparison of biochars derived from wood pellets and pelletized wheat straw as replacements for peat in potting substrates. *Ind. Crop. Prod.* **2013**, *51*, 437–443. [CrossRef]
6. Yu, F.; Steele, P.H.; Gu, M.; Zhao, Y. Using Biochar as Container Substrate for Plant Growth. U.S. Patent 9,359,267, 7 June 2016.
7. Dumroese, R.K.; Heiskanen, J.; Englund, K.; Tervahauta, A. Pelleted biochar: Chemical and physical properties show potential use as a substrate in container nurseries. *Biomass Bioenergy* **2011**, *35*, 2018–2027. [CrossRef]
8. Northup, J. Biochar as a Replacement for Perlite in Greenhouse Soilless Substrates. Master's Thesis, Iowa State Univversity, Ames, IA, USA, 2013.
9. Landis, T.D.; Morgan, N. Growing Media Alternatives for Forest and Native Plant Nurseries. In *National Proceedings: Forest and Conservation Nursery Associations-2008. Proceedings RMRS-P-58*; Dumroese, R.K., Riley, L.E., Eds.; US Department of Agriculture, Forest Service, Rocky Mountain Research Station: Fort Collins, CO, USA, 2009; pp. 26–31.
10. Landis, T.D. Growing media. *Contain. Grow. Media* **1990**, *2*, 41–85.
11. Bohlin, C.; Holmberg, P. Peat: Dominating growing medium in Swedish horticulture. *Acta Hortic.* **2004**, *644*, 177–181. [CrossRef]
12. Fascella, G. Growing substrates alternative to peat for ornamental plants. In *Soilless Culture-Use of Substrates for the Production of Quality Horticultural Crops*; InTech: London, UK, 2015.
13. Rankin, T.; Strachan, I.; Strack, M. Carbon dioxide and methane exchange at a post-extraction, unrestored peatland. *Ecol. Eng.* **2018**, *122*, 241–251. [CrossRef]
14. Carlile, W.; Waller, P. Peat, politics and pressure groups. *Chron. Hortic.* **2013**, *53*, 10–16.
15. Carlile, W.; Cattivello, C.; Zaccheo, P. Organic growing media: Constituents and properties. *Vadose Zone J.* **2015**, *14*. [CrossRef]
16. Wright, R.D.; Browder, J.F. Chipped pine logs: A potential substrate for greenhouse and nursery crops. *HortScience* **2005**, *40*, 1513–1515. [CrossRef]

17. Laird, D.A. The charcoal vision: A win–win–win scenario for simultaneously producing bioenergy, permanently sequestering carbon, while improving soil and water quality. *Agron. J.* **2008**, *100*, 178–181. [CrossRef]
18. Tian, Y.; Sun, X.; Li, S.; Wang, H.; Wang, L.; Cao, J.; Zhang, L. Biochar made from green waste as peat substitute in growth media for *Calathea rotundifola* cv. Fasciata. *Sci. Hortic.* **2012**, *143*, 15–18. [CrossRef]
19. Méndez, A.; Cárdenas-Aguiar, E.; Paz-Ferreiro, J.; Plaza, C.; Gascó, G. The effect of sewage sludge biochar on peat-based growing media. *Biol. Agric. Hortic.* **2017**, *33*, 40–51. [CrossRef]
20. Nieto, A.; Gascó, G.; Paz-Ferreiro, J.; Fernández, J.; Plaza, C.; Méndez, A. The effect of pruning waste and biochar addition on brown peat based growing media properties. *Sci. Hortic.* **2016**, *199*, 142–148. [CrossRef]
21. Zhang, L.; Sun, X.; Tian, Y.; Gong, X. Biochar and humic acid amendments improve the quality of composted green waste as a growth medium for the ornamental plant *Calathea insignis*. *Sci. Hortic.* **2014**, *176*, 70–78. [CrossRef]
22. Headlee, W.L.; Brewer, C.E.; Hall, R.B. Biochar as a substitute for vermiculite in potting mix for hybrid poplar. *Bioenergy Res.* **2014**, *7*, 120–131. [CrossRef]
23. Gvero, P.M.; Papuga, S.; Mujanic, I.; Vaskovic, S. Pyrolysis as a key process in biomass combustion and thermochemical conversion. *Therm. Sci.* **2016**, *20*, 1209–1222. [CrossRef]
24. Lehmann, J.; Rillig, M.C.; Thies, J.; Masiello, C.A.; Hockaday, W.C.; Crowley, D. Biochar effects on soil biota—A review. *Soil Biol. Biochem.* **2011**, *43*, 1812–1836. [CrossRef]
25. Hansen, V.; Müller-Stöver, D.; Ahrenfeldt, J.; Holm, J.K.; Henriksen, U.B.; Hauggaard-Nielsen, H. Gasification biochar as a valuable by-product for carbon sequestration and soil amendment. *Biomass Bioenergy* **2015**, *72*, 300–308. [CrossRef]
26. Bruun, E.W.; Hauggaard-Nielsen, H.; Ibrahim, N.; Egsgaard, H.; Ambus, P.; Jensen, P.A.; Dam-Johansen, K. Influence of fast pyrolysis temperature on biochar labile fraction and short-term carbon loss in a loamy soil. *Biomass Bioenergy* **2011**, *35*, 1182–1189. [CrossRef]
27. Libra, J.A.; Ro, K.S.; Kammann, C.; Funke, A.; Berge, N.D.; Neubauer, Y.; Titirici, M.-M.; Fühner, C.; Bens, O.; Kern, J. Hydrothermal carbonization of biomass residuals: A comparative review of the chemistry, processes and applications of wet and dry pyrolysis. *Biofuels* **2011**, *2*, 71–106. [CrossRef]
28. Kalderis, D.; Kotti, M.; Méndez, A.; Gascó, G. Characterization of hydrochars produced by hydrothermal carbonization of rice husk. *Solid Earth* **2014**, *5*, 477–483. [CrossRef]
29. Wiedner, K.; Rumpel, C.; Steiner, C.; Pozzi, A.; Maas, R.; Glaser, B. Chemical evaluation of chars produced by thermochemical conversion (gasification, pyrolysis and hydrothermal carbonization) of agro-industrial biomass on a commercial scale. *Biomass Bioenergy* **2013**, *59*, 264–278. [CrossRef]
30. Zheng, W.; Sharma, B.; Rajagopalan, N. Using biochar as a soil amendment for sustainable agriculture. 2010. Available online: http://hdl.handle.net/2142/25503 (accessed on 31 January 2019).
31. Roos, C.J. *Clean Heat and Power Using Biomass Gasification for Industrial and Agricultural Projects*; Northwest CHP Application Center: Olympia, WA, USA, 2010.
32. Bridgwater, A.; Meier, D.; Radlein, D. An overview of fast pyrolysis of biomass. *Org. Geochem.* **1999**, *30*, 1479–1493. [CrossRef]
33. Ahmed, I.; Gupta, A. Syngas yield during pyrolysis and steam gasification of paper. *Appl. Energy* **2009**, *86*, 1813–1821. [CrossRef]
34. Fuchs, M.R.; Garcia-Perez, M.; Small, P.; Flora, G. Campfire Lessons: Breaking down the Combustion Process to Understand Biochar Production and Characterization. Available online: https://www.biochar-journal.org/en/ct/47 (accessed on 31 January 2019).
35. Panwar, N.; Kothari, R.; Tyagi, V. Thermo chemical conversion of biomass–eco friendly energy routes. *Renew. Sustain. Energy Rev.* **2012**, *16*, 1801–1816. [CrossRef]
36. Li, J.; Li, Y.; Wu, Y.; Zheng, M. A comparison of biochars from lignin, cellulose and wood as the sorbent to an aromatic pollutant. *J. Hazard. Mater.* **2014**, *280*, 450–457. [CrossRef] [PubMed]
37. Rahman, A.A.; Sulaiman, F.; Abdullah, N. Influence of washing medium pre-treatment on pyrolysis yields and product characteristics of palm kernel shell. *J. Phys. Sci.* **2016**, *27*, 53–75.
38. Jenkins, B.; Bakker, R.; Wei, J. On the properties of washed straw. *Biomass Bioenergy* **1996**, *10*, 177–200. [CrossRef]
39. Boateng, A.; Mullen, C. Fast pyrolysis of biomass thermally pretreated by torrefaction. *J. Anal. Appl. Pyrolysis* **2013**, *100*, 95–102. [CrossRef]

40. Reckamp, J.M.; Garrido, R.A.; Satrio, J.A. Selective pyrolysis of paper mill sludge by using pretreatment processes to enhance the quality of bio-oil and biochar products. *Biomass Bioenergy* **2014**, *71*, 235–244. [CrossRef]
41. Chu, G.; Zhao, J.; Huang, Y.; Zhou, D.; Liu, Y.; Wu, M.; Peng, H.; Zhao, Q.; Pan, B.; Steinberg, C.E. Phosphoric acid pretreatment enhances the specific surface areas of biochars by generation of micropores. *Environ. Pollut.* **2018**, *240*, 1–9. [CrossRef] [PubMed]
42. Zhang, S.; Xiong, Y. Washing pretreatment with light bio-oil and its effect on pyrolysis products of bio-oil and biochar. *RSC Adv.* **2016**, *6*, 5270–5277. [CrossRef]
43. Hina, K.; Bishop, P.; Arbestain, M.C.; Calvelo-Pereira, R.; Maciá-Agulló, J.A.; Hindmarsh, J.; Hanly, J.; Macìas, F.; Hedley, M. Producing biochars with enhanced surface activity through alkaline pretreatment of feedstocks. *Soil Res.* **2010**, *48*, 606–617. [CrossRef]
44. Kołtowski, M.; Oleszczuk, P. Toxicity of biochars after polycyclic aromatic hydrocarbons removal by thermal treatment. *Ecol. Eng.* **2015**, *75*, 79–85. [CrossRef]
45. McCabe, K.G.; Currey, C.J.; Schrader, J.A.; Grewell, D.; Behrens, J.; Graves, W.R. Pelletized soy-based bioplastic fertilizers for container-crop production. *HortScience* **2016**, *51*, 1417–1426. [CrossRef]
46. Fornes, F.; Belda, R.M. Biochar versus hydrochar as growth media constituents for ornamental plant cultivation. *Sci. Agric.* **2018**, *75*, 304–312. [CrossRef]
47. Fornes, F.; Belda, R.M.; Fernández de Córdova, P.; Cebolla-Cornejo, J. Assessment of biochar and hydrochar as minor to major constituents of growing media for containerized tomato production. *J. Sci. Food Agric.* **2017**, *97*, 3675–3684. [CrossRef] [PubMed]
48. Webber, C.L., III; White, P.M., Jr.; Spaunhorst, D.J.; Lima, I.M.; Petrie, E.C. Sugarcane biochar as an amendment for greenhouse growing media for the production of cucurbit seedlings. *J. Agric. Sci.* **2018**, *10*, 104. [CrossRef]
49. Locke, J.C.; Altland, J.E.; Ford, C.W. Gasified rice hull biochar affects nutrition and growth of horticultural crops in container substrates. *J. Environ. Hortic.* **2013**, *31*, 195–202.
50. Cho, M.S.; Meng, L.; Song, J.-H.; Han, S.H.; Bae, K.; Park, B.B. The effects of biochars on the growth of *Zelkova serrata* seedlings in a containerized seedling production system. *For. Sci. Technol.* **2017**, *13*, 25–30. [CrossRef]
51. Housley, C.; Kachenko, A.; Singh, B. Effects of *Eucalyptus saligna* biochar-amended media on the growth of *Acmena smithii*, *Viola* var. *Hybrida*, and *Viola* × *wittrockiana*. *J. Hortic. Sci. Biotechnol.* **2015**, *90*, 187–194. [CrossRef]
52. Sáez, J.; Belda, R.; Bernal, M.; Fornes, F. Biochar improves agro-environmental aspects of pig slurry compost as a substrate for crops with energy and remediation uses. *Ind. Crop. Prod.* **2016**, *94*, 97–106. [CrossRef]
53. Dispenza, V.; De Pasquale, C.; Fascella, G.; Mammano, M.M.; Alonzo, G. Use of biochar as peat substitute for growing substrates of *Euphorbia* × *lomi* potted plants. *Span. J. Agric. Res.* **2017**, *14*, 0908. [CrossRef]
54. Graber, E.R.; Harel, Y.M.; Kolton, M.; Cytryn, E.; Silber, A.; David, D.R.; Tsechansky, L.; Borenshtein, M.; Elad, Y. Biochar impact on development and productivity of pepper and tomato grown in fertigated soilless media. *Plant Soil* **2010**, *337*, 481–496. [CrossRef]
55. Guo, Y.; Niu, G.; Starman, T.; Volder, A.; Gu, M. Poinsettia growth and development response to container root substrate with biochar. *Horticulturae* **2018**, *4*, 1. [CrossRef]
56. Guo, Y.; Niu, G.; Starman, T.; Gu, M. Growth and development of easter lily in response to container substrate with biochar. *J. Hortic. Sci. Biotechnol.* **2018**, *94*, 80–86. [CrossRef]
57. Choi, H.-S.; Zhao, Y.; Dou, H.; Cai, X.; Gu, M.; Yu, F. Effects of biochar mixtures with pine-bark based substrates on growth and development of horticultural crops. *Hortic. Environ. Biotechnol.* **2018**, *59*, 345–354. [CrossRef]
58. Park, J.H.; Choppala, G.K.; Bolan, N.S.; Chung, J.W.; Chuasavathi, T. Biochar reduces the bioavailability and phytotoxicity of heavy metals. *Plant Soil* **2011**, *348*, 439–451. [CrossRef]
59. Dunlop, S.J.; Arbestain, M.C.; Bishop, P.A.; Wargent, J.J. Closing the loop: Use of biochar produced from tomato crop green waste as a substrate for soilless, hydroponic tomato production. *HortScience* **2015**, *50*, 1572–1581. [CrossRef]
60. Nair, A.; Carpenter, B. Biochar rate and transplant tray cell number have implications on pepper growth during transplant production. *HortTechnology* **2016**, *26*, 713–719. [CrossRef]

61. Huang, L. Effects of Biochar and Composts on Substrates Properties and Container-Grown Basil (*Ocimum basilicum*) and Tomato (*Solanum lycopersicum*). Master's Thesis, Texas A&M University, College Station, TX, USA, 2018.
62. Margenot, A.J.; Griffin, D.E.; Alves, B.S.; Rippner, D.A.; Li, C.; Parikh, S.J. Substitution of peat moss with softwood biochar for soil-free marigold growth. *Ind. Crop. Prod.* **2018**, *112*, 160–169. [CrossRef]
63. Bedussi, F.; Zaccheo, P.; Crippa, L. Pattern of pore water nutrients in planted and non-planted soilless substrates as affected by the addition of biochars from wood gasification. *Biol. Fertil. Soils* **2015**, *51*, 625–635. [CrossRef]
64. Awad, Y.M.; Lee, S.-E.; Ahmed, M.B.M.; Vu, N.T.; Farooq, M.; Kim, I.S.; Kim, H.S.; Vithanage, M.; Usman, A.R.A.; Al-Wabel, M. Biochar, a potential hydroponic growth substrate, enhances the nutritional status and growth of leafy vegetables. *J. Clean. Prod.* **2017**, *156*, 581–588. [CrossRef]
65. Nartey, E.G.; Amoah, P.; Ofosu-Budu, G.K.; Muspratt, A.; Kumar Pradhan, S. Effects of co-composting of faecal sludge and agricultural wastes on tomato transplant and growth. *Int. J. Recycl. Org. Waste Agric.* **2017**, *6*, 23–36. [CrossRef]
66. Altland, J.E.; Locke, J.C. Gasified rice hull biochar is a source of phosphorus and potassium for container-grown plants. *J. Environ. Hortic.* **2013**, *31*, 138–144.
67. Altland, J.E.; Locke, J.C. High rates of gasified rice hull biochar affect geranium and tomato growth in a soilless substrate. *J. Plant Nutr.* **2017**, *40*, 1816–1828. [CrossRef]
68. Vaughn, S.F.; Dinelli, F.D.; Jackson, M.A.; Vaughan, M.M.; Peterson, S.C. Biochar-organic amendment mixtures added to simulated golf greens under reduced chemical fertilization increase creeping bentgrass growth. *Ind. Crop. Prod.* **2018**, *111*, 667–672. [CrossRef]
69. Jahromi, N.B.; Walker, F.; Fulcher, A.; Altland, J.; Wright, W.C. Growth response, mineral nutrition, and water utilization of container-grown woody ornamentals grown in biochar-amended pine bark. *HortScience* **2018**, *53*, 347–353. [CrossRef]
70. Fan, R.; Luo, J.; Yan, S.; Zhou, Y.; Zhang, Z. Effects of biochar and super absorbent polymer on substrate properties and water spinach growth. *Pedosphere* **2015**, *25*, 737–748. [CrossRef]
71. Vaughn, S.F.; Eller, F.J.; Evangelista, R.L.; Moser, B.R.; Lee, E.; Wagner, R.E.; Peterson, S.C. Evaluation of biochar-anaerobic potato digestate mixtures as renewable components of horticultural potting media. *Ind. Crop. Prod.* **2015**, *65*, 467–471. [CrossRef]
72. Bilderback, T.E.; Warren, S.L.; Owen, J.S.; Albano, J.P. Healthy substrates need physicals too! *HortTechnology* **2005**, *15*, 747–751. [CrossRef]
73. Noguera, P.; Abad, M.; Puchades, R.; Maquieira, A.; Noguera, V. Influence of particle size on physical and chemical properties of coconut coir dust as container medium. *Commun. Soil Sci. Plant Anal.* **2003**, *34*, 593–605. [CrossRef]
74. Fonteno, W.; Hardin, C.; Brewster, J. *Procedures for Determining Physical Properties of Horticultural Substrates Using the NCSU Porometer*; Horticultural Substrates Laboratory, North Carolina State University: Raleigh, NC, USA, 1995.
75. Méndez, A.; Paz-Ferreiro, J.; Gil, E.; Gascó, G. The effect of paper sludge and biochar addition on brown peat and coir based growing media properties. *Sci. Hortic.* **2015**, *193*, 225–230. [CrossRef]
76. Initiative, I.B. Frequently Asked Questions about Biochar. Available online: https://biochar-international.org/faqs/#1521824701674-1093ca88-fd85 (accessed on 31 January 2019).
77. Khodadad, C.L.; Zimmerman, A.R.; Green, S.J.; Uthandi, S.; Foster, J.S. Taxa-specific changes in soil microbial community composition induced by pyrogenic carbon amendments. *Soil Biol. Biochem.* **2011**, *43*, 385–392. [CrossRef]
78. Lima, I.; Steiner, C.; Das, K. Characterization of designer biochar produced at different temperatures and their effects on a loamy sand. *Ann. Environ. Sci.* **2009**, *3*, 195–206.
79. Li, X.; Shen, Q.; Zhang, D.; Mei, X.; Ran, W.; Xu, Y.; Yu, G. Functional groups determine biochar properties (pH and EC) as studied by two-dimensional 13c nmr correlation spectroscopy. *PLoS ONE* **2013**, *8*, e65949. [CrossRef] [PubMed]
80. Hossain, M.K.; Strezov, V.; Chan, K.Y.; Ziolkowski, A.; Nelson, P.F. Influence of pyrolysis temperature on production and nutrient properties of wastewater sludge biochar. *J. Environ. Manag.* **2011**, *92*, 223–228. [CrossRef] [PubMed]
81. Lawrinenko, M.; Laird, D.A. Anion exchange capacity of biochar. *Green Chem.* **2015**, *17*, 4628–4636. [CrossRef]

82. Altland, J.; Locke, J. Biochar affects macronutrient leaching from a soilless substrate. *HortScience* **2012**, *47*, 1136–1140. [CrossRef]
83. Warnock, D.D.; Lehmann, J.; Kuyper, T.W.; Rillig, M.C. Mycorrhizal responses to biochar in soil–concepts and mechanisms. *Plant Soil* **2007**, *300*, 9–20. [CrossRef]
84. Saito, M. Charcoal as a micro-habitat for VA mycorrhizal fungi, and its practical implication. *Agric. Ecosyst. Environ.* **1990**, *29*, 341–344. [CrossRef]
85. Ezawa, T.; Yamamoto, K.; Yoshida, S. Enhancement of the effectiveness of indigenous arbuscular mycorrhizal fungi by inorganic soil amendments. *Soil Sci. Plant Nutr.* **2002**, *48*, 897–900. [CrossRef]
86. Dubchak, S.; Ogar, A.; Mietelski, J.; Turnau, K. Influence of silver and titanium nanoparticles on arbuscular mycorrhiza colonization and accumulation of radiocaesium in *Helianthus annuus*. *Span. J. Agric. Res.* **2010**, *8*, 103–108. [CrossRef]
87. Conversa, G.; Bonasia, A.; Lazzizera, C.; Elia, A. Influence of biochar, mycorrhizal inoculation, and fertilizer rate on growth and flowering of pelargonium (*Pelargonium zonale* L.) plants. *Front. Plant Sci.* **2015**, *6*, 429. [CrossRef] [PubMed]
88. Masiello, C.A.; Chen, Y.; Gao, X.; Liu, S.; Cheng, H.-Y.; Bennett, M.R.; Rudgers, J.A.; Wagner, D.S.; Zygourakis, K.; Silberg, J.J. Biochar and microbial signaling: Production conditions determine effects on microbial communication. *Environ. Sci. Technol.* **2013**, *47*, 11496–11503. [CrossRef] [PubMed]
89. Belda, R.M.; Lidón, A.; Fornes, F. Biochars and hydrochars as substrate constituents for soilless growth of myrtle and mastic. *Ind. Crop. Prod.* **2016**, *94*, 132–142. [CrossRef]
90. Gu, M.; Li, Q.; Steele, P.H.; Niu, G.; Yu, F. Growth of 'fireworks' gomphrena grown in substrates amended with biochar. *J. Food Agric. Environ.* **2013**, *11*, 819–821.
91. Ain Najwa, K.; Wan Zaliha, W.; Yusnita, H.; Zuraida, A. Effect of different soilless growing media and biochar on growth, yield and postharvest quality of lowland cherry tomato (*Solanum lycopersicum* var. Cerasiforme). *Trans. Malaysian Soc. Plant Physiol.* **2014**, *22*, 53–57.
92. Abubakari, A.-H.; Atuah, L.; Banful, B.K. Growth and yield response of lettuce to irrigation and growth media from composted sawdust and rice husk. *J. Plant Nutr.* **2018**, *41*, 221–232. [CrossRef]
93. Somtrakoon, K.; Chouychai, W. Phytotoxicity of single and combined polycyclic aromatic hydrocarbons toward economic crops. *Russ. J. Plant Physiol.* **2013**, *60*, 139–148. [CrossRef]
94. Huang, X.-D.; Zeiler, L.F.; Dixon, D.G.; Greenberg, B.M. Photoinduced toxicity of PAHs to the foliar regions of *Brassica napus* (canola) and *Cucumbis sativus* (cucumber) in simulated solar radiation. *Ecotoxicol. Environ. Saf.* **1996**, *35*, 190–197. [CrossRef] [PubMed]
95. Hanano, A.; Almousally, I.; Shaban, M.; Moursel, N.; Shahadeh, A.; Alhajji, E. Differential tissue accumulation of 2,3,7,8-tetrachlorinated dibenzo-p-dioxin in *Arabidopsis thaliana* affects plant chronology, lipid metabolism and seed yield. *Bmc Plant Biol.* **2015**, *15*, 193. [CrossRef] [PubMed]
96. John, R.; Ahmad, P.; Gadgil, K.; Sharma, S. Heavy metal toxicity: Effect on plant growth, biochemical parameters and metal accumulation by *Brassica juncea* L. *Int. J. Plant Prod.* **2012**, *3*, 65–76.
97. Peralta, J.; Gardea-Torresdey, J.; Tiemann, K.; Gomez, E.; Arteaga, S.; Rascon, E.; Parsons, J. Uptake and effects of five heavy metals on seed germination and plant growth in alfalfa (*Medicago sativa* L.). *Bull. Environ. Contam. Toxicol.* **2001**, *66*, 727–734. [CrossRef] [PubMed]
98. Lievens, C.; Carleer, R.; Cornelissen, T.; Yperman, J. Fast pyrolysis of heavy metal contaminated willow: Influence of the plant part. *Fuel* **2009**, *88*, 1417–1425. [CrossRef]
99. Bridle, T.; Pritchard, D. Energy and nutrient recovery from sewage sludge via pyrolysis. *Water Sci. Technol.* **2004**, *50*, 169–175. [CrossRef] [PubMed]
100. Jin, J.; Li, Y.; Zhang, J.; Wu, S.; Cao, Y.; Liang, P.; Zhang, J.; Wong, M.H.; Wang, M.; Shan, S.; et al. Influence of pyrolysis temperature on properties and environmental safety of heavy metals in biochars derived from municipal sewage sludge. *J. Hazard. Mater.* **2016**, *320*, 417–426. [CrossRef] [PubMed]
101. Devi, P.; Saroha, A.K. Risk analysis of pyrolyzed biochar made from paper mill effluent treatment plant sludge for bioavailability and eco-toxicity of heavy metals. *Bioresour. Technol.* **2014**, *162*, 308–315. [CrossRef] [PubMed]
102. Buss, W.; Graham, M.C.; Shepherd, J.G.; Mašek, O. Risks and benefits of marginal biomass-derived biochars for plant growth. *Sci. Total Environ.* **2016**, *569*, 496–506. [CrossRef] [PubMed]

103. Shackley, S.; Sohi, S.; Brownsort, P.; Carter, S.; Cook, J.; Cunningham, C.; Gaunt, J.; Hammond, J.; Ibarrola, R.; Mašek, O. *An Assessment of the Benefits and Issues Associated with the Application of Biochar to Soil*; Department for Environment, Food and Rural Affairs, UK Government: London, UK, 2010.
104. McGrath, T.; Sharma, R.; Hajaligol, M. An experimental investigation into the formation of polycyclic-aromatic hydrocarbons (PAH) from pyrolysis of biomass materials. *Fuel* **2001**, *80*, 1787–1797. [CrossRef]
105. Hale, S.E.; Lehmann, J.; Rutherford, D.; Zimmerman, A.R.; Bachmann, R.T.; Shitumbanuma, V.; O'Toole, A.; Sundqvist, K.L.; Arp, H.P.H.; Cornelissen, G. Quantifying the total and bioavailable polycyclic aromatic hydrocarbons and dioxins in biochars. *Environ. Sci. Technol.* **2012**, *46*, 2830–2838. [CrossRef] [PubMed]
106. Wilson, K.; Reed, D. *IBI White Paper—Implications and Risks of Potential Dioxin Presence in Biochar*; International Biochar Initiative: Canandaigua, NY, USA, 2012. Available online: https://www.biochar-international.org/wp-content/uploads/2018/04/IBI_White_Paper-Implications_of_Potential_%20Dioxin_in_Biochar.pdf (accessed on 31 January 2019).
107. Everaert, K.; Baeyens, J. The formation and emission of dioxins in large scale thermal processes. *Chemosphere* **2002**, *46*, 439–448. [CrossRef]
108. Garcia-Perez, M.; Metcalf, J. The Formation of Polyaromatic Hydrocarbons and Dioxins During Pyrolysis: A Review of the Literature with Descriptions of Biomass Composition, Fast Pyrolysis Technologies and Thermochemical Reactions. 2008. Available online: https://research.wsulibs.wsu.edu:8443/xmlui/bitstream/handle/2376/5966/TheFormationOfPolyaromaticHydrocarbonsAndDioxinsDuringPyrolysis.pdf?sequence=1&isAllowed=y (accessed on 31 January 2019).

16

Water-Related Variables for Predicting Yield of Apple under Deficit Irrigation

Riccardo Lo Bianco

Department of Agricultural, Food and Forest Sciences, University of Palermo, Palermo 90128, Italy; riccardo.lobianco@unipa.it; Tel.: +39-091-238-96097

Abstract: Predicting apple yield in relation to tree water use is important for irrigation planning and evaluation. The aim of the present study was to identify measurable variables related to tree water use that could predict final fruit yield of apple trees under different strategies of deficit irrigation. Adult 'Gala' and 'Fuji' apple trees were exposed to conventional irrigation (CI), delivering 100% of crop evapotranspiration; partial root zone drying (PRD), delivering 50% of CI water only on one alternated side of the root-zone; and continuous deficit irrigation (CDI), delivering 50% of CI water on both sides of the root-zone. Integrals of soil (SWD_{int}) and leaf ($LWSD_{int}$) water deficit along with growth and stomatal conductance (Gs_{int}) were calculated across each season and used to estimate total conductance (GS_{tree}) and transpiration (Tr_{tree}) per tree, transpiration efficiency on a fruit (GR_{fruit}/Tr) or tree (GR_{trunk}/Tr) growth basis, and transpiration productivity ($Yield/Tr_{tree}$). 'Fuji' trees had higher $Yield/Tr_{tree}$, but had lower GR_{trunk}/Tr and similar GR_{fruit}/Tr compared to 'Gala' trees. In 'Fuji', CDI reduced yield, trunk growth, leaf hydration, and gas exchange, while in 'Gala', it did not reduce yield and gas exchange. In 'Fuji', a linear combination of GR_{trunk}/Tr, GR_{fruit}/Tr, and Gs_{tree} contributed to predicting yield, with GR_{fruit}/Tr explaining nearly 78% of the model variability. In 'Gala', a linear combination of $LWSD_{int}$ and Gs_{tree} contributed to predicting yield, with Gs_{tree} explaining over 79% of the model variability. These results indicate that measuring tree water status or water use may help predict final apple yields only in those cultivars like 'Gala' that cannot limit dehydration by closing stomates because of carbon starvation. In more vigorous cultivars like 'Fuji', transpiration efficiency based on fruit growth can be a powerful predictor of final yields.

Keywords: leaf water saturation deficit; partial root zone drying; stomatal conductance; transpiration efficiency; transpiration productivity

---

1. Introduction

On cultivated land, it is estimated that environmental stresses significantly limit agricultural production, and global climate changes are constantly increasing such limitations. Environmental stress tolerance is therefore a critical concern for horticulturists if they hope to increase fruit production as population increases. In particular, plant tissue dehydration (drought stress) may cause direct and indirect decreases in fruit quantity and quality. Indeed, drought may also affect photosynthesis and nutrient uptake causing indirect yield reductions.

A great share of the annual precipitation is lost to evapotranspiration (ET, from 70% up to 90% in arid areas) [1]. This fact proves the importance of adequately estimating the ET component of the hydrologic cycle in predicting on-farm irrigation water management and irrigation planning [2], especially if we consider that without ET there is no production [3]. In particular, the transpiration to ET coefficients have been widely used for precise and efficient irrigation management [4].

Crop yield is determined by both available water quantity and plant water use efficiency [5]. At the physiological level, water use efficiency (WUE) can be defined as the ratio between photosynthesis and transpiration, also defined as transpiration efficiency or instantaneous WUE. This WUE is quite difficult to monitor at the whole tree scale, and even more at the orchard level. For horticultural evaluations, WUE can be more easily expressed as fruit yield per unit of irrigation water, or irrigation water productivity (IWP). Recently, many studies have focused on IWP as being directly related to the increase of WUE [6]. In this study, the use of transpiration efficiency or productivity based on the ratios between relative trunk or fruit growth and transpiration rate as well as the ratio between fruit yield and total tree transpiration is proposed.

Several factors can cause variations of WUE in plants, e.g., air humidity, the different carboxylation mechanism of C3 and C4 plants and, in the long period, the losses due to respiration and assimilate partitioning. Indeed, it has been widely demonstrated that it is possible to improve plant carbohydrate distribution towards reproductive structures, such as fruits, by keeping the plants in a state of mild water deficit, in this way controlling the excessive vegetative growth [7]. This concept has represented in the last decades the basis for a long list of trials investigating the outcomes of what was called "regulated deficit irrigation" (RDI) by Chalmers et al. [8] or "controlled deficit irrigation" by English [9] and [10]. As a matter of fact, trials conducted on several crops showed that IWP tends to increase with deficit compared to conventional irrigation [11,12].

Increasing IWP has been more successful in trees than in field crops for several reasons [13]. Fruit quality, for example, strongly affects crop value, but is not associated with biomass production and water use. In addition, tree fruit growth may not be sensitive to water deficit in certain periods and developmental stages [14]. This, in combination with low volume/high frequency irrigation systems, gives the best opportunities to manage fruit trees under controlled water deficit.

A specific deficit irrigation practice that has received particular attention in the last decades and seems to achieve significant water savings with limited information inputs from the grower is partial root zone drying (PRD) [15,16]. With PRD, one half of the root system is cyclically left to dry; roots in drying soil produce chemical signals (abscisic acid, cytokinins, pH changes), which are translocated to the shoots [17] where they induce partial stomatal closure, reduce transpiration, and ultimately increase WUE [15]. Thanks to the well-watered half of the root zone, the effect on plant water potential is minimal [18] and other metabolic and physiological processes associated to water stress are not affected [15,19]. This deficit irrigation technique has produced positive outcomes in a number of fruit species, and in apple, numerous PRD studies have reported significant increases of IWP and even yields similar to those of full irrigated trees [16].

For the reasons above, understanding the transpiration mechanisms of plants and the factors affecting final crop yield under water deficit becomes a priority. Green plants have indeed many structures and control devices, which allow them to function efficiently even in rapidly changing environments. At the leaf level, transpiration is controlled by physiological and structural factors with stomatal aperture and conductance assuming a primary role [20]. Stomatal conductance (g_s) responds to several factors, such as light, CO_2 concentration, vapor pressure deficit, leaf temperature, leaf abscisic acid, and soil water potential. This latter factor influences g_s by a hormonal signal (abscisic acid) originating in the roots, a sort of biological switch when drought occurs [21].

The prediction of apple yield in relation to water requirement or ET is important for irrigation planning and evaluation. Considerable research has led to the development of simple models for predicting mostly yield of field crops from evapotranspiration during the growing season [22–24]. The aim of the present study was to identify measurable variables related to tree water use that could serve for the development of a model to predict final fruit yield of apple trees under deficit irrigation. The same yield predicting variables and models could be useful for fine tuning of deficit irrigation management.

2. Materials and Methods

Data of the present study are further calculations and analysis of measurements reported in Lo Bianco and Francaviglia [25]. The study was conducted in 2008 and 2009 near Caltavuturo (37°49′ N and 850 m a.s.l.), in central Sicily. Plant material consisted of eight-year-old 'Gala' and 'Fuji' apple trees on M.9 rootstock, trained to a central leader, and spaced at 4 m between rows and 1.5 m within rows. Soil type was a sandy clay loam (53.3% sand, 17.6% silt, and 29.1% clay) with pH 7.3 and 1.8% active carbonates, and soil water potential around −17 kPa at field capacity. With the exception of irrigation, all trees received the same cultural practices.

In the field, two nearby rows (one with 36 'Gala' trees, the other with 36 'Fuji' trees) were selected and divided into four blocks, each including three trees per irrigation treatment. Contiguous irrigation treatments on the same row were separated by two buffer trees. In June, three irrigation treatments were imposed: (1) conventional irrigation (CI), delivering 100% of crop evapotranspiration (ET_c); (2) PRD, where trees received 50% of CI water only on one alternated side of the root zone; (3) continuous deficit irrigation (CDI), where trees received 50% of CI water on both sides of the root zone. Wet and dry sides of PRD trees were alternated every 2–3 weeks when soil water potential in the dry side reached values of approximately −100 to −150 kPa.

Weather parameters were monitored with a µMetos weather station (Pessl, Austria) positioned within the experimental plot and used to determine reference evapotranspiration (ET_0) according to the FAO Penman–Monteith method and crop evapotranspiration (Et_c) [26].

Instantaneous vapor pressure deficit (VPD) was calculated from canopy air temperature (in °C) and relative humidity (in %) measured on the same dates and at the same time as stomatal conductance.

Soil water potential was monitored continuously with six Watermark sensors (Irrometer Co., Riverside, CA, USA) directly connected to the weather station. In drip irrigated apple trees, most of the active roots are within the first 60 cm of soil depth. For this reason, Watermark sensors were positioned at a fixed depth of 40 cm and a distance of about 80 cm from emitters and 1 m from the tree trunk in opposite sides of the root-zone. Integrals of soil water deficit (SWD_{int}) across each irrigation season and treatment were calculated as:

$$SWD_{int} = \Sigma_{(1..t)} \mid (SWD_i - SWD_{FC}) \mid, \tag{1}$$

where t is the number of days in the irrigation season, SWD_i are average daily measures of soil water potential, and SWD_{FC} is soil water potential at field capacity. SWD_{int} was used as an indication of soil water deficit accumulated in the root-zone of each treatment during the irrigation periods.

Every two weeks during the irrigation period, at mid-morning two mature, sun-exposed leaves per tree were collected and transported in ice to the laboratory for determination of fresh weight (FW), turgid weight (TW) after rehydrating leaves for 24 h at 8 °C in the dark, and dry weight (DW) after drying leaves at 60 °C to constant weight. Leaf relative water content (RWC) was calculated as [(FW − DW)/(TW − DW)] × 100. Leaf water saturation deficit (LWSD) was calculated as 1 − (RWC/100) and integrated across the irrigation period using the equation proposed by García-Tejero et al. [27] and modified from Myers [28]:

$$LWSD_{int} = \Sigma_{(1..t)} \mid LWD_{i+1} \times (n_{i+1} - n_i) + \frac{1}{2} (LWD_i - LWD_{i+1}) \times (n_{i+1} - n_i) \mid, \tag{2}$$

where t is the number of sampling days, LWD_i and LWD_{i+1} are leaf water deficit values measured on two consecutive sampling days (i and i + 1), and n_{i+1} and n_i the days corresponding to two serial samplings. This variable is the integral of a proportion (0 to 1) and therefore can be considered unitless.

On the same dates, stomatal conductance (g_s) was measured with an AP4 Delta-T porometer (Delta-T Devices, Cambridge, UK) on two leaves similar to those used for RWC measurements. Stomatal conductance was also integrated across the irrigation period according the following equation:

$$Gs_{int} = \Sigma_{(1..t)} \mid g_{i+1} \times (n_{i+1} - n_i) + \frac{1}{2} (g_i - g_{i+1}) \times (n_{i+1} - n_i) \mid, \tag{3}$$

where t is the number of sampling days, g_i and g_{i+1} are leaf stomatal conductance values measured on two consecutive sampling days (i and i + 1), and n_{i+1} and n_i the days corresponding to two serial measurements.

In each year, one fruit per tree was measured bi-weekly in size (height and width) with a digital caliper. Relative seasonal fruit growth was calculated as the total increase in average diameter (mm) divided by the initial diameter of the fruit (mm). Trunk circumference was measured at about 15 cm above the graft union at the beginning and end of the two growing seasons. Trunk cross-section area (TCSA) was derived from trunk circumference and taken as an indicator of apple tree size [29]. Tree growth was calculated as the increase in TCSA divided by the initial TCSA. Total leaf area per tree (LA) was destructively measured on a separate set of trees from the two cultivars and related to TCSA by regression analysis. The function obtained was used to estimate LA from TCSA measurements in the trees in trial.

Integrals of leaf transpiration (Tr) were derived from Gs_{int} and VPD as follows:

$$Tr = Gs_{int} \times (VPD/101.3), \quad (4)$$

Where 101.3 is the barometric pressure in kPa at sea level. Integrals of soil (SWD_{int}) and leaf ($LWSD_{int}$) water deficit along with growth and stomatal conductance (Gs_{int}) were calculated across each season and used to estimate total conductance (GS_{tree}) and transpiration (Tr_{tree}) per tree, transpiration efficiency on a fruit (GR_{fruit}/Tr) or tree (GR_{trunk}/Tr) growth basis, and transpiration productivity ($Yield/Tr_{tree}$). Transpiration efficiency on a per tree growth basis was estimated by dividing trunk growth by Tr (Gr_{trunk}/Tr), transpiration efficiency on a fruit growth basis was estimated by dividing fruit growth by Tr (Gr_{fruit}/Tr), while transpiration productivity was obtained by dividing yield by Tr_{tree} ($Yield/Tr_{tree}$). Transpiration productivity is a very useful measure which is more accurate than IWP and more practical than instantaneous WUE as a trait for improving fruit productivity under limited water resources. Total stomatal conductance per tree was estimated from Gs_{int} and LA, while total transpiration per tree was estimated from Tr and LA.

Data were tested for normal distribution and equal variances and analyzed by analysis of variance and regression procedures using Systat and SigmaPlot software (Systat Software Inc., Richmond, CA, USA). Least squares multiple linear regression with a backward stepwise technique was used to find the best set of variables predicting final apple yield. Means were separated by Tukey's multiple comparison test at $P < 0.05$.

3. Results and Discussion

The relationship between TCSA and LA was described by a non-linear polynomial function, as shown in Figure 1. Canopy and root system size have been linearly related to TCSA in apple [29,30]. In this study with eight-year-old apple trees, the non-linear relationship between TCSA and LA can be explained by canopy size constraints imposed by planting density, tree training form, and pruning. In other words, more vigorous trees (e.g., 'Fuji' trees) were pruned more heavily than weaker trees to remain within the allotted space and avoid competition for light. In this way, canopy and leaf area of trees with different TCSA are brought back to similar sizes determining the observed non-linear relationship between TCSA and LA.

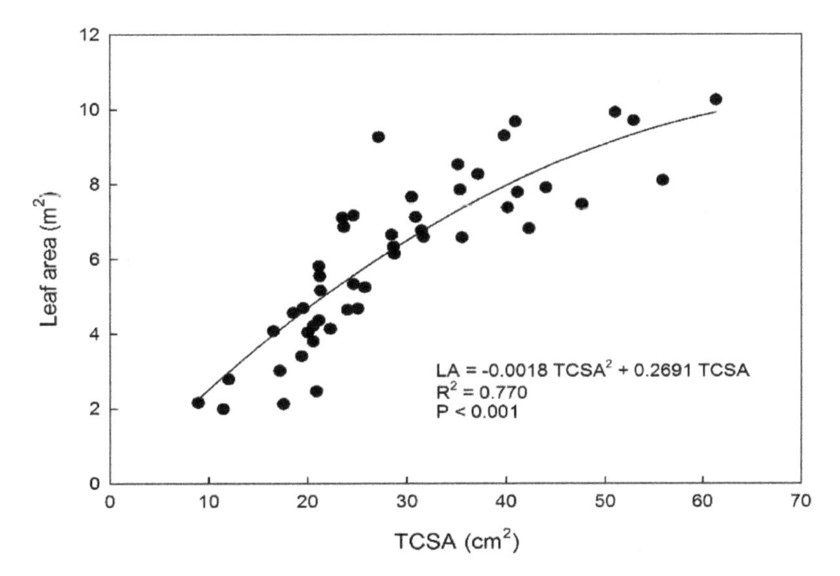

Figure 1. Relationship between trunk cross-section area (TCSA) and total leaf area (LA) in eight-year-old 'Gala' and 'Fuji' apple trees grafted on M.9 rootstock, trained to a central leader, spaced at 4 × 1.5 m, and grown near Caltavuturo, Sicily.

The imposed irrigation treatments effectively determined the expected differences in soil water deficit. Indeed, on average of the two seasons, SWD_{int} was four times higher in CDI (5.62 MPa) than in CI (1.43 MPa) trees; despite the same irrigation volumes, PRD reported intermediate SWD_{int} (4.61 MPa) with 22% lower values than CDI trees. This has been attributed to greater wetted soil surface and consequent soil evaporation in CDI than in PRD in previous studies [25,31,32].

As expected and regardless of irrigation strategy, 'Fuji' trees were bigger, had greater LA, transpired more water, and yielded more fruit than 'Gala' trees, as shown in Tables 1 and 2. On the other hand, transpiration efficiency in terms of whole tree growth was higher in 'Gala' than in 'Fuji' trees; specifically, 'Gala' trees had higher GR_{trunk}/Tr but similar GR_{fruit}/Tr and lower $Yield/Tr_{tree}$ compared to 'Fuji' trees, as shown in Tables 1 and 2, suggesting that they partitioned assimilates mostly to vegetative rather than fruit growth. This may be at least in part due to their smaller size and fewer constraints to acquire soil resources and fill the allotted space compared to 'Fuji' trees.

The two apple cultivars responded differently to soil water deficit and irrigation strategy. In 'Fuji', PRD maintained plant gas exchange, hydration levels, growth, and productivity similar to CI, while CDI induced significant reductions of yield, tree growth, leaf hydration, and gas exchange, as shown in Table 1. In contrast, no yield and gas exchange differences were observed in deficit irrigated 'Gala' trees, while CDI decreased leaf hydration levels and tree growth compared to both CI and PRD, as shown in Table 2. Transpiration productivity was significantly increased by PRD mainly in 'Fuji' while it was reduced by CDI in 'Gala'. Given the milder soil and leaf water deficit induced by PRD compared to CDI, the different responses of the two cultivars to irrigation may be associated with different levels of water stress resistance, with 'Gala' exhibiting lower ability to limit dehydration than 'Fuji'. In other words, both cultivars tend to close stomates and transpire less as soil water deficit progresses, minimizing symptoms of leaf dehydration (isohydric behavior). Yet, 'Gala' showed higher $LWSD_{int}$ (significant leaf dehydration under both PRD and CDI) than 'Fuji' (leaf dehydration only under CDI). This is at least in part due to the larger 'Fuji' root systems which were able to explore more and deeper soil layers and acquire more water and nutrients than 'Gala' roots.

Table 1. Yield, growth (GR), and water-use related variables in 'Fuji' apple trees under conventional irrigation (CI), partial root zone drying (PRD), and continuous deficit irrigation (CDI).

	CI		PRD		CDI		
Yield (kg/tree)	36.3	ab [z]	39.4	a	33.5	b	* [y]
Yield Efficiency (kg cm^{-2})	0.729	ab	0.853	a	0.607	b	**
GR_{trunk} (cm^2 cm^{-2})	0.051	a	0.052	a	0.037	b	*
GR_{fruit} (mm mm^{-1})	0.541		0.545		0.501		ns
Leaf Area (m^2)	9.02		8.70		9.39		ns
Vapor Pressure Deficit (kPa)	158		159		159		ns
$LWSD_{int}$ [x]	10.1	a	10.5	ab	11.1	b	*
Gs_{int} [w] (mol m^{-2} s^{-1})	16.0	a	15.5	a	12.5	b	**
Tr [v] (mol m^{-2} s^{-1})	25.6	a	24.9	a	20.0	b	**
GR_{trunk}/Tr	0.022		0.023		0.019		ns
GR_{fruit}/Tr	0.025		0.025		0.028		ns
Yield/Tr_{tree}	0.177	b	0.210	a	0.191	ab	*
Tr_{tree} [u] (mol s^{-1}/tree)	229		216		188		ns
Gs_{tree} [u] (mol s^{-1}/tree)	143		136		118		ns

z Mean separation within rows by Tukey's multiple comparison test at $P < 0.05$

y Level of statistical significance for irrigation factor from analysis of variance: ns, $P > 0.05$; *, $P < 0.05$; **, $P < 0.01$; ***, $P < 0.001$.

x $LWSD_{int}$ is leaf water saturation deficit integrated across the irrigation period.

w Gs_{int} is stomatal conductance integrated across the irrigation period.

v Tr is transpiration integrated across the irrigation period.

u Seasonal integrals of total conductance (GS_{tree}) and transpiration (Tr_{tree}) per tree.

Table 2. Yield, growth (GR), and water-use related variables in 'Gala' apple trees under conventional irrigation (CI), partial root zone drying (PRD), and continuous deficit irrigation (CDI).

	CI		PRD		CDI		
Yield (kg/tree)	19.3		18.5		17.2		ns [z]
$Yield_{Eff}$ ($kg\ cm^{-2}$)	0.600		0.659		0.537		ns
GR_{trunk} ($cm^2\ cm^{-2}$)	0.082	a [y]	0.071	a	0.053	b	***
GR_{fruit} ($mm\ mm^{-1}$)	0.545		0.568		0.563		ns
Leaf Area (m^2)	6.69		6.31		6.95		ns
Vapor Pressure Deficit (kPa)	169		168		170		ns
$LWSD_{int}$ [x]	7.53	a	8.28	ab	8.93	b	*
Gs_{int} [w] ($mol\ m^{-2}\ s^{-1}$)	13.9		12.9		12.4		ns
Tr [v] ($mol\ m^{-2}\ s^{-1}$)	23.7		21.0		22.0		ns
GR_{trunk}/Tr	0.044	a	0.044	a	0.029	b	*
Gr_{fruit}/Tr	0.028		0.033		0.028		ns
Yield/Tr_{tree}	0.141	ab	0.161	a	0.124	b	*
Tr_{tree} [u] ($mol\ s^{-1}$/tree)	154		131		150		ns
Gs_{tree} [u] ($mol\ s^{-1}$/tree)	90.5		77.4		88.5		ns

z Level of statistical significance for irrigation factor from analysis of variance: ns, $P > 0.05$; *, $P < 0.05$; **, $P < 0.01$; ***, $P < 0.001$.
y Mean separation within rows by Tukey's multiple comparison test at $P < 0.05$
x $LWSD_{int}$ is leaf water saturation deficit integrated across the irrigation period.
w Gs_{int} is stomatal conductance integrated across the irrigation period.
v Tr is transpiration integrated across the irrigation period.
u Seasonal integrals of total conductance (GS_{tree}) and transpiration (Tr_{tree}) per tree.

In 'Fuji', a linear combination of GR_{trunk}/Tr, GR_{fruit}/Tr, and Gs_{tree} contributed to predicting yield of apple trees under soil water deficit, as shown in Table 3. In this cultivar, GR_{fruit}/Tr was the most important variable for predicting yield, explaining nearly 78% of the model variability, while GR_{trunk}/Tr can be considered negligible as it explained only about 3% of the model variability, as shown in Table 3. On the other hand, a linear combination of $LWSD_{int}$ and Gs_{tree} contributed to predicting yield of 'Gala' apple trees under soil water deficit, as shown in Table 4. In this cultivar, Gs_{tree} was the most important variable for predicting yield, explaining over 79% of the model variability, as shown in Table 4. This difference between the two cultivars indicates that fruit yield of 'Gala' trees was more sensitive to stomatal closure compared to 'Fuji' trees. This may be due to differences in tree size and leaf water deficit levels. The increase of $LWSD_{int}$ over the control was indeed greater in 'Gala' (0.75 and 1.4 for PRD and CDI, respectively) than in 'Fuji' (0.4 and 1 for PRD and CDI, respectively). The effect of tree water status on apple yield has been already documented, although other factors like crop load may have stronger effects than water deficit on yield [33]. In addition, the larger 'Fuji' trees may have been less sensitive to stomatal closure than 'Gala' trees because of greater carbon and water storage in permanent structures. In this regard, others have reported contrasting results indicating positive or no effect of tree size and capacitance on water status [34,35], while there is little doubt about the role of permanent structures as carbon reservoirs. In addition to carbon reserves in permanent structures, differences in LA and photosynthetic rates as well as nutrient acquisition may play a significant role in the response of the two cultivars. In the present study, differences in tree size and carbon and water storage may also help explain the higher transpiration productivity in 'Fuji' than in 'Gala' and the major contribution of Gr_{fruit}/Tr to yield prediction in 'Fuji'. $LWSD_{int}$ was a relatively weak yield predictor only in 'Gala', suggesting that even under soil water limiting conditions, factors other than water (e.g., assimilation rate, nutrient status, flower fertility, pollination) are major determinants of apple fruit yield formation and a simple measurement of tree water status may not serve as a solid yield predictor.

Table 3. Multiple linear regression model and parameters contributing to predict yield in 'Fuji' apple trees under deficit irrigation.

Yield = 28.5 + (2774 × GR_{trunk}/Tr) − (295 × GR_{fruit}/Tr) + (0.070 × Gs_{tree})					
N = 65	**R = 0.554**	**R^2 = 0.306**	**SE of Estimate = 8.85**	**P < 0.001**	
Parameters	**Coefficient**	**SE**	**t**	***P***	**% of SSreg**
Constant	28.5	6.74	4.23	<0.001	-
GR_{trunk}/Tr z	2774	973	2.85	0.006	3.1
GR_{fruit}/Tr y	−295	118	−2.51	0.015	77.9
Gs_{tree} x	0.070	0.031	2.26	0.027	19.0

z Trunk growth/transpiration.
y Fruit growth/transpiration.
x Seasonal integral of total conductance (Gs) per tree.

Table 4. Multiple linear regression model and parameters contributing to predict yield in 'Gala' apple trees under deficit irrigation.

Yield = −0.019 + (1.35 × $LWSD_{int}$) + (0.085 × Gs_{tree})					
N = 68	**R = 0.410**	**R^2 = 0.168**	**SE of Estimate = 6.33**	**P = 0.003**	
Parameters	**Coefficient**	**SE**	**t**	***P***	**% of SSreg**
Constant	−0.019	5.50	−0.003	0.997	-
$LWSD_{int}$ z	1.35	0.52	2.59	0.012	20.5
Gs_{tree} y	0.085	0.026	3.23	0.002	79.5

z $LWSD_{int}$ is leaf water saturation deficit integrated across the irrigation period.
y Seasonal integral of total conductance (Gs) per tree.

In conclusion, the more vigorous 'Fuji' trees were more efficient than 'Gala' trees under soil water deficits in terms of yield and transpiration productivity. Our results indicate that measuring tree water status or gas exchange may help predict final apple yields only in those trees and cultivars (like 'Gala' in this study) that are not able to limit dehydration by closing stomates because of carbon starvation. In more vigorous trees and cultivars like 'Fuji', transpiration (or water use) efficiency towards fruit growth seems to be a powerful predictor of final yields.

References

1. Hamon, R.W. Evapotranspiration and water yield predictions. In Evapotranspiration and its role in water resources management. *Conf. Proc. Am. Soc. Agric. Eng.* **1966**, *December*, 8–9.
2. Shockley, D.G. Evapotranspiration and farm irrigation planning and management. In Evapotranspiration and its role in water resources management. *Conf. Proc. Am. Soc. Agric. Eng.* **1966**, *December*, 3–5.
3. Hansen, V.E. Evapotranspiration and water resources management. *Conf. Proc. Am. Soc. Agric. Eng.* **1966**, *December*, 12–13.
4. Kang, S.; Gu, B.; Du, T.; Zhang, J. Crop coefficient and ratio of transpiration to evapotranspiration of winter wheat and maize in a semi-humid region. *Agric. Water Manag.* **2003**, *59*, 239–254. [CrossRef]

5. Xu, L.K.; Hsiao, T.C. Predicted versus measured photosynthetic water-use efficiency of crop stands under dynamically changing field environments. *J. Exp. Bot.* **2004**, *55*, 2395–2411. [CrossRef]
6. Kijne, J.W.; Barker, R.; Molden, D. *Water Productivity in Agriculture: Limits and Opportunities for Improvement*; CABI Publishing: Wallingford, UK, 2003; 332p.
7. Chalmers, D.J.; Mitchell, P.D.; Vanheek, L. Control of peach tree growth and productivity by regulated water supply, tree density and summer pruning. *J. Am. Soc. Hortic. Sci.* **1981**, *106*, 307–397.
8. Chalmers, D.J.; Burge, G.; Jerie, P.H.; Mitchell, P.D. The mechanism of regulation of Bartlett pear fruit and vegetative growth by irrigation withholding and regulated deficit irrigation. *J. Am. Soc. Hortic. Sci.* **1986**, *111*, 904–907.
9. English, M.J. Deficit irrigation: Analytical framework. *J. Irrig. Drain. Eng.* **1990**, *116*, 399–412. [CrossRef]
10. Mitchell, P.D.; Jerie, P.H.; Chalmers, D.J. Effects of regulated water deficits on pear tree growth, flowering, fruit growth and yields. *J. Am. Soc. Hortic. Sci.* **1984**, *109*, 604–606.
11. Zwart, S.J.; Bastiaanssen, W.G. Review of measured crop water productivity values for irrigated wheat, rice, cotton and maize. *Agric. Water Manag.* **2004**, *69*, 115–133. [CrossRef]
12. Fan, T.; Wang, S.; Xiaoming, T.; Luo, J.; Stewart, B.A.; Gao, Y. Grain yield and water use in a long-term fertilization trial in Northwest China. *Agric. Water Manag.* **2005**, *76*, 36–52. [CrossRef]
13. Fereres, E.; Goldhamer, D.A.; Parsons, L.R. Irrigation water management of horticultural crops. *HortScience* **2003**, *38*, 1036–1042.
14. Johnson, R.S.; Handley, D.F. Using water stress to control vegetative growth and productivity of temperate fruit trees. *HortScience* **2000**, *35*, 1048–1050.
15. Dry, P.R.; Loveys, B.R.; Botting, D.G.; Düring, H. Effects of partial root-zone drying on grapevine vigour, yield, composition of fruit and use of water. In Proceedings of the Ninth Australian Wine Industry Technical Conference, Adelaide, South Australia, 16–19 July 1995; pp. 128–131.
16. Lo Bianco, R. Responses of apple to partial root-zone drying. A review. In *Irrigation Management, Technologies, and Environmental Impacts*; Ali, M.H., Ed.; Nova Science Publishers, Inc.: New York, NY, USA, 2013; Chapter 3; pp. 71–86.
17. Dodd, I.C.; Egea, G.; Davies, W.J. Abscisic acid signalling when soil moisture is heterogeneous: Decreased photoperiod sap flow from drying roots limits ABA export to the shoots. *Plant Cell Environ.* **2008**, *31*, 1263–1274. [CrossRef] [PubMed]
18. Gowing, D.J.G.; Davies, W.J.; Jones, H.G. A positive root-sourced signal as an indicator of soil drying in apple, *Malus* × *domestica* Borkh. *J. Exp. Bot.* **1990**, *41*, 1535–1540. [CrossRef]
19. Dry, P.R.; Loveys, B.R.; Düring, H. Partial drying of the root-zone of grape. I. Transient changes in shoot growth and gas exchange. *Vitis* **2000**, *39*, 3–7.
20. Cowan, I.R.; Farquhar, G.D. Stomatal function in relation to leaf metabolism and environment. *Symp. Soc. Exp. Biol.* **1977**, *31*, 471. [PubMed]
21. Davies, W.J.; Zhang, J. Root signals and the regulation of growth and development of plants in drying soil. *Ann. Rev. Plant Biol.* **1991**, *42*, 55–76. [CrossRef]
22. Doorenbos, J.; Kassam, A.H. Yield response to water. In *Irrigation and Drainage Paper No. 33*; FAO: Rome, Italy, 1986.
23. Howel, T.A.; Musick, J.T. Relationship of dry matter production of field crops to water consumption. In *Crop Water Requirements*; Perrier, A., Riou, C., Eds.; INRA: Paris, France, 1985; pp. 247–269.
24. Ouda, S.A.; Khalil, F.A.; Tantawy, M.M. Predicting the impact of water stress on the yield of different maize hybrids. *Res. J. Agric. Biol. Sci.* **2006**, *2*, 369–374.
25. Lo Bianco, R.; Francaviglia, D. Comparative responses of 'Gala' and 'Fuji' apple trees to deficit irrigation: Placement versus volume effects. *Plant Soil* **2012**, *357*, 41–58. [CrossRef]
26. Allen, R.G.; Pereira, L.S.; Raes, D.; Smith, M. Crop evapotranspiration—Guidelines for computing crop water requirements. In *Irrigation and Drainage Paper 56*; FAO: Rome, Italy, 1998.
27. García-Tejero, I.; Romero-Vicente, R.; Jiménez-Bocanegra, J.A.; Martínez-García, G.; Durán-Zuazo, V.H.; Muriel-Fernández, J.L. Response of citrus trees to deficit irrigation during different phenological periods in relation to yield, fruit quality, and water productivity. *Agric. Water Manag.* **2010**, *97*, 689–699. [CrossRef]
28. Myers, B.J. Water stress integral—A link between short term stress and long term growth. *Tree Physiol.* **1988**, *4*, 315–323. [CrossRef]

29. Lo Bianco, R.; Policarpo, M.; Scariano, L. Effect of rootstock vigor and in-row spacing on stem and root growth, conformation, and dry matter distribution of young apple trees. *J. Hortic. Sci. Biotechnol.* **2003**, *78*, 828–836. [CrossRef]
30. Westwood, M.N.; Roberts, A.N. The relationship between trunk cross-sectional area and weight of apple trees. *J. Am. Soc. Hortic. Sci.* **1970**, *95*, 28–30.
31. Marsal, J.; Mata, M.; Del Campo, J.; Arbones, A.; Vallverdú, X.; Girona, J.; Olivo, N. Evaluation of partial root-zone drying for potential field use as a deficit irrigation technique in commercial vineyards according to two different pipeline layouts. *Irrig. Sci.* **2008**, *26*, 347–356. [CrossRef]
32. Mossad, A.; Scalisi, A.; Lo Bianco, R. Growth and water relations of field-grown 'Valencia' orange trees under long-term partial rootzone drying. *Irrig. Sci.* **2018**, *36*, 9–24. [CrossRef]
33. Naor, A.; Naschitz, S.; Peres, M.; Gal, Y. Responses of apple fruit size to tree water status and crop load. *Tree Physiol.* **2008**, *28*, 1255–1261. [CrossRef]
34. Davies, F.S.; Lakso, A.N. Diurnal and seasonal changes in leaf water potential components and elastic properties in response to water stress in apple trees. *Physiol. Plant.* **1979**, *46*, 109–114. [CrossRef]
35. Olien, W.C.; Lakso, A.N. Effect of rootstock on apple (*Malus domestica*) tree water relations. *Physiol. Plant.* **1986**, *67*, 421–430. [CrossRef]

17

Use of Diatomaceous Earth as a Silica Supplement on Potted Ornamentals

Taylor Mills-Ibibofori [1], Bruce Dunn [1,*], Niels Maness [1] and Mark Payton [2]

1 Department of Horticulture & L.A., Oklahoma State University, Stillwater, OK 74078, USA; tsmills@okstate.edu (T.M.-I.); niels.maness@okstate.edu (N.M.)
2 Department of Statistics, Oklahoma State University, Stillwater, OK 74078, USA; mark.payton@okstate.edu
* Correspondence: bruce.dunn@okstate.edu

Abstract: The role of silica as a needed supplement in soilless media is gaining interest. This research studied the effects of diatomaceous earth as a supplement on growth and flower characteristics, physiology, and nutrient uptake in dahlia (*Dahlia Cav.* × *hybrida* 'Dahlinova Montana'), black-eyed Susan (*Rudbeckia hirta* L. 'Denver Daisy'), and daisy (*Gerbera jamesonii* L. 'Festival Light Eye White Shades'). Plants were either well-watered at 10 centibars or water-stressed at 20 centibars. Silicon treatments included top-dressed at 20, 40, 60, and 80 g, or incorporated at 50, 100, 150, and 200 g, in Metro-Mix 360 media without silica plus a control and one treatment of new Metro-Mix 360 with silica already incorporated. Significant effects were seen from diatomaceous earth supplementation, irrigation, and interaction in all plants; growth and flower characteristics, leaf nutrient content, and tolerance to stress were improved by application of diatomaceous earth. An increase in leaf N, P, K, Mg, and Ca was observed for dahlia 'Dahlinova Montana' and black-eyed Susan 'Denver Daisy'. Transpiration was maintained in all three species due to silica supplementation under water-stress. Metro-Mix with silica was similar to the Metro-mix without silica and equivalent to most treatments with supplemental silica for all three species.

Keywords: greenhouse; metro-mix; *Dahlia*; *Rudbeckia*; *Gerbera*

---

1. Introduction

Silicon is the second most abundant element on earth and is present in various forms, including silicon dioxide, also known as silica (Si). In plants, except for members of the family Equisetaceae, Si is a nonessential and beneficial element, meaning that plants can complete their life cycles without the mineral nutrient [1]. However, plants deficient in Si are often weaker structurally and more prone to abnormalities of growth, development, and reproduction. The benefits of Si are mostly evident when plants are under stress conditions [2]. Several studies have shown that plants benefit in many ways from supplemental soluble Si, including greater tolerance of environmental stresses, drought, salinity, mineral toxicity or deficiency, improved growth rates, and resistance to insects and fungi [3–5].

Common use of soilless substrates in greenhouse and nursery production limits the availability of Si to plants [6]. Plants grown with soilless media often appear weaker structurally compared to crops grown in the field [7]. Therefore, adding Si-related compounds as an amendment has been highly recommended. Miyake and Takahashi [8] brought interest to Si nutrition of horticultural crops when they observed Si deficient tomatoes (*Lycopersicon esculentum* Mill.). In the Netherlands, the use of Si supplementation in a hydroponic system was recommended for crops such as cucumber (*Cucumus sativus* L.) and roses (*Rosa hybrida* L.) [9,10]. Roses with Si added to the nutrient formula also showed a decrease in leaf and flower senescence [11]. The shelf life of *Chrysanthemum* L. cut flowers was also extended [12]. Hydroponically-produced gerbera plants supplemented with Si had improved overall crop and flower quality [13].

Other considerations such as solubility, availability, physical properties, and contaminants must be considered before choosing a Si source. Silica is available from natural resources, fertilizers (organic and inorganic), and industrial by-products. Most horticultural studies use Si from by-products such as liquid silicates, slag, and basalt dust [14–16]. Diatomaceous earth (DE) is a sedimentary rock formed from the deposition of silica-rich diatoms. The cell walls of diatoms contain amorphous silica ($SiO_2 \cdot H_2O$). There has been limited research focused on the effects of DE regarding growth and flower characteristics, as well as water-stress related issues in horticultural crops. Most studies utilizing DE focused on retention of water or circulation of oxygen in plant media [17]. However, supplementation of DE has been proven to improve plant growth, quality, and nutrient uptake in agronomic crops, such as rice (*Oryza sativa* L.) [18]. Use of DE to improve plant growth of ornamentals is limited, thus the objectives of this study were to determine the effects of DE as a Si supplement on three potted ornamentals under well-watered and water-stressed conditions.

2. Materials and Methods

2.1. Plant Material and Culture

On 8 May 2015, two 128 plug cell trays of black-eyed Susan (*Rudbeckia hirta* L. 'Denver Daisy'), five 51 plug cell trays of dahlia (*Dahlia Cav.*× *hybrida* 'Dahlinova Montana'), and two 128 plug cell trays of daisy (*Gerbera jamesonii* L. 'Festival Light Eye White Shades') were obtained from Park Seed (Greenwood, SC, USA). Before transplanting, all species were placed on a mist bench. Cuttings and plugs were transplanted into standard 15 cm pots, filled with media (Metro-Mix (MM) 360; Sun Gro Horticulture, Bellevue, WA, USA) that did not contain Si on 28 May 2015, and a single treatment of media (MM + Si; Sun Gro Horticulture, Bellevue, WA, USA) that contained 20 to 50 ppm soluble Si (RESiLIENCE™) derived from wollastonite [19]. A single plant was placed in each pot and plants were grown at the Department of Horticulture and L.A. research greenhouses in Stillwater, OK under natural photoperiods. Temperatures were set at 37 °C during the day and 26 °C during the night.

2.2. Experimental Arrangement

Ten Si treatments were established by adding diatomaceous earth (Perma-Guard, Inc., Kamas, UT, USA). Application of DE included top-dressed (TD) rates at 20, 40, 60, and 80 g, and incorporated (INC) rates at 50, 100, 150, and 200 g. An MM control (using media without DE) and the MM + Si treatments were also included. For each species, there were six pots per Si treatment per irrigation treatment, which served as single pot replicates. Plants were well-watered at 10 centibars or water-stressed at 20 centibars using one drip emitter per pot. Tensiometers (IRROMETER, Riverside, CA, USA) were used to control irrigation by placing a single tensiometer in the middle of a bench in a pot at a depth of 10 cm, which resulted in no leachate. Plant species and Si treatments were randomized within irrigation, which served as blocks.

2.3. Harvesting and Measurements

Data collected on plants included height from the media surface to the tallest opened flower, width (average of two perpendicular measurements), shoot dry weight, number of flowers, flower diameter, and transpiration. Shoot dry weight was determined by cutting the stems at media level and drying for 2 d at 52.2 °C. For foliar nutrient analyses, mature leaves from the middle to upper level of the plant were collected from five plants per Si and irrigation treatment of each species. Soil and leaf nutrient analysis was performed by the Soil, Water, and Forage Analytical Laboratory (SWFAL) at Oklahoma State University, using a LECO TruSpec Carbon and Nitrogen Analyzer (LECO Corporation, St. Joseph, MI, USA). Soil and leaf Si analysis was performed, using the 0.5 M ammonium acetate method [20]. Transpiration was recorded weekly using a LI-1600 Steady State Porometer (LI-COR Inc., Lincoln, NE, USA).

2.4. Statistical Analysis

Pots were arranged in a randomized block design with irrigation serving as the block. Analysis of variance methods (PROC MIXED) were used with a two-factor factorial arrangement, with irrigation and silicon treatment as the factors of interest. Separate analyses were conducted for each of the plant species. When interactions of irrigation and Si treatment were significant, simple effects were reported. Mean separations were determined using a DIFF option in an LSMEANS statement and a SLICE option (when appropriate) and with a 0.05 level of significance.

3. Results

3.1. *Dahlia xhybrida* 'Dahlinova Montana'

A significant interaction of Si treatment with irrigation was seen for transpiration and soil Si (Table 1). Under the well-watered condition, soil Si was greatest when supplemented with 60 and 80 g TD, as well as 100, 150, and 200 g INC (Table 2). Transpiration was greatest in plants under the control as well as 40 and 60 g TD. Under the water-stressed condition, soil Si was greatest for INC plants compared to TD plants and MM + Si plants. Plants treated with 100 g or less DE within the INC treatment had the greatest transpiration.

Table 1. Analysis of variance for growth, flowering, leaf nutrient content, soil silica, and physiology of *Dahlia* × *hybrida* 'Dahlinova Montana' after application of diatomaceous earth (DE) and irrigation controlled with a tensiometer.

Source	Height (cm)	Width (cm)	Shoot Dry Weight (g)	Stem Diameter (cm)	Mean Flower Number	Flower Diameter (cm)
DE Treatment	*^{z}	ns	**	***	ns	ns
Irrigation	****	****	****	***	****	****
DE Treatment × Irrigation	ns	ns	ns	ns	ns	ns

Source	N (%)	P (%)	S (%)	K (%)	Mg (%)	Ca (%)	Na (%)	Si (ppm)	Zn (ppm)	Cu (ppm)	Fe (ppm)	Mn (ppm)	Ni (ppm)
DE Treatment	**	*	*	ns	***	****	ns	****	ns	***	**	****	ns
Irrigation	ns	ns	ns	ns	ns	ns	ns	ns	ns	ns	***	ns	ns
DE Treatment × Irrigation	ns	ns	ns	ns	ns	ns	ns	ns	ns	ns	ns	ns	ns

Source	Transpiration	Soil silica (ppm)
DE Treatment	ns	****
Irrigation	****	****
DE Treatment × Irrigation	****	**

^{z} NS, *, **, ***, **** indicates non-significant or significant at $p \leq 0.05$, 0.01, and 0.001, respectively.

A significant effect of irrigation was seen for all growth and flowering characteristics (Table 1). Well-watered plants had greater height, width, shoot dry weight, mean flower number, stem diameter, and flower diameter, compared to water-stressed plants (Table 3). A significant effect of DE treatment was seen for height, shoot dry weight, and stem diameter (Table 1). Height was greatest for control plants, all TD plants, INC plants at 100 and 200 g, as well as MM + Si plants (Table 4). Shoot dry weight was greatest for control plants, all TD plants, and INC plants at 100 g. Stem diameter was greatest for all TD plants, as well as INC plants at 50 and 100 g.

Table 2. Soil silica (Si), transpiration, and leaf Si affected by interaction of diatomaceous earth treatment with irrigation in *Dahlia* × *hybrida* 'Dahlinova Montana', *Gerbera jamesonii* 'Festival Light Eye White Shades', and *Rudbeckia hirta* 'Denver Daisy'.

		Well-Watered (10 cb)			Water-Stressed (20 cb)		
Cultivar	Application and Rate (g) [z]	Soil Si (ppm)	Transpiration	Leaf Si	Soil Si (ppm)	Transpiration	Leaf Si
Dahlinova Montana	0	47.8 e [y]	8.9 a	x	53.5 bcd	4.8 b	x
	TD 20	56.1 de	6.8 bc	x	49.5 d	4.5 b	x
	TD 40	59.7 bcd	9.0 a	x	51.5 cd	2.9 cd	x
	TD 60	65.8 abc	8.9 a	x	47.8 d	4.2 bc	x
	TD 80	65.1 abc	6.8 bc	x	49.7 d	3.3 cd	x
	INC 50	58.8 cd	7.0 b	x	62.0 ab	4.9 b	x
	INC 100	67.9 ab	5.2 d	x	59.3 abc	6.5 a	x
	INC 150	70.1 a	7.3 b	x	63.1 a	3.2 cd	x
	INC 200	72.7 a	5.7 cd	x	63.6 a	2.9 d	x
	MM + Si	53.1 de	6.7 bc	x	50.3 d	4.6 b	x
Festival Light Eye White Shades	0	43.6 bc	9.0	x	32.7 cd	7.3 abc	x
	TD 20	41.3 c	9.6	x	30.4 d	8.2 ab	x
	TD 40	45.5 bc	9.1	x	35.8 cd	8.5 ab	x
	TD 60	43.0 bc	10.4	x	48.2 b	8.5 ab	x
	TD 80	47.2 bc	9.7	x	47.6 b	8.7 a	x
	INC 50	42.3 c	8.9	x	41.5 bc	6.9 bc	x
	INC 100	50.4 bc	8.3	x	59.5 a	6.1 c	x
	INC 150	52.1 ab	9.2	x	59.4 a	7.0 bc	x
	INC 200	56.9 a	9.4	x	45.3 b	5.8 c	x
	MM + Si	43.6 bc	9.6	x	36.3 cd	7.2 abc	x
Denver Daisy	0	x	x	276.6 bc	x	x	345.7 bc
	TD 20	x	x	281.5 abc	x	x	373.4 bc
	TD 40	x	x	260.2 bc	x	x	357.8 bc
	TD 60	x	x	313.8 abc	x	x	412.0 ab
	TD 80	x	x	261.9 bc	x	x	321.3 bcd
	INC 50	x	x	292.3 abc	x	x	346.3 bc
	INC 100	x	x	377.4 ab	x	x	261.1 cd
	INC 150	x	x	196.4 cd	x	x	208.9 d
	INC 200	x	x	91.4 c	x	x	201.6 d
	MM + Si	x	x	406.5 a	x	x	522.4 a

[z] Top-dressed (TD), Incorporated (INC), and Metro-Mix media (MM). [y] Means (n = 6) with the same letter within the same column and within cultivar are not statistically different at $p \leq 0.05$. [x] Main effects were significant for factors.

Table 3. Growth and flowering characteristics affected by irrigation, controlled by a tensiometer, averaged across diatomaceous earth treatments in *Dahlia* × *hybrida* 'Dahlinova Montana', *Gerbera jamesonii* 'Festival Light Eye White Shades', and *Rudbeckia hirta* 'Denver Daisy'.

Cultivar	Irrigation Rate (cb)	Height (cm)	Width (cm)	Shoot Dry Weight (g)	Stem Diameter (cm)	Mean Flower Number	Flower Diameter (cm)
Dahlinova Montana	10	27.4 a [z]	29.7 a	19.8 a	4.3 a	8.4 a	6.8 a
	20	22.0 b	24.2 b	11.2 b	3.8 b	4.8 b	4.1 b
Festival Light Eye White Shades	10	12.3	23.6 a	6.2	1.1	0.9	1.4
	20	11.2	20.8 b	5.3	0.9	0.6	1.1
Denver Daisy	10	36.8 a	29.9 a	25.0 a	4.6 a	12.6 a	7.9 a
	20	22.0 b	24.2 b	11.2 b	3.7 b	4.8 b	4.1 b

[z] Means (n = 6) with the same letter within the same column and within cultivar are not statistically significant at $p \leq 0.05$.

Table 4. Growth and flowering characteristics affected by diatomaceous earth treatment averaged across irrigation, controlled by a tensiometer, in *Dahlia* × *hybrida* 'Dahlinova Montana', *Gerbera jamesonii* 'Festival Light Eye White Shades', and *Rudbeckia hirta* 'Denver Daisy'.

Cultivar	Application and Rate (g) [z]	Height (cm)	Width (cm)	Shoot Dry Weight (g)	Stem Diameter (cm)	Mean Flower Number	Flower Diameter (cm)
Dahlinova Montana	0	24.9 ab [y]	26.9	15.8 a–d	3.8 bc	7.1	5.7
	TD 20	24.1 ab	29.1	18.9 a	4.5 a	8.0	5.8
	TD 40	24.8 ab	27.9	19.0 a	4.5 a	8.5	5.4
	TD 60	26.9 a	29.3	17.9 ab	4.3 ab	5.7	4.9
	TD 80	26.8 a	26.9	16.2 abc	4.5 a	7.0	5.3
	INC 50	22.2 b	24.3	14.5 bcd	3.9 ac	5.9	4.9
	INC 100	26.6 a	27.5	15.2 a–d	4.3 ab	6.3	6.3
	INC 150	22.1 b	25.9	11.8 d	3.4 c	5.4	5.3
	INC 200	23.9 ab	25.1	12.6 cd	3.4 c	5.5	5.4
	MM + Si	24.9 ab	27.4	13.1 cd	3.8 bc	6.4	5.4
Festival Light Eye White Shades	0	8.9 b [y]	20.8 b	3.9 c	0.9	0.4	0.6
	TD 20	11.1 b	21.7 ab	5.9 bc	1.2	1.1	2.1
	TD 40	11.2 b	22.5 ab	4.9 bc	0.7	0.5	1.1
	TD 60	10.4 b	22.4 ab	5.4 bc	0.6	0.3	0.2
	TD 80	11.8 ab	20.0 b	5.9 bc	0.3	0.8	1.2
	INC 50	12.1 ab	21.6 ab	5.7 bc	1.3	0.6	1.0
	INC 100	15.1 a	25.2 a	8.6 a	1.5	1.2	2.8
	INC 150	15.3 a	24.9 a	7.2 ab	1.5	1.0	2.1
	INC 200	10.8 b	23.1 ab	5.0 bc	1.5	1.0	1.6
	MM + Si	10.8 b	19.5 b	4.9 bc	0.8	0.8	0.6
Denver Daisy	0	30.8 bc [y]	23.0 bc	12.8	4.3	6.4	6.2 b
	TD 20	31.1 bc	26.9 ab	17.6	4.4	10.5	5.4 bc
	TD 40	30.9 bc	26.9 ab	18.1	3.9	9.8	5.5 bc
	TD 60	34.8 abc	27.2 ab	23.4	4.7	11.1	5.9 bc
	TD 80	30.9 bc	28.2 a	24.2	4.5	9.7	6.4 b
	INC 50	32.2 bc	26.1 abc	18.2	4.2	9.6	6.5 ab
	INC 100	46.4 a	27.8 a	19.9	3.8	8.9	5.9 bc
	INC 150	25.3 c	21.9 c	15.5	4.0	7.0	4.9 bc
	INC 200	27.4 bc	23.1 bc	16.1	3.7	8.5	4.5 a
	MM + Si	39.2 ab	29.2 a	20.2	4.3	11.1	8.1 a

[z] Top-dressed (TD), Incorporated (INC), and Metro-Mix media (MM). [y] Means (n = 6) with the same letter within the same column and within cultivar are not statistically different at $p \leq 0.05$.

A significant effect of DE treatment was seen for leaf nutrient content (Table 1). Total nitrogen (N) was greatest for all TD plants and INC plants with rates of 100, 150, and 200 g (Table 5). TD plants at 80 g and INC plants at 100, 150, and 200 g had the greatest values of phosphorus (P). Magnesium (Mg) was greatest for TD plants at 40 and 80 g, as well as INC plants at 50, 100, and 150 g. Calcium (Ca) was greatest for INC plants at 50, 100, and 150 g, as well as MM + Si plants. Sulfur (S) was greatest for all TD plants and INC plants at 100, 150, and 200 g (Table 6). Silica was greatest for control plants, TD plants at 20, 40, and 60 g, INC plants at 100 and 150 g, as well as MM + Si plants. For copper (Cu), the greatest values were seen for TD plants at 40 and 80 g. Iron (Fe) was greatest for TD plants at 40, 60, and 80 g, as well as INC plants at 100 and 150 g. Manganese (Mn) was greatest for INC plants at 150 and 200 g.

Table 5. Leaf macronutrient content affected by diatomaceous earth treatment across irrigation, controlled by a tensiometer, in *Dahlia* × *hybrida* 'Dahlinova Montana', *Gerbera jamesonii* 'Festival Light Eye White Shades', and *Rudbeckia hirta* 'Denver Daisy'.

Cultivar	Application and Rate (g) [z]	N (%)	P (%)	K (%)	Mg (%)	Ca (%)
Dahlinova Montana	0	3.55 bcd [y]	0.29 c	3.51	0.92 bc	1.74 de
	TD 20	3.88 ab	0.33 bc	3.33	0.91 bc	1.81 cd
	TD 40	4.11 ab	0.34 bc	3.19	0.97 ab	1.75 de
	TD 60	4.09 ab	0.33 bc	3.23	0.84 c	1.61 e
	TD 80	3.83 ab	0.34 abc	3.47	0.96 ab	1.85 bcd
	INC 50	3.21 cd	0.29 c	3.48	0.98 ab	1.93 abc
	INC 100	3.75 abcd	0.35 abc	3.54	1.03 a	1.99 ab
	INC 150	3.79 abc	0.38 ab	3.81	0.98 ab	2.00 ab
	INC 200	4.27 a	0.39 a	3.54	0.92 bc	1.88 bcd
	MM + Si	3.19 d	0.33 bc	3.44	0.82 c	2.07 a
Festival Light Eye White Shades	0	2.86 [y]	0.26	2.63	0.62	1.45
	TD 20	2.92	0.31	3.18	0.66	1.55
	TD 40	3.05	0.28	3.30	0.62	1.48
	TD 60	2.93	0.46	3.06	0.77	2.12
	TD 80	3.14	0.54	3.31	0.79	2.02
	INC 50	2.74	0.29	3.10	0.73	1.71
	INC 100	2.82	0.27	3.06	0.62	1.42
	INC 150	2.95	0.26	3.03	0.55	1.26
	INC 200	2.98	0.26	2.97	0.56	1.33
	MM + Si	2.99	0.37	3.10	0.70	1.88
Denver Daisy	0	2.40 de [y]	0.20 e	3.09 d	1.21 a	3.35 b
	TD 20	2.79 cde	0.22 cde	3.31 bcd	1.20 ab	3.33 b
	TD 40	2.98 bc	0.26 bcd	3.53 bcd	1.10 ab	2.89 bc
	TD 60	2.94 bc	0.23 cde	3.25 cd	1.13 ab	3.28 bc
	TD 80	2.87 cd	0.21 de	3.29 bcd	1.12 ab	3.17 bc
	INC 50	2.82 cde	0.24 b–e	3.27 bcd	1.15 ab	3.27 bc
	INC 100	2.96 bc	0.27 abc	3.74 ab	1.12 ab	3.26 bc
	INC 150	3.38 ab	0.28 ab	3.72 abc	1.09 bc	3.26 bc
	INC 200	3.51 a	0.31 a	4.13 a	0.97 cd	2.79 c
	MM + Si	2.37 e	0.21 de	3.14 d	0.96 d	4.01 a
Optimum levels [x]		2.50–4.50	0.20–0.75	1.50–5.50	0.25–1.00	1.00–4.00

[z] Top-dressed (TD), incorporated (INC), and Metro-Mix media (MM). [y] Means (n=6) with the same letter within the same column and within cultivars are not statistically significant at $p \leq 0.05$. [x] According to Kalra (26).

3.2. *Gerbera jamesonii 'Festival Light Eye White Shades'*

A significant interaction of DE with irrigation was seen for soil Si and transpiration (Table 7). Under the well-watered condition, soil Si was greatest when DE was supplemented at 150 and 200 g INC (Table 2). Under the water-stressed condition, soil Si was greatest for 100 and 150 g INC plants. Transpiration was greatest for control plants, TD plants, and MM + Si plants. A significant effect of irrigation was seen for width and leaf nutrient content (Table 7). Well-watered plants had greater widths compared to water-stressed plants (Table 3). Potassium, Ca, sodium (Na), and Mn levels were greater in water-stressed plants compared to well-watered plants (Table 8).

Table 6. Leaf micronutrient and trace element content affected by diatomaceous earth treatment across irrigation, controlled by a tensiometer, in *Dahlia* × *hybrida* 'Dahlinova Montana', *Gerbera jamesonii* 'Festival Light Eye White Shades', and *Rudbeckia hirta* 'Denver Daisy'.

Cultivar	Application and Rate (g) [z]	S (%)	Na (%)	Si (ppm)	Zn (ppm)	Cu (ppm)	Mn (ppm)	Ni (ppm)	Fe (ppm)
Dahlinova Montana	0	0.33 bc [y]	0.02	84.1 a	38.5 a	12.0 cd	151.6 d	0.0	92.2 d
	TD 20	0.39 ab	0.03	69.1 abc	46.6 a	14.4 cd	149.3 d	0.0	106.6 cd
	TD 40	0.39 ab	0.03	77.3 ab	53.6 a	18.2 a	163.9 cd	0.0	232.7 a
	TD 60	0.35 abc	0.02	69.7 abc	39.5 a	14.4 cd	150.2 d	0.0	161.7 a–d
	TD 80	0.41 a	0.02	38.1 c	44.9 a	17.5 ab	188.8 bc	0.0	171.5 abc
	INC 50	0.34 bc	0.02	47.9 bc	45.2 a	14.8 bcd	189.9 bc	0.0	133.8 bcd
	INC 100	0.39 ab	0.03	82.9 a	57.9	15.1 bc	204.6 b	0.0	188.6 ab
	INC 150	0.37 ab	0.03	67.5 abc	42.2	14.1 cd	272.1 a	0.1	163.1 a–d
	INC 200	0.37 ab	0.02	45.8 bc	39.9	13.3 cd	250.4 a	0.0	104.5 cd
	MM + Si	0.31 c	0.03	94.3 a	42.5	11.9 d	194.7 b	0.3	97.1 d
Festival Light Eye White Shades	0	0.40 [y]	0.07	107.6	46.6	24.8	145.4	2.29 a	591.1
	TD 20	0.48	0.08	253.4	56.6	31.5	129.8	0.128 b	392.1
	TD 40	0.39	0.07	185.2	52.8	20.9	129.2	0.002 b	235.6
	TD 60	1.01	0.16	308.0	84.1	133.8	156.9	0.191 b	390.6
	TD 80	1.03	0.14	264.7	115.2	171.8	190.8	0.066 b	566.7
	INC 50	0.42	0.07	223.7	57.9	14.4	165.3	2.12 a	649.9
	INC 100	0.38	0.12	263.2	51.8	20.3	161.9	2.23 a	736.4
	INC 150	0.32	0.07	172.1	43.9	11.6	161.8	0.103 b	305.9
	INC 200	0.31	0.06	163.9	43.6	10.5	219.1	0.131 b	291.1
	MM + Si	0.62	0.08	308.1	79.2	67.9	198.8	1.48 ab	742.7
Denver Daisy	0	0.41 cd	0.02	x	39.7	6.03 de	137.6 bc	0.002	97.9 c
	TD 20	0.49 abc	0.02	x	40.4	8.0 b–e	131.7 bcd	0.002	130.9 bc
	TD 40	0.48 abc	0.02	x	42.7	8.9 b–e	114.4 d	0.002	134.5 bc
	TD 60	0.52 ab	0.04	x	39.8	8.6 a–d	125.7 cd	0.003	132.6 bc
	TD 80	0.44 bcd	0.04	x	33.1	7.5 cde	129.5 cd	0.462	91.7 c
	INC 50	0.55 a	0.02	x	44.2	10.9 ab	137.1 bc	0.533	209.2 ab
	INC 100	0.53 ab	0.03	x	36.8	10.0 abc	152.1 b	0.308	168.8 bc
	INC 150	0.48 abc	0.03	x	37.6	7.6 cde	208.7 a	0.145	139.5 bc
	INC 200	0.50 abc	0.03	x	43.6	11.4 a	203.8 a	0.575	285.2 a
	MM + Si	0.34 d	0.01	x	37.1	5.1e	140.3 bc	0.002	105.9 bc
Optimum levels		0.2–0.8	w	w	27–100	5.0–30.0	20–300	0–5	100–500

[z] Top-dressed (TD) or incorporated (INC). [y] Means (n = 6) with the same letter within the same column and within cultivars are not statistically significant at $p \leq 0.05$. [x] Significant interactions between diatomaceous earth and irrigation reported in another table. [w] Optimum levels not reported.

Table 7. Analysis of variance for growth, flowering, leaf nutrient content, soil silica, and physiology of *Gerbera jamesonii* 'Festival Light Eye White Shades' with application of diatomaceous earth (DE) and irrigation controlled with a tensiometer.

Source	Height (cm)	Width (cm)	Shoot Dry Weight (g)	Stem Diameter (cm)	Mean Flower Number	Flower Diameter (cm)
DE Treatment	* [z]	*	*	ns	ns	ns
Irrigation	ns	**	ns	ns	ns	ns
DE Treatment × Irrigation	ns	Ns	ns	ns	ns	ns

Source	N (%)	P (%)	S (%)	K (%)	Mg (%)	Ca (%)	Na (%)	Si (ppm)	Zn (ppm)	Cu (ppm)	Fe (ppm)
DE Treatment	ns	ns	ns	ns	ns	ns	ns	ns	ns	ns	ns
Irrigation	ns	ns	ns	ns	*	*	**	*	ns	ns	ns
DE Treatment × Irrigation	ns	ns	ns	ns	ns	ns	ns	ns	ns	ns	ns

Source	Transpiration	Soil silica (ppm)
DE Treatment	ns	****
Irrigation	****	*
DE Treatment × Irrigation	**	**

[z] NS, *, **, ***, **** indicates non-significant or significant at $p \leq 0.05$, 0.01, and 0.001, respectively.

Table 8. Leaf nutrient content affected by irrigation, controlled by a tensiometer, across diatomaceous earth treatments in *Dahlia* × *hybrida* 'Dahlinova Montana', *Gerbera jamesonii* 'Festival Light Eye White Shades', and *Rudbeckia hirta* 'Denver Daisy'.

Cultivar	Application and Rate (g) [z]	N (%)	P (%)	S (%)	K (%)	Mg (%)	Ca (%)	Na (%)	Si (ppm)	Zn (ppm)
Dahlinova Montana	10	3.73 [z]	0.35	0.37	3.39	0.93	1.83	0.03	68.1	46.1
	20	3.79	0.33	0.35	3.52	0.94	1.89	0.03	67.2	44.1
Festival Light Eye White Shades	10	2.89 [z]	0.27	0.41	3.01 b	0.61	1.45 b	0.07 b	184.2	51.5
	20	2.99	0.24	0.66	3.14 a	0.72	1.79 a	0.12 a	265.6	74.8
Denver Daisy	10	2.88 [z]	0.25	0.43 b	3.48	1.03 b	2.82 b	0.03	y	37.5
	20	2.92	0.24	0.52 a	3.42	1.18 a	3.70 a	0.22	y	41.5
Optimum levels [x]		2.50–4.50	0.20–0.75	0.25–1.00	1.50–5.50	0.25–1.00	1.00–4.00	w	w	27.0–100.0

[z] Means (n = 6) with the same letter within the same column and within cultivar are not statistically significant $p \leq 0.05$. [y] Significant interaction between diatomaceous earth treatment and irrigation. [x] According to Kalra (26). [w] Optimum levels not reported.

A main effect of DE was seen for height, width, shoot dry weight, and nickel (Ni) content in the leaf tissue (Table 7). Height was greatest for 80 g TD plants, as well as 50, 100, and 150 g INC plants (Table 4). Width was greatest for TD plants at 20, 40, and 60 g, and all INC plants. Shoot dry weight was greatest for INC plants at 100 and 150 g. Nickel was greatest for control plants, INC plants at 50 and 100 g, as well as MM + Si plants (Table 6).

3.3. *Rudbeckia hirta* 'Denver Daisy'

A significant interaction of DE with irrigation was seen for leaf Si content (Table 9). Under the well-watered condition, silica in the leaf was greatest for 20 and 60 g TD plants, 50 and 100 g INC plants, as well as MM + Si plants (Table 2). Under the water-stressed condition, Si in the leaf was greatest for 60 g TD plants and MM + Si plants.

Table 9. Analysis of variance for growth, flowering, leaf nutrient content, soil silica, and physiology of *Rudbeckia hirta* 'Denver Daisy' after application of diatomaceous earth (DE) and irrigation controlled with a tensiometer.

Source	Height (cm)	Width (cm)	Shoot Dry Weight (g)	Stem Diameter (cm)	Mean Flower Number	Flower Diameter (cm)
DE Treatment	* [z]	*	ns	ns	ns	**
Irrigation	**	****	****	***	****	****
DE Treatment × Irrigation	ns	ns	ns	ns	ns	ns

Source	N (%)	P (%)	S (%)	K (%)	Mg (%)	Ca (%)	Na (%)	Si (ppm)	Zn (ppm)	Cu (ppm)	Fe (ppm)	Mn (ppm)	Ni (ppm)
DE Treatment	****	****	***	***	***	**	ns	****	ns	ns	*	****	ns
Irrigation	ns	ns	ns	ns	****	****	ns	****	ns	**	ns	*	ns
DE Treatment × Irrigation	ns	ns	ns	ns	ns	ns	ns	***	ns	ns	ns	ns	ns

Source	Transpiration	Soil silica (ppm)
DE Treatment	ns	****
Irrigation	****	****
DE Treatment × Irrigation	ns	ns

[z] NS, *, **, ***, **** indicates non-significant or significant at $p \leq 0.05$, 0.01, and 0.001, respectively.

A significant effect of irrigation was seen for all growth and flower characteristics, as well as leaf nutrient content, transpiration, and soil Si (Table 9). Plants that were well-watered grew taller and wider, and had greater shoot dry weight, stem diameter, flower number, and flower diameter (Table 3). The nutrients S, Mg, Ca, and Mn were greater in the leaf tissue of water-stressed plants, compared to well-watered plants (Table 8). Soil Si and transpiration were greater in plants under the water-stressed condition, compared to those that were well-watered (Table 10). A main effect of DE treatment was seen for height, width, and flower diameter (Table 9). Height was greatest for 60 g TD plants, 100 g INC plants, and MM + Si plants (Table 4). Width was greatest for all TD plants, INC plants at 50 and 100 g, and MM + Si plants. Flower diameter was greatest for 50 g INC plants and MM + Si plants.

Table 10. Soil silica and transpiration of *Rudbeckia hirta* 'Denver Daisy' affected by irrigation, controlled by a tensiometer, averaged across silica treatments.

Source	Irrigation Rate (cb)	Soil Si (ppm)	Transpiration
Well-watered	10	35.8 b [z]	4.56 b
Water-stressed	20	47.7 a	8.10 a

[z] Means (n = 6) with the same letter within the same column are not statistically significant at $p \leq 0.05$.

4. Discussion

Amending the soilless substrate with varying rates of DE by top-dressing or by incorporating into the substrate increased plant height, width, shoot dry weight, stem diameter, and flower diameter in dahlia 'Dahlinova Montana', daisy 'Festival Light Eye White Shades', and black-eyed Susan 'Denver Daisy', in this study. Several other studies have reported similar benefits of supplemental Si on growth and flowering characteristics. Hwang et al. [21] reported that adding 200 mg L^{-1} of potassium (K) metasilicate increased plant height and shoot dry weight in cut roses. Stem quality was also improved in cut roses when Si was added to a recirculated nutrient solution in a closed hydroponic system [22]. Flower diameter of calibrachoa (*Calibrachoa* × *hybrida* Cerv.), fuchsia (*Fuchsia hybrid* hort. Ex Siebold & Voss), and petunia (*Petunia* × *hybrida* Vilm.) increased when supplemented with a weekly drench of K silicate at 100 mg L^{-1} [23]. Silica supplementation improved growth of two cultivars of French marigolds (*Tagetes patula* L.) by increasing stem diameter, shoots, and dry weights [24]. Growth and biomass parameters were increased in begonia (*Begonia semperflorens* Link et Otto) and pansy (*Viola* × *wittrockiana* Hort.) grown in vitro when supplemented with K silicate [25]. Savvas et al. [13] reported a greater percentage of flowers in hydroponically-grown gerbera (*Gerbera jamesonii*) supplemented with Si.

Amending the soilless substrate with DE increased nutrient content, despite being inert. Based on the analysis of Kalra [26], most nutrients were within the optimum range adequate for plant growth and levels greater than the maximum range were not considered excess or toxic. Nickel concentrations in dahlia 'Dahlinova Montana' were less than the minimum range (Table 4). However, these levels were not considered insufficient because often there are no symptoms to accurately determine Ni deficiency [27]. Epstein [1] has noted that the presence of Si does, in fact, affect absorption and translocation of several macro-nutrients and micro-nutrients. Early studies conducted by Fisher [28] reported that the addition of Si made P more available in barley (*Hordeum vulgare* L.). Mali and Aery [29] found that, in wheat (*Tritium aestivum* L.), potassium uptake was improved even at low concentrations of Si by H-ATPase being activated. Phosphorus and K are essential nutrients for flowering characteristics. Friedman et al. [30] conducted a study on sunflower (*Helianthus annuus* L.) and celosia (*Celosia argentea* L.), and reported that growth and flower parameters were increased when supplemented with an effluent containing high amounts of N, P, and other nutrients. Kamenidou et al. [7–9] also found an increase in N for sunflowers and gerbera, but most of the levels exceeded the optimum range. Nitrogen metabolism is a major factor in stem and leaf growth and too much can delay or prevent flowering. Calcium is part of the structure of cell walls and is necessary for cell growth and division. Ma and Takahashi [31] reported that there was an antagonistic effect between Si and Ca in rice, in which one can decrease the amount of the other. However, our study found the opposite effect, in which MM + Si increased Ca content in dahlia 'Dahlinova Montana' and 'Denver Daisy'. Kamenidou et al. [9] and Savvas et al. [13] also reported that supplemental Si increased Ca within gerbera. There was an increase in metals such as Cu, Fe, and Mn in dahlia 'Dahlinova Montana' and daisy 'Festival Light Eye White Shades', due to DE having trace amounts of these elements [18]. Silica levels in the leaf tissue and media for all the plants were observed in low amounts. Potentially, the plants could be classified as non-accumulators of Si (<0.5%) which has been reported for gerbera [32].

Amending the soilless substrate with varying methods and rates of DE showed mixed results on transpiration in all three species. Improvements in this physiological trait were mostly seen when Si (DE) supplementation interacted with irrigation. Kamenidou et al. [33] found that a foliar spray of Na silicate at 100 mg L^{-1} decreased transpiration in zinnias (*Zinnia* L.). Yoshida and Kitagishi [34] noted that the effects are related to Si being deposited in the cuticular layers of leaves, serving as a barrier which reduces the loss of water. A decrease in transpiration can benefit the floricultural market by improving quality and shelf life of cut flowers [35]. Considering the effect of irrigation, plants under the well-watered condition had greater growth and flowering, which was expected. However, Si is known to maintain the growth and flowering characteristics as well as nutrient levels in water-stressed plants. In Kentucky bluegrass (*Poa pratensis* L.), drought stress hindered physiological and quality attributes, but application of Si alleviated the adverse effects [36].

5. Conclusions

Several growth and flowering characteristics were improved, depending on the rate and application method, by application of DE. Benefits of DE included increased height, width, shoot dry weight, stem, and flower diameter. An increase in nutrients, such as N, P, K, Mg, and Ca, was seen mostly for dahlia 'Dahlinova Montana' and black-eyed Susan 'Denver Daisy'. The adverse effects that typically occur under water-stressed conditions were alleviated and plant quality, as well as transpiration, was maintained in all three plants due to Si supplementation. Silicon is known to play an important role in cell membrane integrity, in which osmosis, photosynthesis, and transpiration all occur. Diatomaceous earth as supplemental Si was beneficial for plant growth, flowering, and nutrient content under both well-watered and water-stressed conditions. For growth and flower characteristics, MM + Si was similar to the control (MM) with no added silica, and equivalent to most treatments with supplemental silica. To conclude, this research supports that DE, one of the many Si sources, is beneficial to plants; however, this is dependent upon species, Si rate, and the method of application. Benefits of DE include an increase in growth parameters, leaf nutrient content, and tolerance to stress, in which plant quality can be maintained. Future studies should further assess the use of DE on other crops and stress conditions.

Author Contributions: Conceptualization, B.D.; methodology, T.M.-I.; formal analysis, M.P.; investigation, T.M.-I.; writing—original draft preparation, T.M.-I.; writing—review and editing, B.D and N.M.; supervision, B.D.

Acknowledgments: We thank Stephen Stanphill for helping with data collection and greenhouse management.

References

1. Epstein, E. The anomaly of silicon in plant biology. *Proc. Natl. Acad. Sci. USA* **1994**, *91*, 11–17. [CrossRef] [PubMed]
2. Ma, J.F.; Yamaji, N. Silicon uptake and accumulation in higher plants. *Plant Sci.* **2006**, *11*, 392–397. [CrossRef] [PubMed]
3. Chérif, M.; Asselin, A.; Bélanger, R.R. Defense responses induced by soluble silicon in cucumber roots infected by *Pythium* spp. *Phytopathology* **1994**, *84*, 236–242. [CrossRef]
4. Ma, J.F. Role of silicon in enhancing the resistance of plants to biotic and abiotic stresses. *Soil Sci. Plant Nutr.* **2004**, *50*, 11–18. [CrossRef]
5. Liang, Y.; Sun, W.; Zhu, Y.G.; Christie, P. Mechanisms of silicon-mediated alleviation of abiotic stresses in higher plants: A review. *Environ. Pollut.* **2006**, *147*, 422–428. [CrossRef] [PubMed]
6. Voogt, W.; Sonneveld, C. *Silicon in Agriculture*; Elsevier: Amsterdam, The Netherlands, 2011.

7. Kamenidou, S.; Cavins, T.J.; Marek, S. Silicon supplements affect horticultural traits of greenhouse-produced ornamental sunflowers. *HortScience* **2008**, *43*, 236–239. [CrossRef]
8. Miyake, Y.; Takahashi, E. Silicon deficiency of tomato plant. *Soil Sci. Plant Nutr.* **1987**, *24*, 175–189. [CrossRef]
9. Kamenidou, S.; Cavins, T.J.; Marek, S. Silicon supplements affect floricultural quality traits and elemental nutrient concentrations of greenhouse produced gerbera. *Sci. Hortic.* **2010**, *123*, 390–394. [CrossRef]
10. De Kreij, C.; Voogt, W.; Baas, R. Nutrient solutions and water quality for soilless cultures. In *Research Station for Floriculture and Glasshouse Vegetables Brochure*; Naaldwijk Office: Naaldwijk, The Netherlands, 1999.
11. Reezi, S.; Babalar, M.; Kalantari, S. Silicon alleviates salt stress, decreases malondialdehyde content and affects petal color of salt stressed cut rose (*Rosa xhybrida* L.) Hot Lady. *Afr. J. Biotechnol.* **2009**, *8*, 1502–1508.
12. Carvalho-Zanao, M.P.; LAZ, J.; Barbosa, J.G.; Grossi, J.A.S.; Ávila, V.T. Yield and shelf life of *Chrysanthemum* in response to the silicon application. *Hortic. Bras.* **2012**, *30*, 403–408. [CrossRef]
13. Savvas, D.; Manos, G.; Kotsiras, A.; Souvaliotis, S. Effects of silicon and nutrient-induced salinity on yield, flower quality and nutrient uptake of gerbera grown in a closed hydroponic system. *J. Appl. Bot. Food Qual.* **2002**, *76*, 153–158.
14. Berthelsen, S.; Noble, A.D.; Kingston, G.; Hurney, A.; Rudd, A.; Garside, A. Improving Yield and ccs in Sugarcane through the Application of Silicon-Based Amendments; Final Report on SRDC Project CLW009. 2003. Available online: http://hdl.handle.net/11079/12957 (accessed on 9 January 2019).
15. Muir, S. Plant-Available Silicon (Si) as A Protectant Against Fungal Diseases in Soil-Less Potting Media. Available online: https://www.ngia.com.au/Story?Action=View&Story_id=1782 (accessed on 9 January 2019).
16. Savant, N.K.; Snyder, G.H.; Datnoff, L.E. Silicon management and sustainable rice production. *Adv. Agron.* **1996**, *58*, 151–199.
17. Meerow, A.W.; Broschat, T.K. Growth of *Hibiscus* in media amended with a ceramic diatomaceous earth granule and treated with a kelp extract. *HortTechnology* **1996**, *6*, 70–73. [CrossRef]
18. Pati, S.; Pal, B.; Badole, S.; Hazra, G.C.; Mandal, B. Effect of silicon fertilization on growth, yield, and nutrient uptake of rice. *Commun. Soil Sci. Plant Anal.* **2016**, *47*, 284–290. [CrossRef]
19. King, P.A.; Reddy, S. Soilless Growth Medium Including Soluble Silicon. Available online: https://patents.google.com/patent/US6074988A/en (accessed on 9 January 2019).
20. Jim, J.W.; Dolda, S.K.; Henderson, R.E. Soil silicon extractability with seven selected extractants in relation to colorimetric and ICP determination. *Soil Sci.* **2004**, *169*, 861–870.
21. Hwang, S.J.; Park, H.M.; Jeong, B.R. Effects of potassium silicate on the growth of miniature rose Pinnochio grown on rockwool and its cut flower quality. *J. Jpn. Soc. Hortic. Sci.* **2005**, *74*, 242–247. [CrossRef]
22. Ehret, D.L.; Menzies, J.G.; Helmer, T. Production and quality of greenhouse roses in recirculating nutrient systems. *Sci. Hortic.* **2005**, *106*, 103–113. [CrossRef]
23. Mattson, N.S.; Leatherwood, W.R. Potassium silicate drenches increase leaf silicon content and affect morphological traits of several floriculture crops grown in a peat-based substrate. *HortScience* **2010**, *45*, 43–47. [CrossRef]
24. Sivanesan, I.; Son, M.S.; Lee, J.P.; Jeong, B.R. Effects of silicon growth of *Tagetes patula* L. 'Boy Orange' and 'Yellow Boy' seedlings cultured in an environment-controlled chamber. *Propag. Ornam. Plants* **2010**, *10*, 136–140.
25. Lim, M.Y.; Lee, E.J.; Jana, S.; Sivanesan, I.; Jeong, B.R. Effect of potassium silicate on growth and leaf epidermal characteristics of Begonia and Pansy grown in vitro. *Hortic. Sci. Technol.* **2012**, *30*, 579–585. [CrossRef]
26. Kalra, Y. *Handbook of Reference Methods for Plant Analysis*; CRC Press: Boston, MA, USA, 1998.
27. Buechel, T. Role of Nickel in Plant Culture. Available online: http://www.pthorticulture.com/en/training-center/role-of-nickel-in-plant-culture/ (accessed on 9 January 2019).
28. Fisher, R.A. A preliminary note on the effect of sodium silicate in increasing the yield of barley. *J. Agric. Sci.* **1929**, *19*, 132–139. [CrossRef]
29. Mali, M.; Avery, N.C. Influence of silicon on growth, relative water contents and uptake of silicon, calcium and potassium in wheat grown in nutrient solution. *J. Plant Nutr.* **2008**, *31*, 1867–1876. [CrossRef]
30. Friedman, H.; Bernstein, N.; Bruner, M.; Rot, I.; Ben-Noon, Z.; Zuriel, A.; Zuriel, R.; Finklestein, S.; Umiel, N.; Hagiladi, A. Application of secondary-treated effluents for cultivation of sunflower (*Helianthus annus* L.) and celosia (*Celosia argentea* L.) as cut flowers. *Sci. Hortic.* **2007**, *115*, 62–69. [CrossRef]

31. Ma, J.F.; Takahashi, E. Interaction between calcium and silicon in water-cultured rice plants. *Plant Soil* **1993**, *148*, 107–113. [CrossRef]
32. Bloodnick, E. Role of Silicon in Plant Culture. Available online: http://www.pthorticulture.com/en/training-center/role-of-silicon-in-plant-culture/ (accessed on 9 January 2019).
33. Kamenidou, S.; Cavins, T.J.; Marek, S. Evaluation of silicon as a nutritional supplement for greenhouse zinnia production. *Sci. Hortic.* **2009**, *119*, 297–301. [CrossRef]
34. Yoshida, S.; Ohnishi, Y.; Kitagishi, K. Chemical forms, mobility and deposition of silicon in rice plant. *Soil Sci. Plant Nutr.* **1962**, *8*, 15–21. [CrossRef]
35. Jana, S.; Jeong, B.R. Silicon: The most under-appreciated element in horticultural crops. *Hortic. Res.* **2014**, *4*, 1–19.
36. Saud, S.; Li, X.; Chen, Y.; Zhang, L.; Fahad, S.; Hussain, S.; Sadiq, A.; Chen, Y. Silicon application increases drought tolerance of Kentucky bluegrass by improving plant water relations and morphophysiological function. *Sci. World J.* **2014**, *23*, 17647–17655. [CrossRef] [PubMed]

18

Effect of LED Lighting and Gibberellic Acid Supplementation on Potted Ornamentals

Taylor Mills-Ibibofori [1], Bruce L. Dunn [1,*], Niels Maness [1] and Mark Payton [2]

1 Department of Horticulture & L.A., Oklahoma State University, Stillwater, OK 74078, USA
2 Department of Statistics, Oklahoma State University, Stillwater, OK 74078, USA
* Correspondence: bruce.dunn@okstate.edu

Abstract: Use of light emitting diode (LED) technology is beginning to replace traditional lighting in greenhouses. This research focused on the effects of LED lighting and gibberellic acid supplementation on growth and flowering of *Dahlia* spp. 'Karma Serena', *Liatris spicata* 'Kobold', and *Lilium asiatic* 'Yellow Cocotte'. Light treatments, used to extend photoperiod, included LED flowering lamps and halogen lamps that emitted a combination of red + far-red + white, red + white, and broad spectrum from late fall to early spring. Gibberellic acid treatments ranged from 40 to 340 mg L^{-1} for Asiatic lily 'Yellow Cocotte', 50 to 250 for gayfeather 'Kobold', and 50 to 150 for dahlia 'Karma Serena'. Results varied within species in response to light and gibberellic acid. A significant interaction of light with gibberellic acid influenced mean flower number and flowering percentage for dahlia 'Karma Serena', while flowering percentage and flower diameter were influenced for Asiatic lily 'Yellow Cocotte'. Effect of light was most significant on growth and flowering measurements, especially for gayfeather 'Kobold' and dahlia 'Karma Serena'. For gayfeather 'Kobold', flowering occurred two weeks earlier under sole LED lighting than under other light treatments and no supplemental light. Although flowering occurred the earliest for dahlia 'Karma Serena' under no supplemental light, plants under light treatments had greater height, width, and shoot weight. Significant effects of gibberellic acid on growth and flowering measurements for dahlia 'Karma Serena' and Asiatic lily 'Yellow Cocotte' were observed for height, width, and flower number.

Keywords: light emitting diodes; GA_3; extended photoperiod; greenhouse

---

1. Introduction

Light is the single most important variable with respect to plant growth and development and is often the most limiting factor in greenhouse production [1]. Therefore, using artificial lighting (AL) or grow lights (GL) in commercial greenhouses is beneficial for plants and growers. Altering photoperiod and increasing light levels are reasons for using these lights. The different lighting sources that growers can use include incandescent (INC) lamps, fluorescent lamps (FL), and high intensity discharge (HID) lamps. Light emitting diodes (LED) are fourth generation lighting sources and are the emerging technology in horticulture [2]. Before choosing a lighting device, several factors, such as efficiency, total energy emissions, life expectancy, and costs need to be considered. In addition, it is important to know the three most important light factors that affect plant growth, which are light quality, light intensity, and light duration [1]. LEDs have proven to be advantageous in all these factors when compared to traditional lighting sources [3].

Energy inputs range from 10% to 30% of total production costs for the greenhouse industry [4]. Thus, any new lighting technology that significantly reduces consumption of electricity for crop lighting, while maintaining or improving crop value is of great interest to growers. Light sources, such as fluorescent, metal halide, high pressure sodium, and incandescent lamps are generally used for plant growth under greenhouse conditions and have been around for half a century. However, these light sources have disadvantages, such as less suitable wavelength for plant growth and limited lifetime of operation. In addition, they require more electricity and produce heat that may injure plant leaves [5].

In the 1990s, light-emitting diodes (LEDs) were investigated for the first time for plant growth and were found to be efficient alternatives to traditional lamps used in lighting systems [6]. Compared with conventional lamps, LEDs are smaller in size and weight, have a long lifetime, low heat emissions, wavelength specificity, and much lower energy consumption [7]. In addition to changes in plant productivity, increased suppression of pathogens has been noted in tomato (*Solanum lycopersicum* L.) and cucumber (*Cucumis sativis* L.) [8]. Physiological and morphological effects of LEDs have been studied in several species, including potato (*Solanum tuberosum* L.), wheat (*Triticum aestivum* L.), lily (*Lilium candidum* L.), lettuce (*Lactuca sativa* L.), spinach (*Spinacia oleracea* L.), strawberry (*Fragaria* × *ananassa* Duchesne), marigold (*Tagetes erecta* L.), chrysanthemum (*Chrysanthemum indicum* L.), and salvia (*Salvia divinorum* Epling and Játiva) using various LED products [9].

Light-emitting diodes have the potential to shorten the crop time, reduce costs, and add new plants for specialty cut flower production during the winter [7]. This light source may also induce greater flowering for winter crops; however, research is limited to propagation, vegetables, and seedling production. Commercial LED fixtures for photoperiodic lighting have been recently developed for flowering applications and are alternatives to INC lamps. Craig and Runkle [10] quantified how red (R) to far-red (FR) ratio of photoperiodic lighting from LEDs influenced flowering and extended the growth of short-day plants. Kohyama [11] investigated the efficacy of commercial LED products developed for flowering applications on long-day plants. Meng and Runkle [12] coordinated grower trials to investigate the efficacy of R + white (W) + FR LEDs to regulate flowering of daylength-sensitive ornamental crops. For some plants, flowering is promoted with a combination of R and FR light [13,14].

Gibberellic acid (GA_3) is a hormone found in plants, which is produced in low amounts. Synthetic GA_3 is commonly used in commercial agriculture. This hormone is very influential and can control plant development, promote growth, and elongate cells. Gibberellic acid can also promote petal growth and enhance other flowering characteristics [15,16]. In certain plant species, GA_3 acts as a mobile signal transmitter for photoperiodic flowering stimulation [17]. For flower induction, soaking bulbs, rhizomes, corms, or spraying the foliage with a GA_3 solution are common applications [18–20]. There are limited but statistically valid interactions between light and GA_3. Both factors are known to have synergistic effects, but mainly on germination of seedlings [21,22]. In certain species, growth and flower initiation are affected by light and GA_3 application [23,24]. More current research needs to be conducted to assess the interaction of light and GA_3 further. Therefore, objectives of this study were to evaluate how gibberellic acid and different combinations of red and far-red light together from LED flowering lamps and halogen lamps, would influence growth and flowering of *Lilium* L., *Dahlia* Cav., and *Liatris* Gaertn. ex Schreb. species.

2. Materials and Methods

2.1. Plant Material and Culture

On 15 September 2015, bulbs of *Lilium asiatic* L. 'Yellow Cocotte' were graded at 16 to 19 cm. Cuttings of *Dahlia* spp. 'Karma Serena', which are short-day plants, arrived 14 October 2015. *Liatris spicata* (L.) Willd. 'Kobold' corms, which are long-day plants, arrived 12 November 2015 and were graded at 8 to 10 cm. Plant materials were obtained from a broker (Gloeckner and Company Incorporated, Harrison, NY, USA). Before transplanting, dahlia 'Karma Serena' cuttings were placed on a mist bench and Asiatic lily 'Yellow Cocotte' were placed in a cooler at 4 °C upon arrival for one month. Gayfeather 'Kobold' corms were immediately treated with GA_3 (Plant Hormones LLC, Auburn, WA, USA). All bulbs and corms were soaked in an aqueous solution of GA_3 for 30 min before being potted. Dahlia leaves were sprayed to glisten once with different rates of GA_3 solution after potting. Tween-20 (Sigma-Aldrich, St. Louis, MO, USA) was also added in the GA_3 solution as a surfactant at a concentration of 0.01%. The GA_3 treatment dates were 24 October 2015, 31 October 2015, and 13 November 2015 for 'Yellow Cocotte ', 'Karma Serena', and 'Kobold', respectively. Dahlia 'Karma Serena', Asiatic lily 'Yellow Cocotte', and liatris 'Kobold' were potted in standard 15 cm pots filled with Metro-Mix 360 media (Sun Gro Horticulture, Bellevue, WA, USA) and were placed in the greenhouses on 16 October 2015, 24 October 2015, and 12 November 2015, respectively.

2.2. *Experimental Arrangement*

The experiment was conducted at four research greenhouses of the Department of Horticulture and L.A. in Stillwater, OK. For each greenhouse, temperatures were set at 23 °C during the day and 18 °C during the night with a photosynthetic photon flux density (PPFD) between 600 to 1200 μmol m^2 s^{-1} and daily light integral of 10–15 mol m^2 d. One light treatment was established in each greenhouse. Light emitting diodes (Philips Green Power Flowering lamps, Amsterdam, The Netherlands) and standard halogen bulbs, which are broad spectrum across the photosynthetically active radiation region, were installed at 0.914 m above the bench area and 0.914 m apart. In the first light treatment, there were 19 14-watt LED R + W + FR flowering lamps (Phillips Lighting, Somerset, NJ, USA) with a spectrum from 420 to 780 nm and peaks at 660 (35%) and 740 (46%). The second light treatment had 11 15-watt LED R + W flowering lamps (Phillips Lighting, Somerset, NJ) with a spectrum from 420 to 720 nm and a peak at 660 (78%) and 12 40-watt halogen bulbs (Osram Sylvania, Wilmington, MA, USA) with a spectrum from 400 to 1200 nm with peaks at 600, 760, and 850 nm with lamps and bulbs installed alternatively. The third light treatment included 23 of the above mentioned 40-watt halogen bulbs, and the fourth treatment did not have lights (control). Plant species and GA_3 rates were randomized within light treatments. Plants were supplemented with seven hours of light after sunset. Before daylight savings time (8 November 2015), lighting was delivered from 1900 to 0200 HR. After daylight savings time, lighting was delivered between 1700 to 2400 HR using timers. A quantum sensor (Spectrum Technologies, Inc., Aurora, IL, USA) measured photosynthetic photon flux density (PPFD) of the LED lamps and halogen bulbs. In each greenhouse where the light was supplemented, measurements were randomly recorded across the bench area and were taken at pot level. The mean photon outputs were 10, 20, and 2 μmol m^{-2} s for LED emitting R + W + FR, LED emitting R + W, and halogen, respectively.

Gibberellic acid rates for gayfeather 'Kobold' were 50, 170, and 250 mg L^{-1} with 12 pots per rate per light treatment. Asiatic lily 'Yellow Cocotte' had rates of 40, 140, and 340 mg L^{-1} with 12 pots per rate per light source. Dahlia 'Karma Serena' rates were 50, 100, and 150 mg L^{-1} with 10 pots per rate per light source. All plants included a controlled rate in which water was used. Plants were watered with drip irrigation as needed. On 23 November 2015, a slow release fertilizer 16-9-12 (3–4 month, Osmocote® Plus, The Scotts Co., Marysville, OH, USA) at a rate of 10 g was added at time of potting and 200 mg L^{-1} 20-10-20 Peat-lite (Jacks, Allentown, PA, USA) water soluble fertilizer was supplemented after three weeks.

2.3. *Harvesting and Measurements*

Data collected from plants included the date of first flower (anthesis), which was only recorded when petals were fully opened. Flower diameter was recorded on 15 November 2015 for dahlia 'Karma Serena' and 22 December 2015 for Asiatic lily 'Yellow Cocotte' using a digital caliper (Tresna Instrument., LTD, Guilin, China). Flowering percent (flowering or not per pot), Number of flowers, plant height (from media surface to tallest flower or bud), and width (average of two perpendicular measurements) were recorded on 18 January 2016 for dahlia 'Karma Serena', 22 Feburary 2016 for Asiatic lily 'Yellow Cocotte', and 27 Feburary 2016 for gayfeather 'Kobold'. Shoot dry weight was recorded on 1 Feburary 2016 for dahlia 'Karma Serena', 29 Feburary 2016 for Asiatic lily 'Yellow Cocotte', and 7 March 2016 for gayfeather 'Kobold' by cutting the stems at the media level, and drying for 3 d at 54.4 °C.

2.4. *Statistical Analysis*

Pots were arranged in a completely randomized design with plant species, GA_3 and light treatments as the specified factors. Data were analyzed with SAS version 9.4 software (SAS Institute, Cary, NC, USA). An analysis of variance methods (PROC MIXED) was used with a two-factor factorial arrangement with light and GA_3 as the factors of interest. For percentage response variables, arcsine square root transformations were used to help normalize the data. Because the levels of the factors changed, separate analyses were conducted for each plant species. When interactions of light with GA_3 were significant, simple effects were reported. Mean separations were determined using protected Fisher-type comparisons (a DIFF option in an LSMEANS statement and a SLICE option when appropriate) and with 0.05, 0.01, 0.001, and 0.0001 levels of significance.

3. Results

3.1. *Liatris spicata 'Kobold'*

A main effect of light was found on all growth measurements, as well as on a number of terminal spikes and days to anthesis (Table 1). Plants under LEDs flowered the earliest, but were not different than halogen or LED + halogen (Table 2). The average number of spikes was greatest with natural light, which was not different than halogen. Plant height and width was greatest under LED and LED + halogen. Shoot dry weight was greatest with halogen lighting. Gibberellic acid rates had a significant effect on plant width, shoot dry weight, and mean spike number (Table 1). For width, plants receiving 0 mg L^{-1} GA_3 were greatest, but were not different from those treated at 50 and 170 mg L^{-1} GA_3 (Table 3). Shoot weight was greatest for 0 mg L^{-1} GA_3, but was not different from 50 and 250 mg L^{-1} GA_3. The average number of spikes was greatest at 250 mg L^{-1} GA_3, but was not different than 0 or 170 mg L^{-1} GA_3.

Table 1. Analysis of variance for growth and flowering measurements of *Liatris spicata* 'Kobold', *Dahlia* spp. 'Karma Serena', and *Lilium asiatica* 'Yellow Cocotte' grown with LED and halogen lights along with multiple rates of gibberellic acid.

Cultivar	Source	Height (cm)	Width (cm)	Shoot Dry Weight (g)	Flowers/Spikes Number [z]	Flower Diameter	Days to Anthesis	Flowering (%)
'Kobold'	Light	**** [y]	****	****	****	- [x]	****	ns
	GA_3	ns	**	*	*	-	ns	ns
	Light × GA_3	ns	ns	ns	ns	-	ns	ns
'Karma Serena'	Light	****	****	****	****	ns	****	ns
	GA_3	****	ns	ns	**	*	ns	ns
	Light × GA_3	ns	ns	ns	*	ns	ns	*
'Yellow Cocotte'	Light	ns	ns	ns	ns	ns	ns	ns
	GA_3	ns	ns	ns	ns	ns	ns	ns
	Light × GA_3	ns	ns	ns	ns	*	ns	*

[z] Number of flowers for 'Karma Serena' and 'Yellow Cocotte', but the number of spikes for 'Kobold'. [y] NS, *, **, ***, **** indicate non-significant or significant at $p \leq 0.05, 0.01, 0.001, 0.0001$, respectively. [x] Data not taken.

Table 2. Growth and flowering measurements of *Liatris spicata* 'Kobold', *Dahlia* spp. 'Karma Serena', and *Lilium asiatica* 'Yellow Cocotte' affected by light averaged across GA_3.

Light Type	Height (cm)	Width (cm)	Shoot Dry Weight (g)	Flower Measurements [z]	Days to Anthesis	Flowering (%)
			'Kobold'			
Control	47.3b [y]	35.2c	13.9b	3.5a	88a	96a
LED	64.7a	49.4a	17.2b	2.3bc	70b	100a
Halogen	52.1b	40.9b	22.0a	3.1ab	73b	98a
LED + Halogen	65.9a	44.9ab	16.8b	1.8c	77ab	98a
			'Karma Serena'			
Control	58.9b	32.5c	9.1d	7.1a	46c	- [x]
LED	67.1b	43.9b	35.0c	7.1a	61b	-
Halogen	95.8a	46.7b	43.6b	8.5a	74a	-
LED + Halogen	85.9a	56.9a	52.9a	7.9a	80a	-
			'Yellow Cocotte'			
Control	45.5a	15.0a	4.0a	2.4a	54a	-
LED	44.5a	19.6a	3.5a	2.0a	47a	-
Halogen	38.4a	16.3a	4.2a	2.0a	43a	-
LED + Halogen	54.1a	19.8a	4.8a	2.1a	55a	-

[z] Mean number of flower spikes for 'Kobold', flower diameter (cm) for 'Karma Serena', and flower number for 'Yellow Cocotte'. [y] Means ($n = 12$ for 'Kobold' and 'Yellow Cocotte'; $n = 10$ for 'Karma Serena') with the same letter within the same column are not statistically significant ($p < 0.05$). [x] Interaction significant for plant measurements.

Table 3. Growth and flowering measurements of *Liatris spicata* 'Kobold', *Dahlia* spp. 'Karma Serena', and *Lilium asiatica* 'Yellow Cocotte' affected by GA_3 averaged across the light.

GA_3 Rate (mg L^{-1})	Height (cm)	Width (cm)	Shoot Dry Weight (g)	Flower Measurements [z]	Days to Anthesis	Flowering (%)
			'Kobold'			
0	59.7a [y]	47.4a	19.8a	2.4ab	77a	98a
50	59.5a	43.2ab	17.6ab	2.3b	76a	94a
170	54.6a	40.6ab	15.2b	2.6ab	78a	100a
250	56.3a	39.3b	17.3ab	3.5a	76a	100a
			'Karma Serena'			
0	65.0b	45.5a	30.5a	8.6a	67a	- [x]
50	81.0a	45.7a	35.8a	7.3ab	62a	-
100	81.3a	45.5a	38.5a	6.8b	64a	-
150	80.3a	43.4a	35.7a	7.8ab	69a	-
			'Yellow Cocotte'			
0	48.5a	19.6a	4.5a	2.1a	- [x]	- [x]
40	47.2a	16.8a	4.4a	2.4a	-	-
140	42.9a	17.3a	3.9a	2.0a	-	-
340	43.4a	17.0a	3.7a	2.0a	-	-

[z] Mean number of flower spikes for 'Kobold', flower diameter (cm) for 'Karma Serena', and flower number for 'Yellow Cocotte'. [y] Means ($n = 12$ for 'Kobold' and 'Yellow Cocotte'; $n = 10$ for 'Karma Serena') with the same letter within the same column are not statistically significant ($p < 0.05$). [x] Interaction significant for plant measurements.

3.2. *Dahlia spp.'Karma Serena'*

There was a significant Light × GA_3 interaction for mean flower number and flowering percentage (Table 1). Flower number within the 50 mg L^{-1} GA_3 rate was greatest for plants under halogen, LED + halogen, and no supplemental light (Table 4). Plants treated with 100 mg L^{-1} GA_3 treatment, no supplemental light, LEDs, and halogen had the greatest number of flowers. The flowering percentage within the 50 and 150 mg L^{-1} GA_3 rates was greatest with no supplemental lighting, halogen, and

LED + halogen light. Plants treated with 100 mg L^{-1} GA_3 treatment, flowering was greatest with natural light, LED, and halogen lighting. The light had a significant effect on height, width, shoot dry weight, and days to anthesis (Table 1). Time to flower was longest under halogen and LED + halogen (Table 2). Height was greatest under halogen, which was not different than LED + halogen. Plant width and shoot dry weight were greatest under LED + halogen. Only height and flower diameter were significantly affected by GA_3 (Table 1). All GA_3 rates produced taller plants compared to no supplemental lighting. No supplemental lighting had the greatest number of flowers though 50 and 150 mg L^{-1} GA_3 were not different (Table 3).

Table 4. Mean flower number and flowering percent of *Dahlia* spp. 'Karma Serena' and *Lilium asiatica* 'Yellow Cocotte' affected by the interaction of light with GA_3.

Plant	Characteristic	Source	GA_3 (mg L^{-1})			
			0	50	100	150
'Karma Serena'	Flower number	Control	3.1c [z]	2.4b	2.3b	2.8a
		LED	2.2c	2.9b	3.1ab	3.1a
		Halogen	6.6a	5.4a	4.4a	3.7a
		LED + Halogen	5.1b	4.5a	4.4a	2.3a
	Flowering percent	Control	100a	100a	100a	100a
		LED	100a	89b	100a	80b
		Halogen	100a	100a	100a	100a
		LED + Halogen	100a	100a	80b	100a
'Yellow Cocotte'			0	40	140	340
	Flower diameter	Control	8.9b	9.2b	9.7a	9.7b
		LED	10.4a	9.8b	10.4a	10.1ab
		Halogen	9.8b	9.4b	10.1a	9.5b
		LED + Halogen	10.5a	10.7a	10.0a	10.9a
	Flowering percent	Control	58b	67b	58bc	75ab
		LED	100a	67b	75a	50bc
		Halogen	75ab	75a	33c	33c
		LED + Halogen	58b	75a	67ab	100a

[z] Means ($n = 10$ for 'Karma Serena'; $n = 12$ for 'Yellow Cocotte') with the same letter within the same column and within plant characteristic are not statistically significant ($p < 0.05$).

3.3. *Lilium Asiatic 'Yellow Cocotte'*

The interaction of Light × GA_3 was seen on flower diameter and flowering percentage (Table 1). Plants treated with 0 mg L^{-1} GA_3 rate, LED and LED + halogen had the greatest flower diameter (Table 4). Plants treated with 40 mg L^{-1} GA_3 rate had the greatest flower diameter under LED + halogen. Plants treated with 340 mg L^{-1} GA_3 rate, plants under LED and LED + halogen had the greatest flower diameters. The flowering percentage was greatest with halogen within the 0 mg L^{-1} GA_3 rate, but was not different from halogen. Plants treated with 40 mg L^{-1} GA_3 rate, plants with halogen and LED + halogen had the greatest flowering percentage. Plants treated with 140 mg L^{-1} GA_3 rate, plants with LED had the greatest flowering percentage, but were not different from LED + halogen. Plants treated with 340 mg L^{-1} GA_3 rate, plants under LED + halogen had the greatest flowering percentage, but were not different from natural lighting. No significant effects were seen by light or GA_3 as main effects on other growth and flowering measurements of 'Yellow Cocotte'.

4. Discussion

The use of LED, LED + halogen, and sole halogen lamps emitting R and FR light effectively promoted growth and flowering in gayfeather 'Kobold' and dahlia 'Karma Serena'. Red light is the most effective at inhibiting flowering in short-day plants (SDP). This was true for dahlia under LED, halogen, and LED + halogen (Table 2). Craig and Runkle [10] reported that flowering in SDPs, such as chrysanthemum (*Chrysanthemum indicum* L.) and dahlia was delayed under incandescent and LED lights. Inhibition of flowering by R light was also seen in cocklebur (*Xanthium strumarium* L.), chrysanthemum, and soybean (*Glycine max* L. Merr.) [25–27]. Delaying flowering in SDPs, such as dahlia especially during the winter months is ideal. During this season, the days are shorter, and the nights are longer. Therefore, SDPs will want to spend photosynthates in the production of reproductive organs, which will result in a lack of growth and development of vegetative parts. Extended growth and greater biomass are promoted under R light, and this was seen for liatris and dahlia under LED flowering lamps and halogen lamps (Table 2). Miyashita et al., [28] noted that R light from LEDs increased shoot length of potato (*Solanum tuberosum* L.) plantlets. Height was also greatest under either LED flowering lamps emitting R + W or R + W + FR, as well as incandescent lamps in ageratum (*Ageratum houstonianum* L.), calibrachoa (*Calibrachoa* x *hybrida* Cerv.), dianthus (*Dianthus* L.), and petunia (*Petunia* x *hybrida* Juss.). Height and shoot dry weight were greatest for salvia (*Salvia splendens* Sellow ex J.A. Schultes) and tomato (*Solanum lyopersicum* L.) under LEDs emitting red [29]. Meng and Runkle [12] reported that the stem length of verbena (*Verbena* x *hybrid* L.) increased under incandescent and LED flowering lamps compared to the control. Dry weight and plant width increased in poinsettia (*Euphorbia pulcherrima* Willd. ex Klotzsch) when grown under supplemental LED lighting emitting R and blue [30]. An increase in all these growth parameters is beneficial for cut flowers.

A combination of R + FR is effective for promoting flowering in long-day plants (LDP). This was true for liatris that were under sole LED lighting emitting R + W + FR (Table 2). Meng and Runkle [12] have also reported that photoperiodic lighting with a mixture of R and FR light from LEDs and incandescent lamps was most effective at promoting flowering in LDPs. The flowering of *Gypsophila paniculata* (L.) 'Baby's Breath' and *Eustoma grandiflorum* (Salisb.) 'Lisianthus' was also promoted under a combination of R and FR light [13,14]. The presence of FR in LED lamps shortened the flowering time and increased number of flowers in petunia. Hastening of flowering, while maintaining plant quality, will decrease the costs of labor and inputs, as well as assure an early market season. Neither R nor FR light from the lamps influenced flowering in Asiatic lily 'Yellow Cocotte'. Bieleski et al. [31] also reported that the use of R light as a night-break was not effective for increasing anthesis or flower bud opening in multiple cultivars of Asiatic lilies. It was also noted that flowering in lilies was more influenced by variations in day-length and not night interruption with supplemental lighting.

Gibberellic acid (GA_3) effectively promoted growth and flowering measurements in gayfeather 'Kobold', lily 'Karma Serena', and Asiatic lily 'Yellow Cocotte'. Previous research has noted the presence and influence of GA_3 in growing tissues, shoot apices, leaves, and flowers [32]. Cell division and expansion are stimulated by GA_3, especially in response to light or darkness [33]. Flower initiation, development, sex expression, and number are also regulated by GA_3 [34]. Bulyalert [35] reported that exogenous applications of GA_3 increased width and height, as well as the flowering percentage in liatris. The significant effect of GA_3 on flower diameter and height in three cultivars of dahlia was not analyzed, but an increase in these features was observed and reported [36]. Flower diameter was also increased in Asiatic hybrid cut lily flowers when treated with GA_3 and a standard preservative [37]. The following studies have reported similar results in other cut flowers. Application of GA_3 promoted shoot elongation in different cultivars of chrysanthemums [38,39]. Foliar application of GA_3 increased stem length in a variety of cut flower cultivars that were field-grown [40]. Bultynck and Lambers [41] reported that the addition of exogenous GA_3 promoted leaf elongation and increased shoot biomass in *Aegilops caudata* (L.) and *Aegilops tauschii* (L.). Pobudkiewicz and Nowak [42] found that flowering size of gerbera (*Gerbera jamesonni* Hooker f.) was enhanced when GA_3 was applied at 200 mg L^{-1}. Mean flower number was increased in philodendron (*Philodendron* Schott) 'Black Cardinal' as GA_3 concentrations increased [43]. Dobrowolska and Janicka [44] also reported that application of GA_3 at a concentration of 10 mg dm^{-3} increased flower number in *Impatiens hawkeri* (L.) 'Riviera Pink'.

Interaction of light with GA_3 effectively promoted growth and flowering measurements of dahlia 'Karma Serena' and Asiatic lily 'Yellow Cocotte'. Yamaguchi and Kamiya [45] have concluded that light and GA_3 are highly interactive and are involved in the same pathways that regulate germination and dormancy. Light and GA_3 are likely interacting with similar pathways regulating growth and flowering. A study reported that cell expansion was promoted in the leaves of dwarf bean (*Phaseolus vulgaris* L.)

and stem elongation was increased in garden peas (*Pisum sativa* L.) when exposed to FR light and saturated with GA_3 [46]. In Kentucky bluegrass (*Poa pratensis* L.), shoot elongation was increased when endogenous levels of GA_3 interacted with light [47]. Williams and Morgan [48] noted that the exposure of GA_3 to FR light hastened flowering in sorghum (*Sorghum bicolor* L.). White et al. [49] reported that although potted greenhouse plants *Aquilegia* × *hybrida* (L.) 'Bluebird' and 'Robin' all flowered when treated with 100 mg L^{-1} exogenous GA_3, there was no synergistic effect with the supplemental lights emitting R and FR. An increase in flower number was also observed, but not due to an interaction of light with GA_3. Another study reported that GA_3 should be applied to plants before cold temperature exposure and light treatments should be applied after cold temperature exposure to improve floral development. There could be even more of an effect between light and GA_3 on lily bulbs based on exposure to cold temperatures before or after as Asiatic lily 'Yellow Cocotte' were the only plants exposed to a cold treatment before applications of GA_3 and light treatments. Possibly, the exposure to cold temperatures before GA_3 treatment contributed to the lack of growth and flowering rates.

5. Conclusions

Light emitting diode flowering lamps are equally effective as halogen lamps at regulating growth and flowering. Although the LED flowering lamps and halogen bulbs have similar light intensity, the energy consumption of LEDs was 14 to 15 watts per lamp, whereas halogen bulbs use considerably more watts per bulb. Not only was there an improvement in energy use, but the quality of plants was maintained and improved with the use of LED flowering lamps. Results of this study and that of many others show that GA_3 also plays an important role in flowering stimulation, as well as plant growth. In addition, light and GA_3 have a synergistic relationship with each other regarding plant and flower development of plants. More research needs to be conducted using an array of LED flowering lamps with different spectrums, and in combination with the plant hormone GA_3 to control plant growth and flowering, as affects are species dependent.

Author Contributions: Conceptualization, B.L.D.; T.M.-I.; methodology, T.M.-I.; formal analysis, M.P.; investigation, T.M.-I.; writing—original draft preparation, T.M.-I.; writing—review and editing, B.L.D and N.M.; supervision, B.L.D.

Acknowledgments: We thank Stephen Stanphill for helping with data collection and greenhouse management.

References

1. Nelson, P. *Greenhouse Operation and Management*, 7th ed.; Pearson: Boston, MA, USA, 2012.
2. Morrow, R. LED lighting in horticulture. *HortScience* **2008**, *43*, 1947–1950. [CrossRef]
3. Bourget, C. An introduction to light-emitting diodes. *HortScience* **2008**, *43*, 1944–1946. [CrossRef]
4. Bessho, M.; Shimizu, K. Latest trends in LED lighting. *Electron. Commum. Jpn.* **2012**, *95*, 315–320. [CrossRef]
5. Singh, D.; Basu, C.; Meinhardt-Wollweber, M.; Roth, B. LEDs for Energy Efficient Greenhouse Lighting. Available online: https://arxiv.org/abs/1406.3016 (accessed on 9 October 2017).
6. Briggs, W.R.; Christie, J.M. Phototropin 1 and phototropin 2: Two versatile plant blue-light receptors. *Trends Plant Sci.* **2002**, *7*, 204–209. [CrossRef]
7. Massa, G.D.; Kim, H.H.; Wheeler, R.M.; Mitchell, C.A. Plant productivity in response to LED lighting. *HortScience* **2008**, *43*, 1951–1956. [CrossRef]
8. Kim, H.H.; Wheeler, R.M.; Sager, J.C.; Yorio, N.C.; Goins, G.D. Light-emitting diodes as an illumination source for plants: A review of research at Kennedy Space Center. *Habitation (Elmsford)* **2005**, *10*, 71–78. [CrossRef] [PubMed]
9. Heo, J.; Lee, C.; Chakrabarty, D.; Paek, K. Growth responses of marigold and salvia bedding plants as affected by monochromic or mixture radiation provided by a light-emitting diode (LED). *Plant Growth Regul.* **2002**, *38*, 225–230. [CrossRef]

10. Craig, D.S.; Runkle, E.S. A moderate to high red to far-red light ratio from light-emitting diodes controls flowering of short-day plants. *J. Am. Soc. Hortic. Sci.* **2013**, *138*, 167–172. [CrossRef]
11. Kohyama, F.; Whitman, C.; Runkle, E.S. Comparing flowering responses of long-day plants under incandescent and two commercial light-emitting diode lamps. *HortTechnology* **2014**, *24*, 490–495. [CrossRef]
12. Meng, Q.; Runkle, E.S. Controlling flowering of photoperiodic ornamental crops with light-emitting diode lamps: A coordinated grower trial. *HortTechnology* **2014**, *24*, 702–711. [CrossRef]
13. Nishidate, K.; Kanayama, Y.; Nishiyama, M.; Yamamoto, T.; Hamaguchi, Y.; Kanahama, K. Far-red light supplemented with weak red light promotes flowering of Gypsophila paniculata. *J. Jpn. Soc. Hortic. Sci.* **2012**, *81*, 198–203. [CrossRef]
14. Yamada, A.; Tanigawa, T.; Suyama, T.; Matsuno, T.; Kunitake, T. Red:Far-red light ratio and far-red light integral promote or retard growth and flowering in Eustoma grandiflorum (Raf.) Shinn. *Sci. Hortic.* **2009**, *120*, 101–106. [CrossRef]
15. Hu, J.; Mitchum, M.G.; Barnaby, N.; Ayele, B.T.; Ogawa, M.; Nam, E.; Lai, W.C.; Hanada, A.; Alonso, J.M.; Ecker, J.R.; et al. Potential sites of bioactive gibberellin production during reproductive growth in Arabidopsis. *Plant Cell* **2008**, *20*, 320–336. [CrossRef] [PubMed]
16. Gupta, R.; Chakrabarty, S.K. Gibberellic acid in plant: Still a mystery unresolved. *Plant Signal. Behav.* **2013**, *8*, 1–5. [CrossRef] [PubMed]
17. Kobayashi, Y.; Weigel, D. Move on up, its time for change: Mobile signals controlling photoperiod-dependent flowering. *Genes Dev.* **2007**, *21*, 2371–2384. [CrossRef] [PubMed]
18. Dennis, D.J.; Doreen, J.; Ohteki, T. Effect of gibberellic acid 'quick-dip' and storage on the yield and quality of blooms from hybrid Zantedeschia tubers. *Sci. Hortic.* **1994**, *57*, 133–142. [CrossRef]
19. Delvadia, D.V.; Ahlawat, T.R.; Meena, B.J. Effect of different GA_3 concentration and frequency on growth, flowering and yield in Gaillardia (Gaillardia pulchella Foug.) cv. Lorenziana. *J. Hortic. Sci.* **2009**, *4*, 81–84.
20. Ranwala, A.P.; Legnani, G.; Reitmeier, M.; Stewart, B.B.; Miller, W.B. Efficacy of plant growth retardants as preplant bulb dips for height control in LA and oriental hybrid lilies. *HortTechnology* **2002**, *12*, 426–431. [CrossRef]
21. Dissanayake, P.; George, D.L.; Gupta, M.L. Effect of light, gibberellic acid and abscisic acid on germination of guayule (Parthenium argentatum Gray) seed. *Ind. Crop Prod.* **2010**, *32*, 111–117. [CrossRef]
22. Toyomasue, T.; Tsuji, H.; Yamane, H.; Nakayama, M.; Yamaguchi, I.; Murofushi, N.; Takahasi, N.; Inoue, Y. Light effect on endogenous levels of gibberellins in photoblastic lettuce seeds. *J. Plant Growth Regul.* **1993**, *12*, 85–90. [CrossRef]
23. Lona, F.; Bocchi, A. Luterferenza dell'acido gibberellieo nell'effecto della lute rossa e rosso-estrema sull'allungamento dell fusto di Perilla ocy~noides L. *L'ateneo Parmense* **1956**, *7*, 645–649.
24. Lockhart, J. A reversal of the light inhibition of pea stem growth by the gibberellins. *Proc. Natl. Acad. Sci. USA* **1956**, *42*, 841–848. [CrossRef]
25. Borthwick, H.A.; Hendricks, S.B.; Parker, M.W. The reaction controlling floral initiation. *Proc. Natl. Acad. Sci. USA* **1952**, *38*, 929–934. [CrossRef]
26. Cathey, H.M.; Borthwick, H.A. Photoreversibility of floral initiation in Chrysanthemum. *Bot. Gaz.* **1957**, *119*, 71–76. [CrossRef]
27. Downs, R.J.; Borthwick, H.A.; Piringer, A.A. Comparison of incandescent and fluorescent lamps for lengthening photoperiods. *Proc. Am. Soc. Hortic. Sci.* **1958**, *71*, 568–578.
28. Miyashita, Y.; Kitaya, Y.; Kubota, C.; Kozai, T.; Kimura, T. Effects of red and far-red light on the growth and morphology of potato plantlets in-vitro: Using light emitting diodes as a light source for micropropagation. *Acta Hortic.* **1995**, *393*, 189–194. [CrossRef]
29. Wollaeger, H.M.; Runkle, E.S. Growth and acclimation of impatiens, salvia, petunia, and tomato seedlings to blue and red light. *HortScience* **2015**, *50*, 522–529. [CrossRef]
30. Bergstrand, K.J.; Asp, H.; Larsson-Jonsson, E.H.; Schussler, H.K. Plant developmental consequences of lighting from above or below in the production of poinsettia. *Eur. J. Hortic. Sci.* **2015**, *80*, 51–55. [CrossRef]
31. Bieleski, R.; Elgar, J.; Heyes, J.; Woolf, A. Flower opening in Asiatic lily is a rapid process controlled by dark-light cycling. *Ann. Bot.* **2000**, *86*, 1169–1174. [CrossRef]
32. Jones, R.L.; Phillips, I.D. Organs of gibberellin synthesis in light-grown sunflower plants. *Plant Physiol.* **1966**, *41*, 1381–1386. [CrossRef]

33. Feng, S.; Martinez, C.; Gusmaroli, G.; Wang, Y.; Zhou, J.; Wang, F. Coordinated regulation of Arabidopsis thaliana development by light and gibberellins. *Nature* **2008**, *451*, 475–479. [CrossRef]
34. Griffiths, J.; Murase, K.; Rieu, I. Genetic characterization and functional analysis of the GID1 gibberellin receptors in Arabidopsis. *Plant Cell* **2006**, *18*, 3399–3414. [CrossRef]
35. Bulyalert, O. Effect of Gibberellic Acid on Growth and Flowering of Liatris Corm (*Liatris spicata*) c.v. Florist Violet Propagated from Seed (Abstract). Available online: http://agris.fao.org/agris-search/search.do?recordID=TH9220028 (accessed on 9 October 2017).
36. Pudelska, K.; Podgajna, E. Decorative value of three dahlia cultivars (Dahlia cultorum Thorsr. et Reis) treated with gibberellin. *Mod. Phytomorphol.* **2013**, *4*, 83–86.
37. Rabiza-Swider, J.; Skutnik, E.; Jedrzejuk, A.; Lukaszewska, A.; Lewandowska, K. The effect of GA_3 and the standard preservative on keeping qualities of cut LA hybrid lily Richmond. *Acta Sci. Pol. Hortorum Cultus.* **2015**, *14*, 51–64.
38. Schmidt, C.; Bellé, A.B.; Nardi, C.; Toledo, A.K. The gibberellic acid (GA_3) in the cut chrysanthemum (*Dedranthema grandiflora* Tzevelev.) Viking: Planting summer/autumn. *Ciência Rural* **2003**, *33*, 267–274. [CrossRef]
39. Zalewska, M.; Żabicka, A.; Wojciechowska, I. The influence of gibberellic acid on the growth and flowering of cascade chrysanthemum cultivars in outside glasshouse. *Zesz. Probl. Post. Nauk. Roln.* **2008**, *525*, 525–533.
40. Bergmann, B.A.; Dole, J.M.; McCall, I. Gibberellic acid shows promise for promoting flower stem length in four field-grown cut flowers. *HortTechnology* **2016**, *26*, 287–292.
41. Bultynck, L.; Lambers, H. Effects of applied gibberellic acid and paclobbutraol on leaf expansion and biomass allocation in two Aegilops species with contrasting leaf elongation rates. *Physiol. Plant.* **2004**, *122*, 143–151. [CrossRef]
42. Pobudkiewicz, A.; Nowak, J. The effect of gibberellic acid on growth and flowering of Gerbera jamesonii Bolus. *Folia Hortic.* **1992**, *4*, 35–42.
43. Chen, J.; Henny, R.J.; McConnell, D.B.; Caldwell, R.D. Gibberellic acid affects growth and flowering of *Philodendron* Black Cardinal. *J. Plant Growth Regul.* **2003**, *41*, 1–6. [CrossRef]
44. Dobrowolska, A.; Janicka, D. The effect of growth regulators on flowering and decorative value of Impatiens hawkeri W. Bull belonging to Riviera group. *Rocz. AR Pozn. Ogrodn.* **2007**, *41*, 35–39.
45. Yamaguchi, S.; Kamiya, Y. Gibberellins and light-simulated seed germination. *J. Plant Growth Regul.* **2001**, *20*, 369–376. [CrossRef]
46. Vince, D. Gibberellic acid and light inhibition of stem elongation. *Planta (Berlin)* **1967**, *75*, 291–308. [CrossRef]
47. Tan, Z.G.; Qian, Y.L. Light intensity affects gibberellic acid content in Kentucky bluegrass. *HortScience* **2003**, *38*, 113–116. [CrossRef]
48. Williams, E.A.; Morgan, P.W. Floral initiation in sorghum hastened by gibberellic acid and far-red light. *Planta* **1979**, *145*, 269–272. [CrossRef]
49. White, J.W.; Chen, H.; Beattie, D.J. Gibberellin, light, and low temperature effects on flowering of Aquilegia. *HortScience* **1990**, *25*, 1422–1424.

Permissions

All chapters in this book were first published by MDPI; hereby published with permission under the Creative Commons Attribution License or equivalent. Every chapter published in this book has been scrutinized by our experts. Their significance has been extensively debated. The topics covered herein carry significant findings which will fuel the growth of the discipline. They may even be implemented as practical applications or may be referred to as a beginning point for another development.

The contributors of this book come from diverse backgrounds, making this book a truly international effort. This book will bring forth new frontiers with its revolutionizing research information and detailed analysis of the nascent developments around the world.

We would like to thank all the contributing authors for lending their expertise to make the book truly unique. They have played a crucial role in the development of this book. Without their invaluable contributions this book wouldn't have been possible. They have made vital efforts to compile up to date information on the varied aspects of this subject to make this book a valuable addition to the collection of many professionals and students.

This book was conceptualized with the vision of imparting up-to-date information and advanced data in this field. To ensure the same, a matchless editorial board was set up. Every individual on the board went through rigorous rounds of assessment to prove their worth. After which they invested a large part of their time researching and compiling the most relevant data for our readers.

The editorial board has been involved in producing this book since its inception. They have spent rigorous hours researching and exploring the diverse topics which have resulted in the successful publishing of this book. They have passed on their knowledge of decades through this book. To expedite this challenging task, the publisher supported the team at every step. A small team of assistant editors was also appointed to further simplify the editing procedure and attain best results for the readers.

Apart from the editorial board, the designing team has also invested a significant amount of their time in understanding the subject and creating the most relevant covers. They scrutinized every image to scout for the most suitable representation of the subject and create an appropriate cover for the book.

The publishing team has been an ardent support to the editorial, designing and production team. Their endless efforts to recruit the best for this project, has resulted in the accomplishment of this book. They are a veteran in the field of academics and their pool of knowledge is as vast as their experience in printing. Their expertise and guidance has proved useful at every step. Their uncompromising quality standards have made this book an exceptional effort. Their encouragement from time to time has been an inspiration for everyone.

The publisher and the editorial board hope that this book will prove to be a valuable piece of knowledge for researchers, students, practitioners and scholars across the globe.

List of Contributors

Ignasi Riera-Vila, Neil O. Anderson, Claire Flavin Hodge and Mary Rogers
Department of Horticultural Sciences, University of Minnesota, Saint Paul, MN 55108, USA

Stefania Toscano and Daniela Romano
Department of Agriculture, Food and Environment (Di3A), Università degli Studi di Catania, Via Valdisavoia 5, 95123 Catania, Italy

Antonio Ferrante
Department of Agricultural and Environmental Sciences, Università degli Studi di Milano, Via Celoria 2, 1-20133 Milano, Italy

Qiansheng Li and Mengmeng Gu
Department of Horticultural Sciences, Texas A & M AgriLife Extension Service, Texas A & M University, College Station, TX 77843, USA

Xiaoqiang Li
Shanghai Jieyou Agriculture Sci & Tech Co., Ltd., Shanghai 201210, China

Bin Tang
Spraying Systems (Shanghai) Co., Shanghai 201611, China

Jennifer K. Boldt and James E. Altland
United States Department of Agriculture, Agricultural Research Service, Application Technology Research Unit, Wooster, OH 44691, USA

Georgios Nikolaou, Nikolaos Katsoulas and Constantinos Kittas
Department of Agriculture Crop Production and Rural Environment, School of Agricultural Sciences, University of Thessaly, Fytokou Str., 38446 Volos, Greece

Damianos Neocleous
Department of Natural Resources and Environment, Agricultural Research Institute, 1516 Nicosia, Cyprus

Luigi Mariani
Lombardy Museum of Agricultural History, via Celoria 2, 20133 Milan, Italy
Department of Agricultural and Environmental Sciences, Università degli Studi di Milano, via Celoria 2, 20133 Milan, Italy

Nadezhda A. Golubkina, Timofey M. Seredin and Marina S. Antoshkina
Agrochemical Research Center, Federal Scientific Center of Vegetable Production, Odintsovo District, Vniissok, Selectsionnaya 14, Moscow 143072, Russia

Olga V. Kosheleva
Federal Research Centre of Nutrition, Biotechnology and Food Safety, Ustinsky pr., 2/14, Moscow 109240, Russia

Gabriel C. Teliban
Department of Horticulture Technology, University of Agriculture Sciences and Veterinary Medicine, 3 M. Sadoveanu, 700490 Iasi, Romania

Gianluca Caruso
Department of Agricultural Sciences, University of Naples Federico II, 80055 Portici, Naples, Italy

William L. Sublett and T. Casey Barickman
North Mississippi Research and Extension Center, Department of Plant and Soil Sciences, Mississippi State University, Verona, MS 38879, USA

Carl E. Sams
Department of Plant Sciences, The University of Tennessee, Knoxville, TN 37996, USA

Marta Guarise
Department of Agriculture and Environmental Science—Production, Landscape, Agroenergy, Università degli Studi di Milano, 20133 Milano, Italy

Gigliola Borgonovo and Angela Bassoli
Department of Food, Environmental and Nutritional Science, Università degli Studi di Milano, 20133 Milano, Italy

Elazar Fallik, Sharon Alkalai-Tuvia, Daniel Chalupowicz and Merav Zaaroor-Presman
Agricultural Research Organization, The Volcani Center, Department of Postharvest Science of Fresh Produce, Rishon Leziyyon 7505101, Israel

Rivka Offenbach, Shabtai Cohen and Effi Tripler
Central and Northern Arava Research and Development, Arava Sapir 8682500, Israel

Angeliki Elvanidi
Department of Agriculture Crop Production and Rural Environment, University of Thessaly, Fytokou Str., 38446 Volos, Greece

Luigi Mariani
Department of Agricultural and Environmental Sciences, Università degli Studi di Milano, via Celoria 2, 20133 Milan, Italy
Lombardy Museum of Agricultural History, via Celoria 2, 20133 Milan, Italy

Filippos Bantis, Athanasios Koukounaras, Anastasios Siomos and Georgios Menexes
School of Agriculture, Aristotle University of Thessaloniki, 54124 Thessaloniki, Greece

Christodoulos Dangitsis and Damianos Kintzonidis
Agris S.A., Kleidi, 59300 Imathia, Greece

Ángela S. Prudencio, Federico Dicenta and Pedro Martínez-Gómez
Department of Plant Breeding, CEBAS-CSIC (Centro de Edafología y Biología Aplicada del Segura-Consejo Superior de Investigaciones Científicas), 30100 Espinardo, Murcia, Spain

Lan Huang
Department of Horticultural Sciences, Texas A&M University, College Station, TX 77843, USA

Riccardo Lo Bianco
Department of Agricultural, Food and Forest Sciences, University of Palermo, Palermo 90128, Italy

Taylor Mills-Ibibofori, Bruce Dunn and Niels Maness
Department of Horticulture & L.A., Oklahoma State University, Stillwater, OK 74078, USA

Mark Payton
Department of Statistics, Oklahoma State University, Stillwater, OK 74078, USA

Index

Printed in the USA
CPSIA information can be obtained
at www.ICGtesting.com
JSHW051413091023
49903JS00006B/410